The Springer Series in Nano-Optics and Nanophotonics provides an expanding selection of research monographs in the area of nano-optics and nanophotonics, science- and technology-based on optical interactions of matter in the nanoscale and related topics of contemporary interest. With this broad coverage of topics, the series is of use to all research scientists, engineers and graduate students who need up-to-date reference books. The editors encourage prospective authors to correspond with them in advance of submitting a manuscript. Submission of manuscripts should be made to the editor-in-chief, one of the editors or to Springer.

Nano-Optics and Nanophotonics

For further volumes:
http://www.springer.com/series/8765

Motoichi Ohtsu

Dressed Photons

Concepts of Light–Matter Fusion Technology

Springer

Motoichi Ohtsu
Department of Electrical Engineering
and Informations Systems
The University of Tokyo
Tokyo
Japan

ISSN 2192-1970 ISSN 2192-1989 (electronic)
ISBN 978-3-642-39568-0 ISBN 978-3-642-39569-7 (eBook)
DOI 10.1007/978-3-642-39569-7
Springer Heidelberg New York Dordrecht London

Library of Congress Control Number: 2013945799

Printed on acid-free paper

Springer is part of Springer Science+Business Media (www.springer.com)

Preface

A dressed photon is a virtual photon that dresses material energy, specifically, the energy of an electron–hole pair, in nanometric space. A quarter of a century has passed since the author pioneered basic research on the dressed photon, and even 20 years have passed since he proposed nanophotonics, which is a novel optical technology exploiting the dressed photon. This technology, which is neither wave-optical technology nor materials technology but a mixture of the two, should be named "Light-Matter Fusion Technology." Although the number of researchers engaged in this technology was quite small in its early stages, it has been rapidly increasing in recent years, and a number of related industries have been born.

In view of the rapid growth of this technical field, the purpose of this book is to disseminate the concepts of the dressed photon. First, Chap. 1 surveys the topics to be discussed in this book. Chapters 2–4 describe the fundamental concepts and theories of dressed photons, using a combination of concepts from optical science, quantum field theory, and condensed matter physics. In Chaps. 5–8, several applications are reviewed. Since the technologies enabling these applications are rapidly progressing, it is recommended that readers refer to the original papers or review articles for details. Finally, Chap. 9 summarizes the topics and presents a future outlook on the field. As supplementary material, Appendices A–H describe related topics and give detailed derivations of the equations appearing in this book.

During the course of establishing the fundamentals and developing applications of dressed photons, the author has gotten a lot of suggestions and comments from leading scientists in the relevant fields of research. Furthermore, fruitful discussions have been held with many young, active scientists, from whom the author has been greatly enlightened.

Since the dressed photon is now being applied to establish generic technologies for constructing infrastructures that will be needed for future society, this book will provide scientific and technical information about dressed photons to scientists, engineers, and students who are and will be engaged in this field.

The author thanks Drs. T. Kawazoe, T. Yatsui, N. Tate, W. Nomura, K. Kitamura (The University of Tokyo), Dr. Naruse (National Institute of Information and Communications Technology), Dr. K. Kobayashi (Yamanashi University), Dr. S. Sangu (Ricoh, Co. Ltd.), and Dr. Y. Tanaka (JFE Steel Corp.) for their collaborations in research on dressed photons. He also extends special thanks to

Drs. M. Tsukada (Tohoku University), H. Hori, and I. Banno (Yamanashi University) for their encouragement throughout the course of the author's research work.

Several application technologies of dressed photons, reviewed in Chaps. 5–8, were developed through academia–industry collaborations under arrangements made by the Specified Nonprofitt Corporation "Nanophotonics Engineering Organization." Finally, the author is grateful to Dr. C. Acheron of Springer–Verlag for his guidance and suggestions throughout the preparation of this book.

May 2013, Tokyo Motoichi Ohtsu

Abeunt studia in mores.
Publius Ovidius Naso, *Heroides*, VI, 83

Contents

1 What is a Dressed Photon? 1
- 1.1 Comparison with Conventional Light 1
- 1.2 Light–Matter Interactions via Dressed Photons 4
- 1.3 Energy Transfer Between Nanomaterials 6
- 1.4 Novel Phenomena Arising from Further Coupling 7
- 1.5 Symbols for Quantum Operators 9
- References 9

2 Physical Picture of Dressed Photons 11
- 2.1 Virtual Photons Dressing Material Energy 11
- 2.2 Range of Interaction Mediated by Dressed Photons 18
  - 2.2.1 Effective Interaction Between Nanomaterials 19
  - 2.2.2 Size-Dependent Resonance and Hierarchy 33
- References 36

3 Energy Transfer and Relaxation by Dressed Photons 37
- 3.1 Coupled States Originating from Two Energy Levels 37
- 3.2 Principles of Dressed-Photon Devices 42
  - 3.2.1 Dressed-Photon Devices Using Two Quantum Dots 43
  - 3.2.2 Dressed-Photon Devices Using Three Quantum Dots 47
- References 56

4 Coupling Dressed Photons and Phonons 59
- 4.1 Novel Molecular Dissociation and the Need for a Theoretical Model 59
  - 4.1.1 Unique Phenomena of Molecular Dissociation by Dressed Photons 59
  - 4.1.2 Lattice Vibrations in the Probe 62
- 4.2 Transformation of the Hamiltonian 67
  - 4.2.1 Diagonalization by Unitary Transformation 67
  - 4.2.2 Physical Picture of the Quasi-Particle 71
  - 4.2.3 The Equilibrium Positions of Atoms 73
- 4.3 Localization Mechanism of Dressed Photons 75
  - 4.3.1 Conditions for Localization 75

4.3.2 Position of Localization 79
4.4 Light Absorption and Emission via Dressed-Photon–Phonons . . . 82
References 88

5 Devices Using Dressed Photons 89
5.1 Structure and Function of Dressed-Photon Devices 89
5.1.1 Devices Utilizing Energy Dissipation 89
5.1.2 Devices in Which Coupling with Propagating Light is Controlled 115
5.2 Characteristics of Dressed-Photon Devices 117
5.2.1 Low Energy Consumption 118
5.2.2 Tamper-Resistance 125
5.2.3 Skew Resistance 126
5.2.4 Autonomy in Energy Transfer 127
References 134

6 Fabrication Using Dressed Photons 137
6.1 Molecular Dissociation by Dressed-Photon–Phonons 137
6.1.1 Comparison Between Experiments and Theories 137
6.1.2 Deposition by Molecular Dissociation 144
6.2 Lithography Using Dressed-Photon–Phonons 147
6.3 Fabrication by Autonomous Annihilation of Dressed-Photon–Phonons 160
6.3.1 Smoothing a Material Surface by Etching 160
6.3.2 Repairing Scratches on a Substrate Surface by Deposition 166
6.3.3 Other Related Methods 168
References 169

7 Energy Conversion Using Dressed-Photons 171
7.1 Conversion From Optical to Optical Energy 171
7.1.1 Multi-Step Excitation 176
7.1.2 Non-Degenerate Excitation and Applications 184
7.2 Conversion From Optical to Electrical Energy 190
7.2.1 Multi-Step Excitation and Autonomous Fabrication. 191
7.2.2 Wavelength Selectivity and Light Emission 195
7.3 Conversion From Electrical to Optical Energy 200
7.3.1 Autonomous Device Fabrication 201
7.3.2 Device Operation 203
7.3.3 Applications to Other Related Devices 208
References 213

8 Spatial Features of the Dressed-Photon and its Mathematical Scientific Model 215
8.1 Hierarchy 215
8.1.1 Hierarchical Memory. 216
8.1.2 Hierarchy Based on the Constituents of Nanomaterials. 219
8.1.3 Hierarchy and Local Energy Dissipation 221
8.1.4 Applications Exploiting the Differences Between Propagating Light and Dressed Photons 223
8.2 Conversion From an Electric Quadrupole to an Eelectric Dipole 227
8.3 Probe-Free Methods 230
8.3.1 Magnified Transcription of the Spatial Distribution of the Interaction 230
8.3.2 Spatial Modulation of the Energy Transfer Between Quantum Dots 231
8.4 Mathematical Scientific Model 233
8.4.1 Formation of Nanomaterials 235
8.4.2 Statistical Modeling of Morphology 240
References 245

9 Summary and Future Outlook 247
9.1 Summary 247
9.2 Future Outlook 250
References 251

Appendix A: Multipolar Hamiltonian 253

Appendix B: Elementary Excitation and Exciton-Polariton 259

Appendix C: Projection Operator and Effective Interaction Operator 265

Appendix D: Transformation from Photon Base to Polariton Base 275

Appendix E: Derivation of the Equations for Size-Dependent Resonance 279

Appendix F: Energy States of a Semiconductor Quantum Dot 283

Appendix G: Solutions of the Quantum Master Equations for the Density Matrix Operators 295

Appendix H: Derivation of Equations in Chap. 4 301

Index 317

Chapter 1
What is a Dressed Photon?

Incipe quidquid agas, pro toto est prima operis pars.
Decimus Magnus Ausonius, *Idyllia*, XII

To start with our description of the dressed photon (DP) and its applications, the present chapter surveys some common concepts that are used throughout this book.

1.1 Comparison with Conventional Light

First, try to answer the following three simple questions.

[Question 1]
After the end of a glass fiber is sharpened to form a nanometer-sized apex, its tapered side surface is coated with an opaque film, leaving the apex of the fiber uncoated to form a nanometer-sized aperture. Such a sharpened glass fiber is called a fiber probe (probe), which has been frequently used in the technical fields to be covered in this book. Now, assume that this probe is placed in a vacuum chamber, as shown in Fig. 1.1. The chamber is filled with a low-pressure gas, whose molecules dissociate by absorbing ultraviolet light. Visible light is injected into the probe from its tail.

Now, the question is: Do the freely moving gas molecules in this chamber dissociate when they arrive at the apex of the probe?

[Question 2]
As an example, consider optical lithography, which is a popular method of forming a fine pattern on the surface of a crystal substrate, such as silicon (Si) (Fig. 1.2). Here, the crystal surface to be patterned is coated with a photo-resist film, whose structure changes due to a photo-induced chemical reaction caused by the absorption of ultraviolet light. A photo-mask with a nanometer-sized aperture is placed on the photoresist film, and the photo-mask surface is irradiated with visible light.

M. Ohtsu, *Dressed Photons*, Nano-Optics and Nanophotonics,
DOI: 10.1007/978-3-642-39569-7_1, © Springer-Verlag Berlin Heidelberg 2014

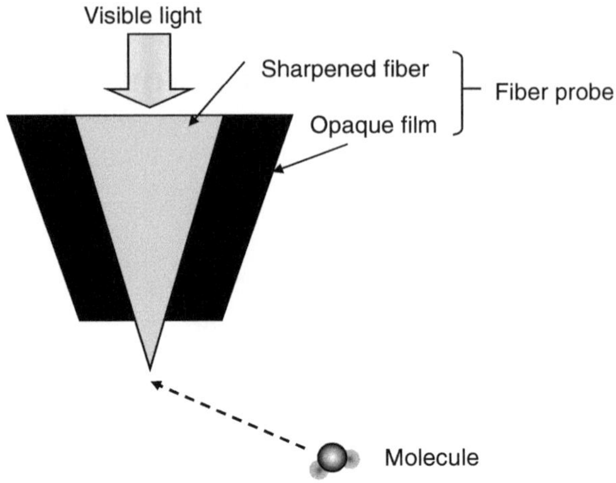

Fig. 1.1 Dissociation of molecules by using visible light and a fiber probe with a nanometer-sized apex

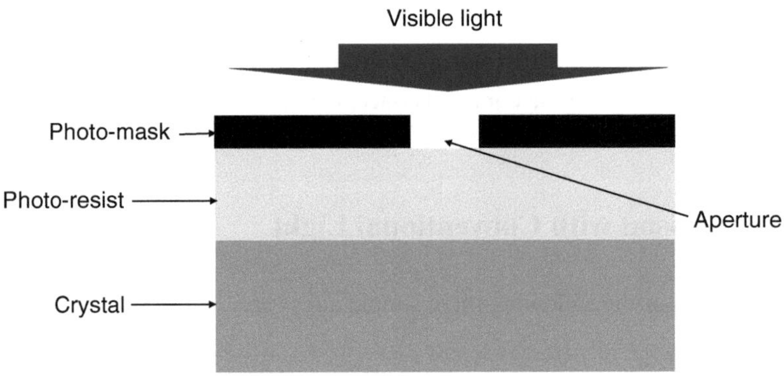

Fig. 1.2 Optical lithography using visible light and a photo-mask with a nanometer-sized aperture

Now, the question is: Is the aperture pattern of the photo-mask transcribed to the surface of the photo-resist film?

[Question 3]
Is it possible to fabricate a light emitting diode by using a bulk Si crystal?

The answers to these three questions must be "No!", as a natural consequence of the principles of conventional optical and materials sciences. The reasons are as follows:

[Question 1]
There are two reasons: (1) This probe works as a cut-off optical waveguide to visible light because the aperture size is much smaller than its wavelength. Therefore, the molecules in the proximity of the probe apex are not illuminated by the light.

(2) Even if they were illuminated, they would not dissociate because they do not absorb visible light, whose photon energy is much lower than that of ultraviolet light.

[Question 2]
There are two reasons, which are almost equivalent to those of Question 1: (1) The visible light does not transmit through the photo-mask because the size of the aperture is much smaller than its wavelength, meaning that the photo-resist is not illuminated by the light. (2) Even if it were illuminated, no chemical reaction would be induced because the photo-resist film does not absorb visible light, whose photon energy is much lower than that of ultraviolet light.

[Question 3]
Since Si is an indirect transition-type semiconductor: Electrons have to transition from the conduction band to the valence band in order to emit light. However, in the case of an indirect transition-type semiconductor, the wave-numbers (momenta) of the electrons at the bottom of the conduction band and at the top of the valence band are different. Therefore, for electron–hole recombination, a phonon is required in order to satisfy the momentum conservation law. In other words, an electron–phonon interaction is required. However, the probability of this interaction is low, resulting in a low interband transition probability.

In reality, however, the answers to these questions have already turned out to be "Yes!", with the advent of a novel optical science that overturns long-held beliefs in conventional optical and materials sciences. One objective of this book is to describe the reasons why the answers have turned out to be "Yes!".

To answer "Yes!" to Questions 1 and 2, one has to assume that a minute light field is generated on a nanometer-sized material (called a nanomaterial; for example, the probe apex or the photo-mask aperture), and furthermore, that its energy is as high as that of ultraviolet light. To answer "Yes!" to Question 3, one has to assume that this minute light field is generated also in the Si crystal, which assists the electrons in order to satisfy the momentum conservation law. The details of the reasons for the affirmative answers to Questions 1–3 will be described in Sects. 6.1, 6.2, and 7.3, respectively, from which it will be found that this minute light field is nothing more than the dressed photon (DP). By exploiting the interaction between the DP and the nanomaterial, a novel and innovative optical technology, called "dressed-photon technology", has emerged.

Table 1.1 classifies several optical technologies according to the combination of light and material used. Among them, the ones using propagating light are nothing more than conventional optical technology, which is also called photonics. Although some recent optical methods exploit nanomaterials, they still remain in the category of conventional optical technology because propagating light is used. Thus, their answers to Questions 1–3 are still "No!".

The conventional optical technologies in the right column in this table exploit the wave nature of propagating light. One way to distinguish the differences between

Table 1.1 Classification of optical technologies according to the combination of light and materials used

	Nanometric dressed photon	Macroscopic propagating light
Nanometric material	Dressed-photon technology (Nanophotonics)	Plasmonics, metamaterials, and photonic crystals
Macroscopic material	———	Conventional optical technology

technology that exploits DPs and technology that exploits propagating light is to examine whether the momenta of the particles involved in the light–matter interaction (photons, electrons, and phonons) are conserved or not: The uncertainty relation $\Delta k \cdot \Delta x \geq 1$ holds between the uncertainty Δk of the wave-number k of the light (the photon momentum) and that Δx of its position x. In the case of the DP, since $\Delta x \ll \lambda$ holds because its size is smaller than the optical wavelength λ, one derives $\Delta k \gg k$ from the uncertainty relation . This means that the wave-number and the momentum are uncertain and non-conserved. In other words, the dispersion relation, i.e., the relation between the wave-number (momentum) and the energy, cannot be used for analyzing phenomena in which the DP is involved. Accordingly, the refractive index, representing the phase-delayed feature of the optical response of the material, cannot be the fundamental physical quantity.

On the other hand, in the case of propagating light, since $\Delta k \ll k$ holds because $\Delta x \gg \lambda$, the uncertainties of the wave-number and the momentum are negligible, and these quantities are conserved. Thus, the wave-number, momentum, dispersion relation, and refractive index are allowed to be used to describe phenomena in which propagating light is involved. Although the recently developed areas of plasmonics, metamaterials, and photonic crystals employ sub-wavelength–sized materials, they are unrelated to dressed-photon technology because they use propagating light and rely on the dispersion relation.

1.2 Light–matter Interactions via Dressed Photons

Conventional optical technology has relied on materials science and technology to explore and develop novel materials. By processing these materials, optical devices have been constructed for efficiently emitting, detecting, or modulating propagating light. In other words, conventional optical technology has used propagating light merely as a tool instead of exploring new types of light. In contrast, dressed-photon technology was born as a result of exploring a new type of light, that is, the DP. Since conventional classical and quantum theories of light cannot be directly applied to describe the DP, novel concepts and theoretical bases are required.

In the conventional quantum theory of light, the concept of a photon was established by quantizing the electromagnetic field of light that propagates through macroscopic free space whose size is larger than the wavelength. A photon corresponds to an electromagnetic mode in a virtual cavity defined in free space. Since a photon is massless, it is difficult to express its wave function in a coordinate representation in order to draw a picture of the photon as a spatially localized point particle like an electron. Thus, interactions between photons and electrons in a nanometric space must be carefully investigated. For this investigation, this book always pays attention to whether the light field is nanometric or macroscopic in size, an approach that has never been described in conventional textbooks on light. The main scope of conventional optics textbooks has been a comparison between the classical and quantum features of light.

For describing a light field in a nanometric space, the energy transfer between two nanomaterials and detection of the transferred energy are formulated by assuming that the nanomaterials are arranged in close proximity to each other and illuminated by propagating light (Sect. 2.1). Although the separation between the two nanomaterials is much shorter than the optical wavelength, it is sufficiently long to prevent electron tunneling. As a result, the energy is transferred not by a tunneled electron but by some sorts of optical interactions between the two nanomaterials.

A serious problem, however, is that a virtual cavity cannot be defined in a subwavelength–sized nanometric space, unlike the conventional quantum theory of light. In order to solve this problem, an infinite number of electromagnetic modes, with an infinite number of frequencies, polarization states, and energies, must be assumed. In parallel with this assumption, infinite numbers of energy levels must also be assumed for the electrons and holes. As a result of these assumptions, the DP is found to be a that dresses the material energy, i.e., the energy of the electron–hole pair. The interaction between the two nanomaterials can be represented by energy transfer due to the annihilation of a DP from the first nanomaterial and its creation on the second nanomaterial.

The DP field is modulated temporally and spatially. The temporal modulation feature is represented by an infinite number of modulation sidebands, i.e., an infinite series of photon eigen-energies. As a result of the dressing mentioned above, the electron–hole pair also dresses the photon energy, with the result that its eigen-energy exhibits a similar modulation feature. Consequently, a dual relation of the modulation is established between the photon and the electron–hole pair.

The DP has always been named the "optical near field" in the author's early study [1–3]. Although this name appropriately represented the spatial features of the DP, it was not sufficient to represent the detailed interactions with nanomaterials, nor to convey a physical picture of the dressing the energy of the electron–hole pair. Thus, the name "optical near field" has been replaced with "DP" in order to express the detailed interaction explicitly and to provide a clearer physical picture.

Nanophotonics, an innovative optical technology based on the DP, was first proposed by the author and has led to the development of a variety of applications [4–7]. A recent trend in optical technology is the tendency to accept even plasmonics, metamaterials, and photonic crystals (refer to the right column in Table 1.1) into the category of Nanophotonics. However, since they use propagating light and, thus, are unrelated to the DP from the viewpoint of light–matter interactions in nanometric space, the original field of Nanophotonics has now been renamed as "dressed-photon technology" to avoid confusion.

The actual nanomaterials used for dressed-photon technology are buried in or fixed on a crystal substrate and are illuminated by light. That is, the actual nanomaterials are always surrounded by a macroscopic system composed of macroscopic materials and electromagnetic fields. Therefore, the contribution from the macroscopic system must be included in the analysis of the energy transfer between the nanomaterials, which is not straightforward (Sect. 2.2). Furthermore, it is also difficult to define a virtual cavity for a nanometric system surrounded by a macroscopic system. In order to avoid these difficulties, a novel theory was established to describe the "effective interaction" between nanomaterials mediated by the DP. This interaction is also called a "near-field optical interaction", whose energy is expressed by a Yukawa function. Its magnitude rapidly decreases with increasing separation between the two nanomaterials, whose decay length is equivalent to the size of the nanomaterial. The spatially modulated feature of the DP field is represented by this Yukawa function, and a unique spatial feature of the interaction, named "hierarchy", appears due to the size-dependent spatial modulation. This hierarchy has been applied to information security systems (Sects. 8.1–8.3).

1.3 Energy Transfer Between Nanomaterials

Semiconductor nanomaterials, called quantum dots (QDs), have often been used for dressed-photon technology. Novel states of coupling between QDs have been found through study of the interactions between electrons, holes, and excitons in QDs in order to describe the energy transfer mediated by DPs and subsequent relaxation between closely spaced QDs. In the case of DP-mediated interactions, the long-wavelength approximation is invalid because the range of interaction is as short as the size of the QDs. This suggests that an electric dipole transition that has been forbidden in the case of propagating light excitation is allowed in the case of DP excitation. This means that the electric dipole-forbidden energy levels in QDs can be exploited for nanometer-sized photonic devices (DP devices; Chap. 5). The advantage of using these forbidden energy levels is that the contribution of the propagating light to the DP device operation can be excluded in order to avoid malfunction of the device.

In a DP device, DPs mediate the transfers of the energies of the electrons, holes, and excitons in a nanomaterial to an adjacent nanomaterial, and uni-directional energy transfer is realized by the subsequent energy dissipation via interactions with phonons in the heat bath, destroying quantum coherence (Sect. 3.2). As a result, signal transmission becomes possible from one nanomaterial to the other, guaranteeing reliable operation of the DP device.

1.4 Novel Phenomena Arising from Further Coupling

In actual materials, such as semiconductors, the contribution of the crystal lattice also needs to be included in the theoretical model of the DP. By doing so, it has been found that the DP interacts with phonons, i.e., quanta of normal modes of the crystal vibration. As a result of this interaction, a novel quasi-particle is generated on the surface of a nanomaterial (Chap. 4), and the energy of this quasi-particle can transfer to the adjacent nanomaterial, where it induces a novel photo-chemical reaction (Chap. 6). Here, since translational symmetry is broken due to the finite size of the nanomaterial, the momentum (or wave-number) of the quasi-particle has a large uncertainty and is non-conserved, as was the case of the DP itself. Furthermore, in a finite-sized nanomaterial, it is possible to generate multi-mode coherent phonons as a result of the DP–phonon interaction.

The quasi-particle generated by the DP–phonon interaction is called a dressed-photon–phonon (DPP), which is a DP dressing the energy of the multi-mode coherent phonon. As was the case of the DP, the DPP field is temporally and spatially modulated. As a result of the temporal modulation, the DPP gains an infinite number of modulation sidebands. As a dual relation of this modulation, the electron–hole pair dresses the energies of the photon and phonon, which means that the eigen-energy of the electron–hole pair in the nanomaterial is modulated. Furthermore, as a result of spatial modulation, the DPP field leaks out from the nanomaterial surface with a spatial extent as short as the size of the nanomaterial, as was the case of the DP field. Therefore, the DPP can transfer energy from one nanomaterial to another if these two nanomaterials are in close proximity to each other. Since the transferred energy can modulate the electron–hole pair in the second nanomaterial, the quantum state of the second nanomaterial has to be represented by the direct product of the electronic state and phonon state for estimating the magnitude of the transferred energy. The phenomena induced by the DPP have enabled emission and absorption of photons in a material even when their energies are lower than the bandgap energy E_g of electrons in the material. By utilizing these unique phenomena, a novel technology has been developed for up-converting the optical and electrical energies (Chap. 7).

In the conventional interaction between propagating light and a material, the quantum state represented by the above-mentioned direct product was not required because only the electric dipole-allowed transitions have been involved. As a result, in contrast to the interaction mediated by the DPP, the phenomena originating from this interaction are induced only when the photon energy of the propagating light is larger than E_g. In other words, light with a photon energy smaller than E_g (i.e., with a wavelength longer than the cut-off wavelength $\lambda_c = E_g/hc$) is not emitted or absorbed by the material.

Conventional optical technologies, i.e., photonics, in the right column of Table 1.1, have used a variety of materials to construct devices that emit or absorb propagating light. Nanotechnology has rapidly progressed in recent years, producing several nanomaterials such as carbon nanotubes (CNTs) and QDs. When they are used in a laser, for example, a large number of CNTs or QDs are provided in the laser cavity to be used as light emitting media. Although they contribute to improving some aspects of the laser oscillation performance as compared with using a conventional macroscopic material, these devices are nothing more than one class of laser. That is, since this technology improves some aspects of the optical device performance, the innovation brought about by this technology is regarded merely as a "quantitative innovation".

Photonics developed rapidly after the advent of lasers in the 1960s and matured in the 1990s. However, it has become difficult to meet the requirements of increasing speed/capacity in optical information transmission, increasing density in optical information storage, and increasing resolution in optical fabrication, that are demanded in order to construct infrastructures for future society. Furthermore, since rare or toxic materials have been used for optical devices, it is difficult to solve the problems related to resource conservation and environmental protection.

The principal cause of these problems is that propagating light has been used. Stated another way, spatial averaging–the so-called long-wavelength approximation–has been used for light–matter interactions. This averaging, or approximation, involves diffraction of light, which limits miniaturization of devices, acting as a barrier to increasing the optical storage density, increasing the resolution of optical fabrication, and so on–a problem known as the diffraction limit. Furthermore, only the electric dipole-allowed transitions have been exploited because of the long-wavelength approximation. That is, only materials compatible with these types of transitions have been explored and used to construct optical devices. In other words, conventional optical technology has been based on the principle of wave optics, where light and matter are dealt with separately. On the other hand, since dressed-photon technology uses DPs, i.e., virtual photons dressing material energy, it can be called "light–matter fusion technology", which fuses light and matter instead of dealing with them separately and independently. The principle of the DP has given birth to novel optical functions that break through these technical limits, including the diffraction limit. An essential innovation brought about by such breakthroughs is called a "qualitative innovation".

In order to answer the questions "What is the dressed photon?" and "What is its contribution to optical science and technology?", this book reviews theoretical bases established by combining the concepts of optical science, quantum field theory, and condensed matter physics. On these bases, all of the answers to Questions 1–3 in Sect. 1.1 turn out to be "Yes!". Furthermore, these principles have been already applied to a variety of technologies, including optical devices, optical fabrication, energy conversion, and optical information security systems, allowing us to break through the limits faced by conventional optical technology and realize qualitative innovations. As a by-product, quantitative innovations have also been realized. dressed-photon technology is now progressing rapidly, establishing new bases of optical technology.

Table 1.2 Annihilation operators used in this book*

Symbol for operator (equation number in which these operators appear for the first time)	Quasi-particle
$\hat{a}$ (2.1)	Photon
$\tilde{a}$ (2.20)	Dressed photon
$\hat{b}$ (2.1)	Electron–hole pair (Exciton)
$\hat{c}$ (4.16)	Phonon
$\hat{e}$ (2.3)	Electron
$\hat{h}$ (2.3)	Hole
$\hat{\alpha}$ (4.30a)	Dressed-photon–phonon
$\hat{\xi}$ (2.27)	Exciton polariton

*The corresponding creation operator, i.e., the Hermitian conjugate of the annihilation operator, is expressed by adding the superscript † to each symbol in this table

1.5 Symbols for Quantum Operators

For the readers' convenience, Table 1.2 summaries the annihilation operators of photons and other relevant particles, which will appear in the following chapters.

References

1. M. Ohtsu (ed.), *Near-Field Nano/Atom Optics and Technology* (Springer-Verlag, Berlin, 1998)
2. M. Ohtsu, H. Hori, *Near-Field Nano-Optics* (Kluwer Academic, New York, 1999)
3. M. Ohtsu, K. Kobayashi, *Optical Near Fields* (Springer, Berlin, 2004)
4. M. Ohtsu (ed.), *Progress in Nano-Electro-Optics I - VII* (Springer-Verlag, Berlin, 2003–2010)
5. M. Ohtsu, K. Kobayashi, T. Kawazoe, T. Yatsui, M. Naruse, *Principles of Nanophotonics* (CRC Press, Boca Raton, 2008)
6. M. Ohtsu (ed.), *Nanophotonics and Nanofabrication* (Wiley-VHC, Weinheim, 2009)
7. M. Ohtsu (ed.), *Progress in Nanophotonics I* (Springer, Berlin, 2011)

Chapter 2
Physical Picture of Dressed Photons

Veritatis simplex oratio est.
Lucius Annaeus Seneca, *Epistulae*, XLIX,12

This chapter derives a physical picture of the dressed photon (DP) by analyzing light–matter interactions in nanometric space, in which an infinite number of photon modes and an infinite number of energy levels of electron–hole pairs are involved. Based on this physical picture, energy transfer between two nanomaterials under light illumination and detection of this energy transfer are described. Furthermore, an effective energy of the interaction between the nanomaterials is derived by noting that the actual nanometric system in which the DP is generated is surrounded by a macroscopic light–matter system.

2.1 Virtual Photons Dressing Material Energy

In the conventional quantum theory of light, the concept of a photon is the quantization of an electromagnetic field that propagates through macroscopic free space having a size greater than the wavelength [1]. A photon corresponds to an electromagnetic mode in a virtual cavity defined in free space for quantizing light. Since a photon is massless, it is difficult to construct a wave function in a coordinate representation that gives a picture of the photon as a spatially localized point particle like an electron [2]. However, if a detector–an atom in the simplest case–is placed near the light source to absorb a photon in an area whose linear dimension is much smaller than the wavelength of light, it would be possible to detect the photon with a spatial resolution equivalent to the size of the detector [3, 4].

In order to understand the properties of dressed photons (DPs), this chapter investigates the interactions between photons and electrons in a nanometric space by taking the above-mentioned properties of photons into account. For this investigation, multiple nanomaterials, arranged in close proximity to each other and illuminated by propagating light, are considered. By considering the case of just two nanomaterials

M. Ohtsu, *Dressed Photons*, Nano-Optics and Nanophotonics,
DOI: 10.1007/978-3-642-39569-7_2,

for simplicity, the energy transfer between them and detection of this energy transfer are formulated. The present discussion deals with the case where the separation between the two nanomaterials is much shorter than the optical wavelength but sufficiently long to prevent electron tunneling. Therefore, the energy is transferred not by electron tunneling but by the electromagnetic interaction. This section describes the physical picture of photons that mediate this interaction and energy transfer.

It should be noted that the two nanomaterials and the light cannot be treated independently. This is because the nanomaterials emit or absorb virtual photons driven by fluctuations in the electromagnetic field, e.g, zero-point fluctuations of the vacuum. These absorption and emission processes, which have been called virtual processes, violate the energy conservation law; however, they are consistent with the Heisenberg uncertainty principle. As a result of these virtual processes, nanomaterials are covered with a cloud of virtual photons, and the clouds of virtual photons on the two nanomaterials spatially overlap each other.

When the nanomaterials are in excited states, conventional theories such as perturbation theories cannot be employed for describing the interaction because real photons (propagating light) are emitted in addition to the virtual photons. Although several theoretical attempts have been made to describe this interaction, no sufficiently accurate theory, including a proper description of the relaxation process, has been established. This section presents a novel theory for virtual photons around nanomaterials, even in the excited states. An advantage of this theory is that the energy transfer between the nanomaterials can be described by the emission or absorption of DPs, as will be described later. In the following parts of this section, several properties of photons generated around nanomaterials will be discussed by analyzing the interactions between photons, electrons, and positive holes in the nanomaterials. A serious problem, however, is that a virtual cavity cannot be defined in a sub-wavelength sized nanometric space, unlike the conventional quantum theory of light. In order to solve this problem, an infinite number of electromagnetic modes, with infinite frequencies, polarization states, and energies, must be assumed. Due to this assumption, an infinite number of energy states must be also assumed for the electrons and holes.

Under illumination with propagating light having photon energy $\hbar\omega_o$ the interaction between a photon and an electron–hole pair in nanometric space can be described by a multipolar Hamiltonian (refer to Appendix A)

$$\hat{H} = \sum_{k\lambda} \hbar\omega_k \hat{a}^{\dagger}_{k\lambda}\hat{a}_{k\lambda} + \sum_{\alpha>F,\ \beta<F} \left(E_\alpha - E_\beta\right) \hat{b}^{\dagger}_{\alpha\beta}\hat{b}_{\alpha\beta} + \hat{H}_{\mathrm{int}}. \tag{2.1}$$

The first term represents the photon energy generated in the nanometric space, which is given by the sum of the infinite number of photon modes with angular frequency $\omega_{\boldsymbol{k}}$, polarization state λ, and energy $\hbar\omega_{\boldsymbol{k}}$.[1] Here, the subscript $\boldsymbol{k}$ represents the

[1] This is different from the case of an exciton-polariton in a macroscopic material (refer to Appendix B). That is, since a virtual cavity can be defined in the case of a super-wavelength sized macroscopic space, it is sufficient to write the single mode photons in the first term of Eq. (B.1) in Appendix B,

wave-vector, and $\hat{a}_{k\lambda}$ and $\hat{a}^{\dagger}_{k\lambda}$ are its annihilation and creation operators, respectively, which satisfy a commutation relation

$$\left[\hat{a}_{k\lambda}, \hat{a}^{\dagger}_{k'\lambda'}\right] = \delta_{kk'}\delta_{\lambda\lambda'} \tag{2.2}$$

where $\delta_{kk'}$ and $\delta_{\lambda\lambda'}$ are Kronecker deltas.

The second term represents the energy of the electron–hole pair, which is also given by the sum of the energies of the electron–hole pairs of the infinite number of energy states, identified by the subscripts α and β. The energy difference $E_\alpha - E_\beta$ represents the bandgap energy in the case of a semiconductor, and F represents the Fermi energy level. The operators

$$\hat{b}_{\alpha\beta} = S\hat{e}_\alpha \hat{h}_\beta, \tag{2.3a}$$
$$\hat{b}^{\dagger}_{\alpha\beta} = S^*\hat{e}^{\dagger}_\alpha \hat{h}^{\dagger}_\beta, \tag{2.3b}$$

respectively represent the simultaneous annihilation and creation of the electron and hole; i.e., they represent the annihilation and creation operators of the electron–hole pair. Here, S and S^* are complex numbers for representing the time inversion symmetry and its complex conjugate, respectively. Their absolute values are unity. $\hat{e}_\alpha$ and $\hat{e}^{\dagger}_\alpha$ are the annihilation and creation operators of the electron in the energy level α. $\hat{h}_\beta$ and $\hat{h}^{\dagger}_\beta$, on the other hand, are those of the hole in the energy level β. Since the electron–hole pair is a boson, its operators $\hat{b}_{\alpha\beta}$ and $\hat{b}^{\dagger}_{\alpha\beta}$ satisfy the commutation relation

$$[\hat{b}_{\alpha\beta}, \hat{b}^{\dagger}_{\alpha'\beta'}] = \delta_{\alpha\alpha'}\delta_{\beta\beta'}. \tag{2.3c}$$

The third term of Eq. (2.1) represents the energy of the interaction between the photon and the electron–hole pair, which is given by

$$\hat{H}_{\text{int}} = -\int \hat{\psi}^{\dagger}(\boldsymbol{r})\boldsymbol{p}(\boldsymbol{r})\hat{\psi}(\boldsymbol{r}) \cdot \hat{\boldsymbol{D}}^{\perp}(\boldsymbol{r})dv, \tag{2.4}$$

where $\boldsymbol{p}(\boldsymbol{r})$ is an electric dipole moment. $\hat{\psi}(\boldsymbol{r})$ is an annihilation operator for the field of the electron–hole pair, which is expressed as

$$\hat{\psi}(\boldsymbol{r}) = \sum_{\alpha>F} \varphi_{e\alpha}(\boldsymbol{r})\,\hat{e}_\alpha + \sum_{\beta<F} \varphi_{h\beta}(\boldsymbol{r})\,\hat{h}_\beta \tag{2.5}$$

by using the state functions $\varphi_{e\alpha}(\boldsymbol{r})$ and $\varphi_{h\beta}(\boldsymbol{r})$ of the electron and hole. The field creation operator $\hat{\psi}^{\dagger}(\boldsymbol{r})$ is the Hermitian conjugate of Eq. (2.5). When these field

Footnote 1 continued
whose angular frequency is equal to ω_0 of the incident light. In contrast, in a sub-wavelength sized nanometric space, an infinite number of photon modes is needed in the first term of Eq. 2.1.

operators are inserted into Eq. (2.4), only the operators of simultaneous annihilation, $\hat{e}_\alpha \hat{h}_\beta$, or creation, $\hat{e}^\dagger_\alpha \hat{h}^\dagger_\beta$, of the electron and hole are retained, and Eqs. (2.3a)–(2.3c) are used. $\hat{\boldsymbol{D}}^\perp(\boldsymbol{r})$ is the transverse component of the electric displacement operator of the incident photon, which is perpendicular to the wave–vector $\boldsymbol{k}$.[2] This operator is expressed as

$$\hat{\boldsymbol{D}}^\perp(\boldsymbol{r}) = i \sum_{\boldsymbol{k}} \sum_{\lambda=1}^{2} N_{\boldsymbol{k}} \boldsymbol{e}_{\boldsymbol{k}\lambda}(\boldsymbol{k}) \left\{ \hat{a}_{\boldsymbol{k}\lambda}(\boldsymbol{k}) e^{i\boldsymbol{k}\cdot\boldsymbol{r}} - \hat{a}^\dagger_{\boldsymbol{k}\lambda}(\boldsymbol{k}) e^{-i\boldsymbol{k}\cdot\boldsymbol{r}} \right\}, \tag{2.6}$$

where plane waves were used for the mode functions. Here, $N_{\boldsymbol{k}}$ and $\boldsymbol{e}_{\boldsymbol{k}\lambda}(\boldsymbol{k})$ are a proportionality constant and the unit vector along the direction of polarization, respectively. Inserting Eqs. (2.5) and (2.6) into Eq. (2.4) yields the interaction Hamiltonian

$$\begin{aligned}
\hat{H}_{\text{int}} = & -i \sum_{\boldsymbol{k}\lambda} N_{\boldsymbol{k}} \sum_{\alpha>F,\, \beta<F} \int \left(\varphi^*{}_{h\beta} \varphi_{e\alpha} (\boldsymbol{p} \cdot \boldsymbol{e}_{\boldsymbol{k}\lambda}(\boldsymbol{k})) \hat{b}_{\alpha\beta} \right. \\
& \left. + \varphi^*{}_{e\alpha} \varphi_{h\beta} (\boldsymbol{p} \cdot \boldsymbol{e}_{\boldsymbol{k}\lambda}(\boldsymbol{k})) \hat{b}^\dagger_{\alpha\beta} \right) \left[\hat{a}_{\boldsymbol{k}\lambda} e^{i\boldsymbol{k}\cdot\boldsymbol{r}} - \hat{a}^\dagger_{\boldsymbol{k}\lambda} e^{-i\boldsymbol{k}\cdot\boldsymbol{r}} \right] dv \\
= & -i \sum_{\boldsymbol{k}\lambda} N_{\boldsymbol{k}} \sum_{\alpha>F,\, \beta<F} \left\{ \int \varphi^*{}_{h\beta} \varphi_{e\alpha} e^{i\boldsymbol{k}\cdot\boldsymbol{r}} (\boldsymbol{p} \cdot \boldsymbol{e}_{\boldsymbol{k}\lambda}(\boldsymbol{k})) \hat{b}_{\alpha\beta} \hat{a}_{\boldsymbol{k}\lambda} \right. \\
& + \int \varphi^*{}_{e\alpha} \varphi_{h\beta} e^{i\boldsymbol{k}\cdot\boldsymbol{r}} (\boldsymbol{p} \cdot \boldsymbol{e}_{\boldsymbol{k}\lambda}(\boldsymbol{k})) \hat{b}^\dagger_{\alpha\beta} \hat{a}_{\boldsymbol{k}\lambda} \\
& - \int \varphi^*{}_{h\beta} \varphi_{e\alpha} e^{-i\boldsymbol{k}\cdot\boldsymbol{r}} (\boldsymbol{p} \cdot \boldsymbol{e}_{\boldsymbol{k}\lambda}(\boldsymbol{k})) \hat{b}_{\alpha\beta} \hat{a}^\dagger{}_{\boldsymbol{k}\lambda} \\
& \left. - \int \varphi^*{}_{e\alpha} \varphi_{h\beta} e^{-i\boldsymbol{k}\cdot\boldsymbol{r}} (\boldsymbol{p} \cdot \boldsymbol{e}_{\boldsymbol{k}\lambda}(\boldsymbol{k})) \hat{b}^\dagger_{\alpha\beta} \hat{a}^\dagger_{\boldsymbol{k}\lambda} \right\} dv.
\end{aligned} \tag{2.7}$$

By representing the spatial Fourier transform of the electric dipole moment as

$$\rho_{\beta\alpha\lambda}(\boldsymbol{k}) = \int \varphi^*{}_{h\beta}(\boldsymbol{r}) \varphi_{e\alpha}(\boldsymbol{r}) (\boldsymbol{p}(\boldsymbol{r}) \cdot \boldsymbol{e}_{\boldsymbol{k}\lambda}(\boldsymbol{k})) e^{i\boldsymbol{k}\cdot\boldsymbol{r}} dv, \tag{2.8a}$$

$$\rho^*{}_{\beta\alpha\lambda}(\boldsymbol{k}) = \int \varphi^*{}_{e\alpha}(\boldsymbol{r}) \varphi_{h\beta}(\boldsymbol{r}) (\boldsymbol{p}(\boldsymbol{r}) \cdot \boldsymbol{e}_{\boldsymbol{k}\lambda}(\boldsymbol{k})) e^{-i\boldsymbol{k}\cdot\boldsymbol{r}} dv, \tag{2.8b}$$

$$\rho_{\alpha\beta\lambda}(\boldsymbol{k}) = \int \varphi^*{}_{e\alpha}(\boldsymbol{r}) \varphi_{h\beta}(\boldsymbol{r}) (\boldsymbol{p}(\boldsymbol{r}) \cdot \boldsymbol{e}_{\boldsymbol{k}\lambda}(\boldsymbol{k})) e^{i\boldsymbol{k}\cdot\boldsymbol{r}} dv, \tag{2.9a}$$

$$\rho^*{}_{\alpha\beta\lambda}(\boldsymbol{k}) = \int \varphi^*{}_{h\beta}(\boldsymbol{r}) \varphi_{e\alpha}(\boldsymbol{r}) (\boldsymbol{p}(\boldsymbol{r}) \cdot \boldsymbol{e}_{\boldsymbol{k}\lambda}(\boldsymbol{k})) e^{-i\boldsymbol{k}\cdot\boldsymbol{r}} dv, \tag{2.9b}$$

[2] The transverse component $\boldsymbol{F}^\perp$ of a vector field $\boldsymbol{F}(\boldsymbol{r})$ is defined by $\nabla \cdot \boldsymbol{F}^\perp = 0$. The longitudinal component $\boldsymbol{F}^\parallel$, on the other hand, is by $\nabla \times \boldsymbol{F}^\parallel = 0$.

Eq. (2.7) are transformed to

$$\hat{H}_{\text{int}} = -i \sum_{\boldsymbol{k}\lambda} N_{\boldsymbol{k}} \sum_{\alpha>F,\ \beta<F} \left\{ \rho_{\beta\alpha\gamma}(\boldsymbol{k}) \hat{b}_{\alpha\beta} \hat{a}_{\boldsymbol{k}\lambda} + \rho_{\alpha\beta\lambda}(\boldsymbol{k})\, \hat{b}^{\dagger}_{\alpha\beta} \hat{a}_{\boldsymbol{k}\lambda} \right.$$
$$\left. - \rho^{*}{}_{\alpha\beta\lambda}(\boldsymbol{k})\, \hat{b}_{\alpha\beta} \hat{a}^{\dagger}_{\boldsymbol{k}\lambda} - \rho^{*}{}_{\beta\alpha\lambda}(\boldsymbol{k})\, \hat{b}^{\dagger}_{\alpha\beta} \hat{a}^{\dagger}_{\boldsymbol{k}\lambda} \right\}$$
$$= -i \sum_{\boldsymbol{k}\lambda} N_{\boldsymbol{k}} \sum_{\alpha>F,\ \beta<F} \left\{ \left[\rho_{\beta\alpha\gamma}(\boldsymbol{k})\, \hat{b}_{\alpha\beta} + \rho_{\alpha\beta\lambda}(\boldsymbol{k})\, \hat{b}^{\dagger}{}_{\alpha\beta} \right] \hat{a}_{\boldsymbol{k}\lambda} \right.$$
$$\left. - \left[\rho^{*}{}_{\alpha\beta\lambda}(\boldsymbol{k})\, \hat{b}_{\alpha\beta} + \rho^{*}{}_{\beta\alpha\lambda}(\boldsymbol{k})\, \hat{b}^{\dagger}{}_{\alpha\beta} \right] \hat{a}^{\dagger}_{\boldsymbol{k}\lambda} \right\}. \tag{2.10}$$

Furthermore, by using

$$\hat{\gamma}_{\alpha\beta\lambda}(\boldsymbol{k}) = \rho^{*}{}_{\alpha\beta\lambda}(\boldsymbol{k})\, \hat{b}_{\alpha\beta} + \rho^{*}{}_{\beta\alpha\lambda}(\boldsymbol{k})\, \hat{b}^{\dagger}_{\alpha\beta}, \tag{2.11}$$
$$\hat{\gamma}^{\dagger}_{\alpha\beta\lambda}(\boldsymbol{k}) = \rho_{\alpha\beta\lambda}(\boldsymbol{k})\, \hat{b}^{\dagger}_{\alpha\beta} + \rho_{\beta\alpha\lambda}(\boldsymbol{k})\, \hat{b}_{\alpha\beta}, \tag{2.12}$$

Eq. (2.10) is simply expressed as

$$\hat{H}_{\text{int}} = -i \sum_{\boldsymbol{k}\lambda} N_{\boldsymbol{k}} \sum_{\alpha>F,\beta<F} \left(\hat{\gamma}^{\dagger}_{\alpha\beta\lambda}(\boldsymbol{k}) \hat{a}_{\boldsymbol{k}\lambda} - \hat{\gamma}_{\alpha\beta\lambda}(\boldsymbol{k}) \hat{a}^{\dagger}_{\boldsymbol{k}\lambda} \right). \tag{2.13}$$

After inserting Eq. (2.13) into Eq. (2.1), a unitary transform operator

$$\hat{U} = e^{\hat{S}}, \tag{2.14a}$$

is used for diagonalizing the total Hamiltonian $\hat{H}$. Here, the relation

$$\hat{U}^{\dagger} = \hat{U}^{-1} \tag{2.14b}$$

holds, and $\hat{S}$ is an anti-Hermitian operator, which is defined by

$$\hat{S} = -i \sum_{\boldsymbol{k}\lambda} N_{\boldsymbol{k}} \sum_{\alpha>F,\beta<F} \left(\hat{\gamma}^{\dagger}_{\alpha\beta\lambda}(\boldsymbol{k})\, \hat{a}_{\boldsymbol{k}\lambda} + \hat{\gamma}_{\alpha\beta\lambda}(\boldsymbol{k})\, \hat{a}^{\dagger}_{\boldsymbol{k}\lambda} \right) \tag{2.15}$$

satisfying the relation $\hat{S} = -\hat{S}^{\dagger}$. Applying Eq. (2.14a) to $\hat{H}$ in Eq. (2.1) yields the diagonalized Hamiltonian

$$\tilde{H} = \hat{U}^{-1} \hat{H} \hat{U} = \sum_{\boldsymbol{k}\lambda} \sum_{\alpha>F,\beta<F} \left[\hbar\omega'_{\boldsymbol{k}} \tilde{a}^{\dagger}_{\boldsymbol{k}\lambda} \tilde{a}_{\boldsymbol{k}\lambda} + \left(E'_{\alpha} - E'_{\beta} \right) \tilde{b}^{\dagger}_{\alpha\beta} \tilde{b}_{\alpha\beta} \right]. \tag{2.16}$$

This equation means that new normal vibrations with angular frequencies $\omega'_{\boldsymbol{k}}$ and $(E'_{\alpha} - E'_{\beta})/\hbar$ are created as a result of the coupling of the two harmonic oscillators in

Eq. (2.1), which have the angular frequencies $\omega_{\boldsymbol{k}}$ and $(E_\alpha - E_\beta)/\hbar$. In the first term of Eq. (2.16), $\hbar\omega'_{\boldsymbol{k}}$, $\tilde{a}_{\boldsymbol{k}\lambda}$, and $\tilde{a}^\dagger_{\boldsymbol{k}\lambda}$ represent the eigenenergy, annihilation operator, and creation operator of a new quantum, respectively. In the second term, $E'_\alpha - E'_\beta$, $\tilde{b}_{\alpha\beta}$, and $\tilde{b}^\dagger_{\alpha\beta}$ represent those of another new quantum. The annihilation operator $\tilde{a}_{\boldsymbol{k}\lambda}$ in the first term can be derived by using the formula

$$\tilde{a}_{\boldsymbol{k}\lambda} = \hat{U}^{-1}\hat{a}_{\boldsymbol{k}\lambda}\hat{U} = \hat{a}_{\boldsymbol{k}\lambda} + \left[\hat{a}_{\boldsymbol{k}\lambda}, \hat{S}\right] + \frac{1}{2!}\left[\left[\hat{a}_{\boldsymbol{k}\lambda}, \hat{S}\right], \hat{S}\right] + \cdots \tag{2.17}$$

Since the relation

$$\begin{aligned}\left[\hat{a}_{\boldsymbol{k}\lambda}, \hat{S}\right] &= \left[\hat{a}_{\boldsymbol{k}\lambda}, -i\sum_{\boldsymbol{k}\lambda} N_{\boldsymbol{k}} \sum_{\alpha>F,\beta<F}\left(\hat{\gamma}_{\alpha\beta\lambda}(\boldsymbol{k})\,\hat{a}^\dagger_{\boldsymbol{k}\lambda}\right)\right]\\ &= -iN_{\boldsymbol{k}} \sum_{\alpha>F,\beta<F} \hat{\gamma}_{\alpha\beta\gamma}(\boldsymbol{k})\\ &= -iN_{\boldsymbol{k}} \sum_{\alpha>F,\beta<F}\left(\rho^*{}_{\alpha\beta\lambda}(\boldsymbol{k})\,\hat{b}_{\alpha\beta} + \rho^*{}_{\beta\alpha\lambda}(\boldsymbol{k})\,\hat{b}^\dagger_{\alpha\beta}\right)\end{aligned} \tag{2.18}$$

is derived by using the commutation relation of Eq. (2.2), inserting Eq. (2.18) into Eq. (2.17) and retaining the term of the lowest order of $\hat{S}$ gives

$$\tilde{a}_{\boldsymbol{k}\lambda} = \hat{a}_{\boldsymbol{k}\lambda} - iN_{\boldsymbol{k}} \sum_{\alpha>F,\beta<F}\left(\rho^*{}_{\alpha\beta\lambda}(\boldsymbol{k})\,\hat{b}_{\alpha\beta} + \rho^*{}_{\beta\alpha\lambda}(\boldsymbol{k})\,\hat{b}^\dagger_{\alpha\beta}\right). \tag{2.19}$$

Similarly,

$$\tilde{a}^\dagger_{\boldsymbol{k}\lambda} = \hat{U}^{-1}\hat{a}^\dagger_{\boldsymbol{k}\lambda}\hat{U} = \hat{a}^\dagger_{\boldsymbol{k}\lambda} + \left[\hat{a}^\dagger_{\boldsymbol{k}\lambda}, \hat{S}\right] + \frac{1}{2!}\left[\left[\hat{a}^\dagger{}_{\boldsymbol{k}\lambda}, \hat{S}\right], \hat{S}\right] + \cdots, \tag{2.20}$$

$$\begin{aligned}\left[\hat{a}^\dagger_{\boldsymbol{k}\lambda}, \hat{S}\right] &= \left[\hat{a}^\dagger_{\boldsymbol{k}\lambda}, -i\sum_{\boldsymbol{k}\lambda} N_{\boldsymbol{k}} \sum_{\alpha>F,\beta<F}\left(\hat{\gamma}^\dagger_{\alpha\beta\lambda}(\boldsymbol{k})\,\hat{a}_{\boldsymbol{k}\lambda}\right)\right]\\ &= iN_{\boldsymbol{k}} \sum_{\alpha>F,\beta<F} \hat{\gamma}^\dagger_{\alpha\beta\lambda}(\boldsymbol{k})\\ &= iN_{\boldsymbol{k}} \sum_{\alpha>F,\beta<F}\left(\rho_{\alpha\beta\lambda}(\boldsymbol{k})\,\hat{b}^\dagger_{\alpha\beta} + \rho_{\beta\alpha\lambda}(\boldsymbol{k})\,\hat{b}_{\alpha\beta}\right),\end{aligned} \tag{2.21}$$

giving

$$\tilde{a}^\dagger_{\boldsymbol{k}\lambda} = \hat{a}^\dagger_{\boldsymbol{k}\lambda} + iN_{\boldsymbol{k}} \sum_{\alpha>F,\beta<F}\left(\rho_{\alpha\beta\lambda}(\boldsymbol{k})\,\hat{b}^\dagger_{\alpha\beta} + \rho_{\beta\alpha\lambda}(\boldsymbol{k})\,\hat{b}_{\alpha\beta}\right). \tag{2.22}$$

The right-hand side of Eq. (2.19) indicates that the operator $\hat{b}_{\alpha\beta}$ of the electron–hole pair is added to the photon operator $\hat{a}_{\boldsymbol{k}\lambda}$. This means that the field represented by the operator $\tilde{a}_{\boldsymbol{k}\lambda}$ on the left-hand side is not the free photon but the photon dressing the material energy, i.e., the energy of the electron–hole pair. Finally, annihilation and creation operators of the DP are obtained by summing $\tilde{a}_{\boldsymbol{k}\lambda}$ and $\tilde{a}^{\dagger}_{\boldsymbol{k}\lambda}$ with respect to $\boldsymbol{k}$ and λ, i.e., $\sum_{\boldsymbol{k}\lambda} \tilde{a}_{\boldsymbol{k}\lambda}$ and $\sum_{\boldsymbol{k}\lambda} \tilde{a}^{\dagger}_{\boldsymbol{k}\lambda}$. The quantities $\rho_{\alpha\beta\lambda}(\boldsymbol{k})$, $\rho^*{}_{\alpha\beta\lambda}(\boldsymbol{k})$, $\rho_{\beta\alpha\lambda}(\boldsymbol{k})$, and $\rho^*{}_{\beta\alpha\lambda}(\boldsymbol{k})$ in the second and third terms of Eqs. (2.19) and (2.22) indicate that the DP is generated as a result of the interaction between the photon and the electron–hole pair. Also, the extent of the interaction range is given by the spatial distributions of $\varphi^*{}_{e\alpha}(\boldsymbol{r})\,\varphi_{h\beta}(\boldsymbol{r})$ and $\varphi^*{}_{h\beta}(\boldsymbol{r})\,\varphi_{e\alpha}(\boldsymbol{r})$ in $\rho_{\alpha\beta\lambda}(\boldsymbol{k})$ and $\rho_{\beta\alpha\lambda}(\boldsymbol{k})$ of Eqs. (2.8a), (2.8b), (2.9a), and (2.9b). However, a more detailed estimation of the interaction range is required because $\varphi_{e\alpha}(\boldsymbol{r})$ and $\varphi_{h\beta}(\boldsymbol{r})$ are nothing more than the state functions of the electron and hole, respectively, whose penetration length outside the nanomaterial surface is very short. A detailed discussion of the interaction range will be given in Sect. 2.2, in which the interaction range is shown to be equivalent to the size of the nanomaterial. That is, the effect of the DP spreads throughout a nanometric space whose volume is equivalent to that of the nanomaterial. In the process of this derivation, a physical picture of the virtual photons will be given in relation to the energy conservation law.

By using the DP annihilation and creation operators, the interaction between the two nanomaterials can be represented by the energy transfer, i.e, the annihilation of a DP from the first nanomaterial and its creation on the second nanomaterial, or in other words, emission and absorption of the DP. Here, since the range of interaction mediated by the DP is equivalent to the size of the nanomaterial, emission and absorption of the DP is realized by keeping the separation between the two nanomaterials as short as their sizes. This energy transfer can be regarded as tunneling of the DP. However, if the separation is too short, electrons can tunnel from one nanomaterial to the other, which goes against the goal of realizing pure optical devices (reviewed in Chap. 5) and should be avoided.

Annihilation and creation operators $\tilde{b}_{\alpha\beta}$ and $\tilde{b}^{\dagger}_{\alpha\beta}$ in Eq. (2.16) are also derived in a similar way and are expressed as

$$\tilde{b}_{\alpha\beta} = \hat{U}^{-1}\hat{b}_{\alpha\beta}\hat{U} = \hat{b}_{\alpha\beta} - i\sum_{\boldsymbol{k}\lambda}\left(\rho_{\alpha\beta\lambda}(\boldsymbol{k})\,\hat{a}_{\boldsymbol{k}\lambda} + \rho^*_{\beta\alpha\lambda}(\boldsymbol{k})\,\hat{a}^{\dagger}_{\boldsymbol{k}\lambda}\right), \qquad (2.23a)$$

$$\tilde{b}^{\dagger}_{\alpha\beta} = \hat{U}^{-1}\hat{b}^{\dagger}_{\alpha\beta}\hat{U} = \hat{b}^{\dagger}_{\alpha\beta} - i\sum_{\boldsymbol{k}\lambda}\left(\rho_{\beta\alpha\lambda}(\boldsymbol{k})\,\hat{a}_{\boldsymbol{k}\lambda} + \rho^*_{\alpha\beta\lambda}(\boldsymbol{k})\,\hat{a}^{\dagger}_{\boldsymbol{k}\lambda}\right). \qquad (2.23b)$$

By summing with respect to α and β, the operators $\sum_{\alpha>F,\beta<F} \tilde{b}_{\alpha\beta}$ and $\sum_{\alpha>F,\beta<F} \tilde{b}^{\dagger}_{\alpha\beta}$ represent the electron–hole pair dressing the photon energy. However, since the separation between the two nanomaterials is sufficiently long to avoid electron tunneling, the interaction between the two nanomaterials is described by the absorption and

emission of the DP, and, thus, the operators $\sum_{k\lambda} \tilde{a}_{k\lambda}$ and $\sum_{k\lambda} \tilde{a}^{\dagger}_{k\lambda}$ are used in the following discussions.

Unlike an exciton-polariton in a macroscopic material (refer to Appendix B), since the wavenumber, momentum, and resulting phase are not conserved in the nanometric space, annihilation ($\sum_{k\lambda} \tilde{a}_{k\lambda}$) and creation ($\sum_{k\lambda} \tilde{a}^{\dagger}_{k\lambda}$) of the DP do not occur temporally or spatially in an anti-correlated manner to the annihilation and creation of the electron–hole pair ($\sum_{\alpha>F,\beta<F} \tilde{b}_{\alpha\beta}$ and $\sum_{\alpha>F,\beta<F} \tilde{b}^{\dagger}_{\alpha\beta}$). That is, their wave amplitudes in the classical picture are represented not by simple sinusoidal functions (see the end of Appendix B) but by temporally and spatially modulated function. The angular frequencies ω'_{k} in the first term of Eq. (2.16) are for such a modulation sideband component of the photon. As a dual relation, in the second term, the eigenenergy $E'_{\alpha} - E'_{\beta}$ is also for the modulation sideband of the electron–hole pair.

2.2 Range of Interaction Mediated by Dressed Photons

The end of the previous section described the sidebands of the eigenenergy, generated as a result of the temporal modulation feature of the DP field. As the next step, this section describes the spatial modulation feature. Although, for simplicity, the nanomaterials are assumed to be isolated from the outside in the previous section, actual nanomaterials are surrounded by, for example, a macroscopic substrate on which they are fixed or a macroscopic host crystal in which they are buried. Furthermore, the applied propagating light is also a macroscopic electromagnetic field. In short, since actual nanomaterials are always surrounded by a macroscopic system composed of macroscopic materials and electromagnetic fields, the contribution from the macroscopic system must be considered in analyzing the interaction between the nanomaterials and the resultant energy transfer. This section discusses the electromagnetic interaction between such nanomaterials surrounded by a macroscopic system. It should be noted that it is also difficult to define a virtual cavity for quantization because the nanomaterials are in a nanometric system, and, more complicatedly, they couple with the surrounding macroscopic material. This section describes in detail the spatial modulation feature of DPs by solving these complicated problems [5].

In order to derive the effective interaction energy between the two nanomaterials surrounded by the macroscopic system (see Fig. 2.1), the interaction is renormalized by using the projection operator method (refer to Appendix D). This effective interaction is called a near-field optical interaction [6, 7]. As a result of renormalization, the energy of the near-field optical interaction between the two nanomaterials as a function of their separation r will be given by Eq. (2.76); i.e., it is also expressed by a Yukawa function

$$V_{eff} = \frac{e^{-r/a}}{r}, \tag{2.24}$$

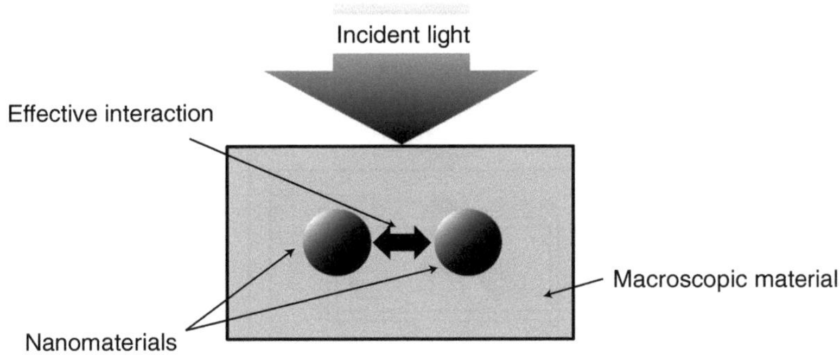

Fig. 2.1 Effective interaction between two nanomaterials surrounded by macroscopic materials and electromagnetic fields

where a represents the interaction range. This range is equivalent to the size of the nanomaterial and does not depend on the wavelength of the applied incident light. This means that the DP is localized on the surface of the nanomaterial, and therefore, the electromagnetic interaction between the nanomaterials is interpreted as originating from the DP energy transfer between them [8–10]. In the following subsections, an equation representing the near-field optical interaction is derived by using an effective interaction operator $\hat{V}_{eff}$ (refer to Eq. (C.25) of Appendix C) based on the projection operator method [6, 11].

2.2.1 *Effective Interaction Between Nanomaterials*

The light–matter interaction system under consideration here is composed of two subsystems, as schematically shown in Fig. 2.1. One is a nanometric subsystem composed of two nanomaterials. The other is a macroscopic material system including the incident light, whose size is much larger than the wavelength of the incident light. They interact with each other as shown in Fig. 2.2.

Here, the nanometric subsystem is called a relevant subsystem n, and the macroscopic subsystem an irrelevant subsystem M. Since the interaction induced in subsystem n is to be analyzed, it is essential to renormalize the effects originating from subsystem M in a consistent and systematic way. The following description reviews a formulation based on the projection operator method.

(a) Bare interaction and effective interaction

The operator $\hat{V}$ of the interaction between nanomaterials s and p surrounded by subsystem M is expressed in the multipolar formalism under the electric dipole approximation as

$$\hat{V} = -\frac{1}{\varepsilon_0}\left\{\hat{\boldsymbol{p}}_s \cdot \hat{\boldsymbol{D}}^{\perp}(\boldsymbol{r}_s) + \hat{\boldsymbol{p}}_p \cdot \hat{\boldsymbol{D}}^{\perp}\left(\boldsymbol{r}_p\right)\right\}, \tag{2.25}$$

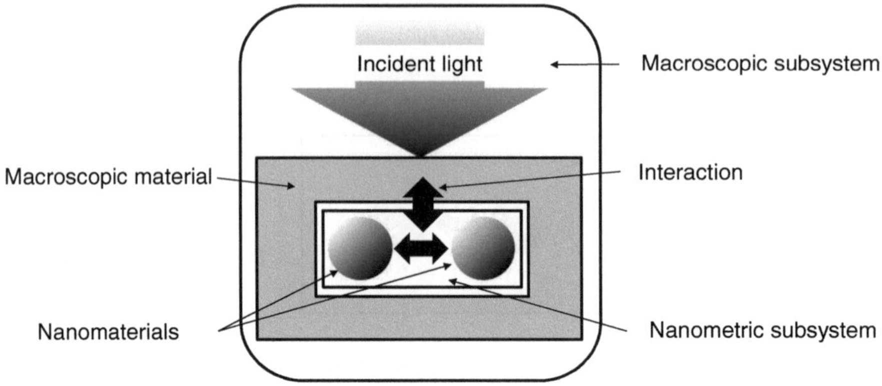

Fig. 2.2 Nanometric and macroscopic subsystems

(refer to Eq. (A.23) in Appendix A). Here, $\hat{\boldsymbol{p}}_\alpha$ $(\alpha = s, p)$ is an electric dipole operator. The positions of the nanomaterials s and p are denoted by $\boldsymbol{r}_s$ and $\boldsymbol{r}_p$, respectively. The transverse component of the electric displacement operator of the incident light is expressed as $\hat{\boldsymbol{D}}^{\perp}(\hat{\boldsymbol{r}})$, which has been given in Eq. (2.6).

Exciton–polariton states are employed as a basis to describe subsystem M (refer to Appendix D) because the incident light reaches and excites the nanomaterials s and p after propagating through the macroscopic material. For this purpose, after the annihilation and creation operators in Eq. (2.6) are rewritten using the exciton–polariton operators $\hat{\xi}(\boldsymbol{k})$ and $\hat{\xi}^{\dagger}(\boldsymbol{k})$, they are inserted into Eq. (2.25). Using the electric dipole operator defined by

$$\hat{\boldsymbol{p}}_\alpha = \left\{\hat{b}(\boldsymbol{r}_\alpha) + \hat{b}^{\dagger}(\boldsymbol{r}_\alpha)\right\} \boldsymbol{p}_\alpha, \tag{2.26}$$

with the annihilation and creation operators $\hat{b}(\boldsymbol{r}_\alpha)$ and $\hat{b}^{\dagger}(\boldsymbol{r}_\alpha)$ of the electron–hole pair (exciton) in subsystem n and the transition dipole moments $\boldsymbol{p}_\alpha$ yields the bare interaction operator in the exciton–polariton picture:

$$\hat{V} = -i \sum_{\alpha=s}^{p} \sum_{\boldsymbol{k}} \sqrt{\frac{\hbar}{2\varepsilon_0 V}} \left(\hat{b}(\boldsymbol{r}_\alpha) + \hat{b}^{\dagger}(\boldsymbol{r}_\alpha)\right) \left(K_\alpha(\boldsymbol{k})\,\hat{\xi}(\boldsymbol{k}) - K_\alpha^{*}(\boldsymbol{k})\,\hat{\xi}^{\dagger}(\boldsymbol{k})\right). \tag{2.27}$$

Here, ε_0 is the dielectric constant in vacuum, and V is the volume of the virtual cavity in subsystem M for describing the exciton-polariton. $K_\alpha(\boldsymbol{k})$ and $K_\alpha^{*}(\boldsymbol{k})$ are the coupling coefficients between subsystem n and the exciton-polariton in subsystem M, given by (refer to Eq. (D.18) in Appendix D)

$$K_\alpha(\boldsymbol{k}) = \sum_{\lambda=1}^{2} \left(\boldsymbol{p}_\alpha \cdot \boldsymbol{e}_\lambda(\boldsymbol{k})\right) f(k)\, e^{i\boldsymbol{k}\cdot\boldsymbol{r}_\alpha}, \tag{2.28}$$

where $\boldsymbol{e}_\lambda(\boldsymbol{k})$ is a unit vector along the direction of polarization of light. The wavenumber dependence of $f(k)$ will be rewritten as Eq. (2.70) (refer to Eq. (D.19) in Appendix D).

In order to derive the effective interaction energy, it is preferable to express $|\psi\rangle$ for the total system in terms of a basis whose degree of freedom is as low as possible. For this purpose, the total system is divided into two functional spaces. One is called the relevant P space. The other is the complementary space to the P space, which is called Q space. The next step is to evaluate the energy of the effective interaction exerted in the P space after tracing out the exciton-polariton degree of freedom:

$$V_{eff} = \langle\phi_{Pf}|\,\hat{V}_{eff}\,|\phi_{Pi}\rangle\,. \tag{2.29}$$

The two states $|\phi_{Pi}\rangle$ and $\langle\phi_{Pf}|$ on the right-hand side of this equation are the initial and final states in the P space before and after the interaction, respectively, both of which are eigenstates of the unperturbed Hamiltonian. The effective interaction operator $\hat{V}_{eff}$ on the right-hand side of this equation is expressed by using the bare interaction operator $\hat{V}$ of Eq. (2.27):

$$\hat{V}_{eff} = \sum_j \hat{P}\hat{V}\hat{Q}\,|\phi_{Qj}\rangle\langle\phi_{Qj}|\,\hat{Q}\hat{V}\hat{P}\left(\frac{1}{E^0_{Pi}-E^0_{Qj}}+\frac{1}{E^0_{Pf}-E^0_{Qj}}\right), \tag{2.30}$$

(refer to Eq. (C.44) in Appendix C). Using this expression, Eq. (2.29) is given by

$$V_{eff} = \sum_j \langle\phi_{Pf}|\,\hat{P}\hat{V}\hat{Q}\,|\phi_{Qj}\rangle\langle\phi_{Qj}|\,\hat{Q}\hat{V}\hat{P}\,|\phi_{Pi}\rangle\left(\frac{1}{E^0_{Pi}-E^0_{Qj}}+\frac{1}{E^0_{Pf}-E^0_{Qj}}\right). \tag{2.31}$$

Here, $\hat{P}$ is the projection operator, which is defined by using the basis $\{|\phi_{Pj}\rangle\}$ in the P space as

$$\hat{P} = \sum_j |\phi_{Pj}\rangle\langle\phi_{Pj}|, \tag{2.32}$$

(refer to Eq. (C.3) in Appendix C).
This projection operator transforms the arbitrary state $|\psi\rangle$ into the P space spanned by the basis $\{|\phi_{Pj}\rangle\}$. The complimentary operator $\hat{Q}$ is defined by using the basis $\{|\phi_{Qj}\rangle\}$ in the Q space as

$$\hat{Q} = \sum_j |\phi_{Qj}\rangle\langle\phi_{Qj}|. \tag{2.33}$$

The eigenenergies of the initial and final states in the P space are denoted by E^0_{Pi} and E^0_{Pf}, respectively. The eigenenergy of the intermediate state in the Q space is E^0_{Qj}. On the right-hand side of Eq. (2.30), the bare interaction operator $\hat{V}$ is placed between the projection operators $\hat{P}$ and $\hat{Q}$, which represents the screening effect by the P and Q spaces. Furthermore, the right-hand side of Eq. (2.31) represents the virtual transition from the initial state $|\phi_{Pi}\rangle$ in the P space to the intermediate state $|\phi_{Qj}\rangle$ in the Q space, and the successive virtual transition from this intermediate state $|\phi_{Qj}\rangle$ to the final state $|\phi_{Pf}\rangle$ in the P space.

(b) Magnitude of effective interaction energy

As the initial state $|\phi_{Pi}\rangle$ in the P space, it is assumed that the electron–hole pairs (excitons) in the nanomaterials s and p in subsystem n are in their excited and ground states $|s_{ex}\rangle$ and $|p_g\rangle$, respectively. In addition, the exciton-polariton states, which are used as bases to describe subsystem M, are in their vacuum state $|0_{(M)}\rangle$. Therefore, the initial state in the P space is expressed as

$$|\phi_{Pi}\rangle = |s_{ex}\rangle |p_g\rangle \otimes |0_{(M)}\rangle, \tag{2.34}$$

where the symbol $\otimes$ represents the direct product.[3] By the energy transfer from the nanomaterial s to nanomaterial p as a result of the interaction, the electron–hole pairs (excitons) in the nanomaterials s and p transition to the ground and excited states $|s_g\rangle$ and $|p_{ex}\rangle$, respectively. The exciton–polariton in subsystem M is also in the vacuum state $|0_{(M)}\rangle$. Therefore, the final state $|\phi_{Pf}\rangle$ in the P space is expressed as

$$|\phi_{Pf}\rangle = |s_g\rangle |p_{ex}\rangle \otimes |0_{(M)}\rangle. \tag{2.35}$$

The basis $\{|\phi_{Pj}\rangle\}$ in the P space is spanned by the two states given by Eqs. (2.34) and (2.35). Other states cannot be employed for the basis because they violate the energy conservation law in subsystem n in the process of interaction.

On the other hand, as shown in Fig. 2.3, the complimentary Q space is composed of many states, including states that violate the energy conservation law in subsystem n, and therefore, they are not employed for the basis of the P space. That is, the basis $\{|\phi_{Qj}\rangle\}$ of the Q space is spanned by

$$|\phi_{Q1n}\rangle = |s_g\rangle |p_g\rangle \otimes |n_{(M)}\rangle, \tag{2.36a}$$

and

$$|\phi_{Q2n}\rangle = |s_{ex}\rangle |p_{ex}\rangle \otimes |n_{(M)}\rangle. \tag{2.36b}$$

Here, $|n_{(M)}\rangle$ represents the state in which $n_{(M)}$ quanta of exciton-polaritons exist in subsystem M. However, as will be described immediately after presenting Eq. (2.43), since only the state $|1_{(M)}\rangle$ gives a nonzero value of the effective interaction energy,

[3] It is convenient to define the direct product by using matrices because a matrix is commonly used in quantum mechanics to represent an operator and a state. For example, in the case of the 2×2 matrices $\boldsymbol{A} = \begin{pmatrix} a_{11} & a_{12} \\ a_{21} & a_{22} \end{pmatrix}$ and $\boldsymbol{B} = \begin{pmatrix} b_{11} & b_{12} \\ b_{21} & b_{22} \end{pmatrix}$, their direct product $\boldsymbol{C} = \boldsymbol{A} \otimes \boldsymbol{B}$ is defined by $\boldsymbol{C} = \begin{pmatrix} a_{11}\boldsymbol{B} & a_{12}\boldsymbol{B} \\ a_{21}\boldsymbol{B} & a_{22}\boldsymbol{B} \end{pmatrix}$, where $a_{ij}\boldsymbol{B} = \begin{pmatrix} a_{ij}b_{11} & a_{ij}b_{12} \\ a_{ij}b_{21} & a_{ij}b_{22} \end{pmatrix}$, $(i, j) = 1, 2$.

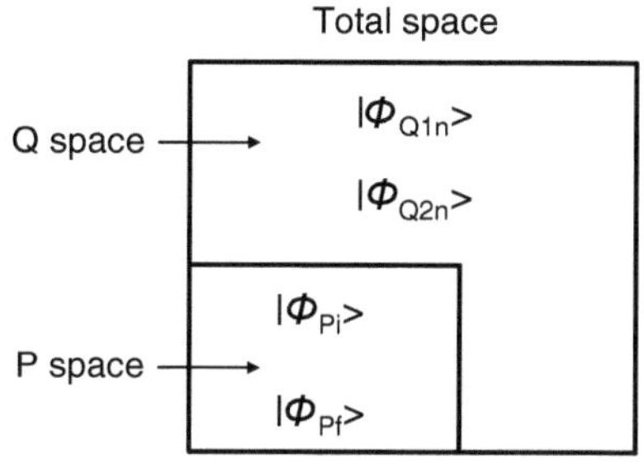

Fig. 2.3 P space and Q space

the states with $|1_{(M)}\rangle$ are extracted from Eqs. (2.36a) and (2.36b) and expressed as $|\phi_{Q11}\rangle$ and $|\phi_{Q21}\rangle$. Furthermore, for consistency of expression with Eq. (2.30), $|\phi_{Q11}\rangle$ and $|\phi_{Q21}\rangle$ are written as

$$|\phi_{Q1}\rangle = |s_g\rangle |p_g\rangle \otimes |1_{(M)}\rangle, \tag{2.37a}$$

and

$$|\phi_{Q2}\rangle = |s_{ex}\rangle |p_{ex}\rangle \otimes |1_{(M)}\rangle. \tag{2.37b}$$

By noting that the relations

$$\hat{P} |\phi_{Pj}\rangle = |\phi_{Pj}\rangle \tag{2.38}$$

and

$$\hat{Q} |\phi_{Qjn}\rangle = |\phi_{Qjn}\rangle \qquad (j = 1, 2) \tag{2.39}$$

hold due to Eqs. (2.32) and (2.33), one can derive

$$\langle\phi_{Qjn}| \hat{Q}\hat{V}\hat{P} |\phi_{Pi}\rangle = \langle\phi_{Qjn}| \hat{V} |\phi_{Pi}\rangle \tag{2.40}$$

and

$$\langle\phi_{Pf}| \hat{P}\hat{V}\hat{Q} |\phi_{Qjn}\rangle = \langle\phi_{Pf}| \hat{V} |\phi_{Qjn}\rangle. \tag{2.41}$$

When the bare interaction operator $\hat{V}$ of Eq. (2.27) is inserted into these equations, the annihilation and creation operators $\hat{b}(\boldsymbol{r}_\alpha)$ and $\hat{b}^\dagger(\boldsymbol{r}_\alpha)$ of the electron–hole pair (exciton) apply only to $|n_{(M)}\rangle$ ($n_{(M)} = 0, 1, 2, \ldots$), and those of the exciton-polariton, $\hat{\xi}(\boldsymbol{k})$ and $\hat{\xi}^\dagger(\boldsymbol{k})$ apply only to $|s_g\rangle$, $|s_{ex}\rangle$, $|p_g\rangle$, and $|p_{ex}\rangle$. From Eqs. (2.34)–(2.37b) one can derive

$$\langle\phi_{Q1n}| \hat{V} |\phi_{Pi}\rangle$$
$$= -i \sum_{\alpha=s}^{p} \sum_{\boldsymbol{k}} \sqrt{\frac{\hbar}{2\varepsilon_0 V}} \langle s_g| \langle p_g| \otimes \langle n_{(M)}| \left(\hat{b}(\boldsymbol{r}_\alpha) + \hat{b}^\dagger(\boldsymbol{r}_\alpha)\right)$$

$$\times \left(K_\alpha(\boldsymbol{k})\,\hat{\xi}(\boldsymbol{k}) - K_\alpha^*(\boldsymbol{k})\,\hat{\xi}^\dagger(\boldsymbol{k}) \right) |s_{ex}\rangle\, |p_g\rangle \otimes |0_{(M)}\rangle$$

$$= -i \sum_{\boldsymbol{k}} \sqrt{\frac{\hbar}{2\varepsilon_0 V}}\, K_s(\boldsymbol{k}) \tag{2.42}$$

and

$$\langle \phi_{Pf}|\, \hat{V}\, |\phi_{Q1n}\rangle$$

$$= -i \sum_{\alpha=s}^{p} \sum_{\boldsymbol{k}} \sqrt{\frac{\hbar}{2\varepsilon_0 V}}\, \langle s_g|\, \langle p_{ex}| \otimes \langle 0_{(M)}| \left(\hat{b}(\boldsymbol{r}_\alpha) + \hat{b}^\dagger(\boldsymbol{r}_\alpha) \right)$$

$$\times \left(K_\alpha(\boldsymbol{k})\,\hat{\xi}(\boldsymbol{k}) - K_\alpha^*(\boldsymbol{k})\,\hat{\xi}^\dagger(\boldsymbol{k}) \right) |s_g\rangle\, |p_g\rangle \otimes |n_{(M)}\rangle$$

$$= i \sum_{\boldsymbol{k}} \sqrt{\frac{\hbar}{2\varepsilon_0 V}}\, K_p^*(\boldsymbol{k})\,. \tag{2.43}$$

Here, only the term $n_{(M)} = 1$ gives a nonzero value in the second rows of Eqs. (2.42) and (2.43) because $\langle 0_{(M)}| \left(\hat{b}(\boldsymbol{r}_\alpha) + \hat{b}^\dagger(\boldsymbol{r}_\alpha) \right) |n_{(M)}\rangle = 0\ (n_{(M)} \neq 1)$, which leads to the third rows. Therefore, the $(j = 1)$-th term on the right-hand side of Eq. (2.31) is expressed as

$$\langle \phi_{Pf}|\, \hat{P}\hat{V}\hat{Q}\, |\phi_{Q1}\rangle \langle \phi_{Q1}|\, \hat{Q}\hat{V}\hat{P}\, |\phi_{Pi}\rangle \left(\frac{1}{E_{Pi}^0 - E_{Q1}^0} + \frac{1}{E_{Pf}^0 - E_{Q1}^0} \right)$$

$$= \sum_{\boldsymbol{k}} \frac{\hbar}{2\varepsilon_0 V} K_s(\boldsymbol{k})\, K_p^*(\boldsymbol{k}) \left(\frac{1}{E_{Pi}^0 - E_{Q1}^0} + \frac{1}{E_{Pf}^0 - E_{Q1}^0} \right). \tag{2.44}$$

In order to similarly express the $(j = 2)$-th term on the right-hand side of Eq. (2.31), the relations

$$\langle \phi_{Q2n}|\, \hat{V}\, |\phi_{Pi}\rangle$$

$$= -i \sum_{\alpha=s}^{p} \sum_{\boldsymbol{k}} \sqrt{\frac{\hbar}{2\varepsilon_0 V}}\, \langle s_{ex}|\, \langle p_{ex}| \otimes \langle n_{(M)}| \left(\hat{b}(\boldsymbol{r}_\alpha) + \hat{b}^\dagger(\boldsymbol{r}_\alpha) \right)$$

$$\times \left(K_\alpha(\boldsymbol{k})\,\hat{\xi}(\boldsymbol{k}) - K^*{}_\alpha(\boldsymbol{k})\,\hat{\xi}^\dagger(\boldsymbol{k}) \right) |s_{ex}\rangle\, |p_g\rangle \otimes |0_{(M)}\rangle$$

$$= i \sum_{\boldsymbol{k}} \sqrt{\frac{\hbar}{2\varepsilon_0 V}}\, K_p^*(\boldsymbol{k}) \tag{2.45}$$

and

$$\begin{aligned}&\left\langle \phi_{Pf}\right| \hat{V} \left|\phi_{Q2n}\right\rangle \\ &= -i\sum_{\alpha=s}^{p}\sum_{k}\sqrt{\frac{\hbar}{2\varepsilon_0 V}}\left\langle s_g\right| \left\langle p_{ex}\right| \otimes \left\langle 0_{(M)}\right| \left(\hat{b}\left(\boldsymbol{r}_\alpha\right)+\hat{b}^{\dagger}\left(\boldsymbol{r}_\alpha\right)\right) \\ &\left(K_\alpha\left(\boldsymbol{k}\right)\hat{\xi}\left(\boldsymbol{k}\right)-K^{*}{}_\alpha\left(\boldsymbol{k}\right)\hat{\xi}^{\dagger}\left(\boldsymbol{k}\right)\right)\left|s_{ex}\right\rangle\left|p_{ex}\right\rangle \otimes \left|n_{(M)}\right\rangle \\ &= -i\sum_{k}\sqrt{\frac{\hbar}{2\varepsilon_0 V}}K_s\left(\boldsymbol{k}\right)\end{aligned} \tag{2.46}$$

are used, which are similar to Eqs. (2.42) and (2.43), respectively. Then, the $(j = 2)$-th term is expressed as

$$\begin{aligned}&\left\langle \phi_{Pf}\right| \hat{P}\hat{V}\hat{Q}\left|\phi_{Q2}\right\rangle\left\langle \phi_{Q2}\right| \hat{Q}\hat{V}\hat{P}\left|\phi_{Pi}\right\rangle\left(\frac{1}{E^0_{Pi}-E^0_{Q2}}+\frac{1}{E^0_{Pf}-E^0_{Q2}}\right) \\ &= \sum_{\boldsymbol{k}}\frac{\hbar}{2\varepsilon_0 V}K_s\left(\boldsymbol{k}\right)K_p^{*}\left(\boldsymbol{k}\right)\left(\frac{1}{E^0_{Pi}-E^0_{Q2}}+\frac{1}{E^0_{Pf}-E^0_{Q2}}\right).\end{aligned} \tag{2.47}$$

By summing Eqs. (2.44) and (2.47), Eq. (2.31) is rewritten as

$$\begin{aligned}V_{eff}\left(s\rightarrow p\right) = \sum_{\boldsymbol{k}}\frac{\hbar}{2\varepsilon_0 V}K_s\left(\boldsymbol{k}\right)K_p^{*}\left(\boldsymbol{k}\right)&\left(\frac{1}{E^0_{Pi}-E^0_{Q1}}+\frac{1}{E^0_{Pf}-E^0_{Q1}}\right. \\ &\left.+\frac{1}{E^0_{Pi}-E^0_{Q2}}+\frac{1}{E^0_{Pf}-E^0_{Q2}}\right).\end{aligned} \tag{2.48}$$

Here V_{eff} on the left-hand side of Eq. (2.31) was rewritten as $V_{eff}\left(s\rightarrow p\right)$ in order to represent that energy is transferred from the nanomaterial s to the nanomaterial p. By replacing the sum for the wave-vector $\boldsymbol{k}$ by the integral $\frac{V}{(2\pi)^3}\int_0^\infty d\boldsymbol{k}$, the symbol V is eliminated from this equation, yielding

$$\begin{aligned}V_{eff}\left(s\rightarrow p\right) = \frac{\hbar^2}{(2\pi)^3\varepsilon_0}\int_0^\infty d\boldsymbol{k}K_s\left(\boldsymbol{k}\right)K_p^{*}\left(\boldsymbol{k}\right)&\left(\frac{1}{E^0_{Pi}-E^0_{Q1}}+\frac{1}{E^0_{Pf}-E^0_{Q1}}\right. \\ &\left.+\frac{1}{E^0_{Pi}-E^0_{Q2}}+\frac{1}{E^0_{Pf}-E^0_{Q2}}\right).\end{aligned} \tag{2.49}$$

By denoting the eigenenergies of the states $\left|s_g\right\rangle$, $\left|s_{ex}\right\rangle$, $\left|p_g\right\rangle$, and $\left|p_{ex}\right\rangle$ in the nanomaterials s and p by $E_{s,g}$, $E_{s,ex}$, $E_{p,g}$, and $E_{p,ex}$, and by denoting the eigenenergy of the state $\left|1_{(M)}\right\rangle$ in the exciton-polariton by $E(k)$, one obtains

$$E^0_{Pi} - E^0_{Q1} = \left(E_{s.ex} + E_{p,g}\right) - \left(E_{s,g} + E_{p,g} + E(k)\right)$$
$$= \left(E_{s.ex} - E_{s,g}\right) - E(k) = -\left(E(k) - E_s\right), \tag{2.50a}$$
$$E^0_{Pi} - E^0_{Q2} = \left(E_{s.ex} + E_{p,g}\right) - \left(E_{s,ex} + E_{p,ex} + E(k)\right)$$
$$= -\left(E_{p.ex} - E_{p,g}\right) - E(k) = -\left(E(k) + E_p\right), \tag{2.50b}$$
$$E^0_{Pf} - E^0_{Q1} = \left(E_{s.g} + E_{p,ex}\right) - \left(E_{s,g} + E_{p,g} + E(k)\right)$$
$$= \left(E_{p.ex} - E_{p,g}\right) - E(k) = -\left(E(k) - E_p\right), \tag{2.50c}$$
$$E^0_{Pf} - E^0_{Q2} = \left(E_{s.g} + E_{p,ex}\right) - \left(E_{s,ex} + E_{p,ex} + E(k)\right)$$
$$= -\left(E_{s.ex} - E_{s,g}\right) - E(k) = -\left(E(k) + E_s\right). \tag{2.50d}$$

Here, the difference $E_{\alpha,ex} - E_{\alpha,g}$ between the eigenenergies of the excited state ($E_{\alpha,ex}$) and ground state ($E_{\alpha,g}$) was replaced with the transition energy E_α ($\alpha = s, p$). Inserting these into Eq. (2.49) yields

$$V_{eff}(s \to p) = -\frac{\hbar^2}{(2\pi)^3 \varepsilon_0} \int_0^\infty d\boldsymbol{k} K_s(\boldsymbol{k}) K_p^*(\boldsymbol{k}) \left(\frac{1}{E(k) - E_s} + \frac{1}{E(k) - E_p} + \frac{1}{E(k) + E_p} + \frac{1}{E(k) + E_s} \right) \tag{2.51}$$

By exchanging the subscripts of the nanomaterials s and p in order to assume the initial and final states

$$|\phi_{Pi}\rangle = |s_g\rangle |p_{ex}\rangle \otimes |0_{(M)}\rangle, \tag{2.52}$$

and

$$|\phi_{Pf}\rangle = |s_{ex}\rangle |p_g\rangle \otimes |0_{(M)}\rangle \tag{2.53}$$

the energy $V_{eff}(p \to s)$ transferred from the nanomaterial p to the nanomaterial s can be derived in the same manner as above, and is given by

$$V_{eff}(p \to s) = -\frac{\hbar^2}{(2\pi)^3 \varepsilon_0} \int_0^\infty d\boldsymbol{k} K_p(\boldsymbol{k}) K_s^*(\boldsymbol{k}) \left(\frac{1}{E(k) - E_p} + \frac{1}{E(k) - E_s} + \frac{1}{E(k) + E_s} + \frac{1}{E(k) + E_p} \right). \tag{2.54}$$

Furthermore, inserting Eq. (2.28) into Eqs. (2.51) and (2.54) and summing them gives

$$V_{eff}(\boldsymbol{r}) = -\frac{\hbar^2}{(2\pi)^3} \varepsilon_0 \sum_{\lambda=1}^{2} \int_0^\infty f^2(k)\, d\boldsymbol{k} \left(\boldsymbol{p}_s \cdot \boldsymbol{e}_\lambda(\boldsymbol{k})\right) e^{i\boldsymbol{k}\cdot(\boldsymbol{r}_s - \boldsymbol{r}_p)} \left(\boldsymbol{p}_p \cdot \boldsymbol{e}_\lambda(\boldsymbol{k})\right)$$
$$\times \left(\frac{1}{E(k) - E_s} + \frac{1}{E(k) - E_p} + \frac{1}{E(k) + E_p} + \frac{1}{E(k) + E_s} \right)$$

$$= -\frac{\hbar^2}{(2\pi)^3 \varepsilon_0} \sum_{\lambda=1}^{2} \sum_{\alpha=s}^{p} \int_0^{\infty} \left(\boldsymbol{p}_s \cdot \boldsymbol{e}_\lambda(\boldsymbol{k})\right) \left(\boldsymbol{p}_p \cdot \boldsymbol{e}_\lambda(\boldsymbol{k})\right) f^2(k)$$

$$\times \left(\frac{1}{E(k) + E_\alpha} + \frac{1}{E(k) - E_\alpha}\right) e^{i\boldsymbol{k}\cdot\boldsymbol{r}} d\boldsymbol{k}, \tag{2.55}$$

where $\boldsymbol{r} = \boldsymbol{r}_s - \boldsymbol{r}_p$.

(c) Summation and integration for deriving a Yukawa function

(i) Summation with respect to the polarization states

Summation with respect to polarization state λ in Eq. (2.55) gives

$$\sum_{\lambda=1}^{2} \left(\boldsymbol{p}_p \cdot \boldsymbol{e}_\lambda(\boldsymbol{k})\right) \left(\boldsymbol{p}_s \cdot \boldsymbol{e}_\lambda(\boldsymbol{k})\right) = \sum_{\lambda=1}^{2} \sum_{i,j=1}^{3} \left(p_{pi} e_{\lambda i}(\boldsymbol{k})\right) \left(p_{sj} e_{\lambda j}(\boldsymbol{k})\right). \tag{2.56}$$

Here, by denoting the j-th component of the unit vector $\boldsymbol{u}_k = \boldsymbol{k}/k$ by u_{kj},

$$\sum_{\lambda=1}^{3} e_{\lambda i}(\boldsymbol{k}) e_{\lambda j}(\boldsymbol{k}) = \sum_{\lambda=1}^{2} e_{\lambda i}(\boldsymbol{k}) e_{\lambda j}(\boldsymbol{k}) + e_{3i}(\boldsymbol{k})\, e_{3j}(\boldsymbol{k})$$

$$= \sum_{\lambda=1}^{2} e_{\lambda i}(\boldsymbol{k}) e_{\lambda j}(\boldsymbol{k}) + u_{ki} u_{kj} = \delta_{ij} \tag{2.57}$$

holds. Transferring the term $u_{ki} u_{kj}$ to the right-hand side of this equation gives

$$\sum_{\lambda=1}^{2} e_{\lambda i}(\boldsymbol{k}) e_{\lambda j}(\boldsymbol{k}) = \delta_{ij} - u_{ki} u_{kj}, \tag{2.58}$$

and therefore, inserting this equation into Eq. (2.56) gives

$$\sum_{\lambda=1}^{2} \left(\boldsymbol{p}_p \cdot \boldsymbol{e}_\lambda(\boldsymbol{k})\right) \left(\boldsymbol{p}_s \cdot \boldsymbol{e}_\lambda(\boldsymbol{k})\right) = \sum_{i,j=1}^{3} p_{pi} p_{sj} \left(\delta_{ij} - u_{ki} u_{kj}\right). \tag{2.59}$$

Equation (2.59) is inserted into Eq. (2.55), and then the integral with respect to the azimuth angles ϑ and ϕ of $\boldsymbol{k}$ is taken, where

$$d\boldsymbol{k} = k^2 dk d\Omega = k^2 dk \sin\vartheta d\vartheta d\phi \tag{2.60}$$

holds. As the first step of this integration, the integral $\int u_{ki} u_{kj} e^{i\boldsymbol{k}\cdot\boldsymbol{r}} d\Omega$ is taken: expressing the j-th component of $\boldsymbol{k}$ as k_j, gives $\nabla_i \nabla_j e^{i\boldsymbol{k}\cdot\boldsymbol{r}} = -k_i k_j e^{i\boldsymbol{k}\cdot\boldsymbol{r}}$ because $\nabla_j e^{i\boldsymbol{k}\cdot\boldsymbol{r}} = i k_j e^{i\boldsymbol{k}\cdot\boldsymbol{r}}$, and thus, $\nabla_i \nabla_j e^{i\boldsymbol{k}\cdot\boldsymbol{r}}$

$= -k^2 u_{ki} u_{kj} e^{i\boldsymbol{k}\cdot\boldsymbol{r}}$ is obtained because $k_j = k u_{kj}$. It follows that

$$\int u_{ki} u_{kj} e^{i\boldsymbol{k}\cdot\boldsymbol{r}} d\Omega = -\frac{1}{k^2} \nabla_i \nabla_j \int e^{i\boldsymbol{k}\cdot\boldsymbol{r}} d\Omega. \tag{2.61a}$$

Furthermore, by noting that

$$\int e^{i\boldsymbol{k}\cdot\boldsymbol{r}} d\Omega = \int_0^{2\pi} \int_{-1}^{1} e^{ikr\cos\vartheta} d(\cos\vartheta)\, d\phi = \frac{2\pi}{ikr}\left(e^{ikr} - e^{-ikr}\right), \tag{2.61b}$$

the right-hand side of Eq. (2.61a) is transformed to

$$-\frac{1}{k^2} \nabla_i \nabla_j \int e^{i\boldsymbol{k}\cdot\boldsymbol{r}} d\Omega = -\frac{2\pi}{ik^3} \nabla_i \nabla_j \left(\frac{e^{ikr} - e^{-ikr}}{r}\right). \tag{2.62}$$

As the next step, by using this equation and noting that Eq. (2.61b) holds, the integral in Eq. (2.55) is taken over the azimuth angles to derive

$$\begin{aligned}
&\int \left(\delta_{ij} - u_{ki} u_{kj}\right) e^{i\boldsymbol{k}\cdot\boldsymbol{r}} d\Omega \\
&= \delta_{ij} \frac{2\pi}{ik}\left(\frac{e^{ikr} - e^{-ikr}}{r}\right) + \frac{2\pi}{ik^3} \nabla_i \nabla_j \left(\frac{e^{ikr} - e^{-ikr}}{r}\right) \\
&= 2\pi[\delta_{ij} \frac{\left(e^{ikr} - e^{-ikr}\right)}{ikr} + \left(\delta_{ij} - 3u_{ri} u_{rj}\right) \\
&\quad \times \left\{\frac{\left(e^{ikr} + e^{-ikr}\right)}{k^2 r^2} - \frac{\left(e^{ikr} - e^{-ikr}\right)}{ik^3 r^3}\right\} - \frac{\left(e^{ikr} - e^{-ikr}\right)}{ikr} u_{ri} u_{rj}],
\end{aligned} \tag{2.63}$$

where u_{ri} is the i-th component of the unit vector $\boldsymbol{u_r} = \boldsymbol{r}/r$. As a result of the above summation and integration, the effective interaction energy is expressed as[4]

$$\begin{aligned}
V_{\mathit{eff}}(r) = &-\frac{\hbar^2}{(2\pi)^2 \varepsilon_0} \int_{-\infty}^{\infty} k^2 dk\, f^2(k) \sum_{\alpha=s}^{p} \left(\frac{1}{E(k) + E_\alpha} + \frac{1}{E(k) - E_\alpha}\right) \\
&\times \left\{ \left(p_s \cdot p_p\right) e^{ik\cdot r} \left(\frac{1}{ikr} + \frac{1}{k^2 r^2} - \frac{1}{ik^3 r^3}\right)\right.
\end{aligned}$$

[4] The range of the integration with respect to the absolute value ($k(= |\boldsymbol{k}|)$) is set as $(-\infty, \infty)$ in this equation. This is because the range of integration of $-e^{-ikr}/r$ is changed to $(-\infty, 0)$ by changing k to $-k$ when the right-hand side of Eq. (2.63) is integrated over $(0, \infty)$. As a result of this change, $e^{-i\boldsymbol{k}\cdot\boldsymbol{r}}$ in Eq. (2.63) is eliminated, and, thus, Eq. (2.64) contains $e^{i\boldsymbol{k}\cdot\boldsymbol{r}}$ only.

$$- \left(\boldsymbol{p}_s \cdot \boldsymbol{u}_r\right) \left(\boldsymbol{p}_p \cdot \boldsymbol{u}_r\right) e^{i\boldsymbol{k}\cdot\boldsymbol{r}} \left(\frac{1}{ikr} + \frac{3}{k^2r^2} - \frac{3}{ik^3r^3}\right) \Bigg\} \tag{2.64}$$

(ii) Averaging over the azimuth angle of the electric dipole moment
In order to average $\left(\boldsymbol{p}_s \cdot \boldsymbol{u}_r\right) \left(\boldsymbol{p}_p \cdot \boldsymbol{u}_r\right)$ on the right-hand side of Eq. (2.64) over the azimuth angles θ and φ of $\boldsymbol{r}$, for simplicity and without loss of generality, the electric dipoles $\boldsymbol{p}_s$ and $\boldsymbol{p}_p$ are assumed to be parallel to each other. Under this assumption, one obtains

$$\left\langle \left(\boldsymbol{p}_s \cdot \boldsymbol{u}_r\right) \left(\boldsymbol{p}_p \cdot \boldsymbol{u}_r\right) \right\rangle_{\theta,\varphi} = \frac{p_s p_p}{4\pi} \int_0^{2\pi} d\varphi \int_0^{\pi} \cos^2\theta \sin\theta d\theta = \frac{p_s p_p}{3}, \tag{2.65}$$

which means that the average is one-third of $\left(\boldsymbol{p}_s \cdot \boldsymbol{p}_p\right)$. By inserting this into Eq. (2.64), only the term $1/r$ is retained on the right-hand side, whereas $1/r^2$ and $1/r^3$ in the first and second terms cancel each other out. As a result, the effective interaction energy is expressed as

$$\begin{aligned} V_{eff}(\boldsymbol{r}) = & -\frac{2\hbar^2 p_s p_p}{3(2\pi)^2\varepsilon_0} \int_{-\infty}^{\infty} k^2 dk f^2(k) \\ & \times \sum_{\alpha=s}^{p} \left(\frac{1}{E(k) + E_\alpha} + \frac{1}{E(k) - E_\alpha}\right) \frac{e^{i\boldsymbol{k}\cdot\boldsymbol{r}}}{ikr}. \end{aligned} \tag{2.66}$$

(iii) Integration over the wavenumber
The eigenenergy E_α of the nanomaterial $\alpha\,(= s, p)$ is expressed as $E_\alpha = p_\alpha^2/2m_\alpha$, where p_α and m_α are the momentum and effective mass of the exciton, respectively. Inserting $p_\alpha = h/a_\alpha$ into this expression yields

$$E_\alpha = \frac{1}{2m_\alpha}\left(\frac{h}{a_\alpha}\right)^2, \tag{2.67}$$

where a_α is the size of the nanomaterial. On the other hand, the energy $E(k)$ is assumed to follow the dispersion relation

$$E(k) = E_m + \frac{(\hbar k)^2}{2m_{pol}} \tag{2.68}$$

(refer to Fig. B.1 of Appendix B), where m_{pol} and E_m are the effective mass of the exciton-polariton and the eigenenergy of the exciton of subsystem M, respectively. In the case of a semiconductor, for example, E_m corresponds to the bandgap energy E_g. The nanomaterials in subsystem n are excited by propagating light whose photon energy is adjusted to be lower than E_m in order to avoid absorption by the macroscopic subsystem M. By adjusting the photon energy in this way, the light propagates through subsystem M without attenuating its power and successfully reaches subsystem n. Under this transparent situation, E_m can be excluded from Eq. (2.68), and the energy of the exciton-polariton of subsystem M, contributing the effective interaction energy, is simply expressed as

$$E\left(k\right)=\frac{(\hbar k)^2}{2m_{pol}}. \tag{2.69}$$

By using this equation, the term $f(k)$ in Eq. (2.28) is given by

$$f\left(k\right)=\frac{ck\sqrt{\frac{\hbar}{2m_{pol}}}}{\sqrt{\frac{\hbar^2}{2m_{pol}^2}k^2-c^2}} \tag{2.70}$$

where c is the speed of light in vacuum. By using Eqs. (2.69) and (2.66) is expressed as

$$\begin{aligned}V_{eff}\left(\boldsymbol{r}\right)=&-\frac{2\hbar^2 p_s p_p}{3(2\pi)^2\varepsilon_0}\int_{-\infty}^{\infty}k^2 dk f^2\left(k\right)\sum_{\alpha=s}^{p}\frac{2m_{pol}}{\hbar^2}\\&\times\left\{\frac{1}{(k+i\Delta_{\alpha+})\,(k-i\Delta_{\alpha+})}+\frac{1}{(k+i\Delta_{\alpha-})\,(k-i\Delta_{\alpha-})}\right\}\frac{e^{i\boldsymbol{k}\cdot\boldsymbol{r}}}{ikr}\\&\equiv\sum_{\alpha=s}^{p}\left[V_{eff,\alpha+}\left(\boldsymbol{r}\right)+V_{eff,\alpha-}\left(\boldsymbol{r}\right)\right]\end{aligned} \tag{2.71}$$

where

$$\Delta_{\alpha\pm}\equiv\frac{1}{\hbar}\sqrt{2m_{pol}\left(\pm E_\alpha\right)}. \tag{2.72}$$

After taking the complex integral over k, by noting the pole of the first order $k=i\Delta_{\alpha\pm}$ and by rewriting $f(k)$ in Eq. (2.71) as $f\left(i\Delta_{\alpha\pm}\right)$, $V_{eff,\alpha+}\left(\boldsymbol{r}\right)$ and $V_{eff,\alpha-}\left(\boldsymbol{r}\right)$ in the third row of Eq. (2.71) are expressed as

$$V_{eff,\alpha\pm}\left(\boldsymbol{r}\right)=\mp\frac{p_s p_p}{3\left(2\pi\right)\varepsilon_0}W_{\alpha\pm}(\Delta_{\alpha\pm})^2\frac{e^{-\Delta_{\alpha\pm}r}}{r}, \tag{2.73}$$

where

$$W_{\alpha\pm}\equiv\frac{m_{pol}c^2}{\left(m_{pol}c^2\pm E_\alpha\right)}. \tag{2.74}$$

Inserting Eqs. (2.73) and (2.74) into Eq. (2.71) yields

$$V_{eff}\left(\boldsymbol{r}\right)=-\frac{p_s p_p}{3\left(2\pi\right)\varepsilon_0}$$

$$\times \sum_{\alpha=s}^{p} \left[W_{\alpha+}(\Delta_{\alpha+})^2 \frac{e^{-\Delta_{\alpha+} r}}{r} - W_{\alpha-}(\Delta_{\alpha-})^2 \frac{e^{-\Delta_{\alpha-} r}}{r} \right], \tag{2.75}$$

which is the expression for the effective interaction energy, representing the spatial modulation feature of the DP. As a result of this interaction, the exciton in the nanomaterial emits propagating light after time γ_{rad}^{-1}, which is the inverse of the radiative relaxation rate γ_{rad} (explained in Sect. 3.2.1), depending on the structure and size of the nanomaterial. Then, the emitted propagating light, i.e., the scattered light, can be detected in the far field.

Equation (2.75) is composed of two terms. By noting Eq. (2.72), it is found that the first term is merely the Yukawa function

$$Y(\Delta_{\alpha+}) = \frac{\exp\left(-2\pi\sqrt{\frac{m_{pol}}{m_\alpha}}\frac{r}{a_\alpha}\right)}{r}, \tag{2.76}$$

as has been shown in Eq. (2.24), which was derived because $\Delta_{\alpha+}$ takes a real number. The value of this function decreases rapidly with increasing r. The interaction range, i.e., a in Eq. (2.24), is found to be $(a_\alpha/2\pi)\sqrt{m_\alpha/m_{pol}}$ from this equation, which is proportional to the size a_α of the nanomaterial α. Thus, Eq. (2.76) means that the electromagnetic field exists on the surface of the nanomaterial α within a range that depends on the material size. In other words, the nanomaterial α is covered with a localized electromagnetic field cloud. The effective interaction energy mediated by the DP is represented by this function $Y(\Delta\alpha_+)$.

The second term is given by the function

$$Y(\Delta_{\alpha-}) = \frac{\exp\left(-i2\pi\sqrt{\frac{m_{pol}}{m_\alpha}}\frac{r}{a_\alpha}\right)}{r}, \tag{2.77}$$

which was derived because $\Delta_{\alpha-}$ takes an imaginary number. The numerator of this equation sinusoidally oscillates with period $\lambda_\alpha = a_\alpha\sqrt{m_\alpha/m_{pol}}$ by varying r. This means that Eq. (2.77) represents a spherical wave with wavelength λ_α. However, this is not the propagating lightwave to be detected in the far field, and furthermore, the wavelength λ_α is not correlated with the wavelength of the propagating light incident on the nanomaterial α. Equation (2.77) originates from the fact that no boundary conditions are set on the nanomaterial. That is, the spatial distribution of the electromagnetic field is, in general, determined by the boundary conditions if the wavelength is sufficiently short. However, because no boundary conditions are set in the present case, Eq. (2.77) means that an oscillatory electromagnetic field leaks out from the surface of the nanomaterial. It is expected that this leaking field can be eliminated by using a more detailed theoretical model with accurate boundary conditions in the future.

Now, Eqs. (2.76) and (2.77) are compared from the viewpoint of the energy transfer between the subsystems n and M. Equation (2.51) means that Eq. (2.76) originates from the transition from the initial state $|\phi_{Pi}\rangle$ of Eq. (2.34) to the final state $|\phi_{Pf}\rangle$ of Eq. (2.35) via the intermediate state $|\phi_{Q2}\rangle$ of Eq. (2.37b), as schematically illustrated in Fig. 2.4a. Here, in the initial state $|\phi_{Pi}\rangle$, the nanomaterial p is in the ground state $|p_g\rangle$, and the exciton-polariton in subsystem M is in the vacuum state $|0_{(M)}\rangle$. First, by the transition from the initial state $|\phi_{Pi}\rangle$ to the intermediate state $|\phi_{Q2}\rangle$, the nanomaterial p is excited to the excited state $|p_{ex}\rangle$, and one exciton-polariton is generated in subsystem M; i.e., it is excited to the state $|1_{(M)}\rangle$. Therefore, this transition violates the energy conservation law because the two subsystems increase their energies simultaneously. Next, by the transition from the intermediate state $|\phi_{Q2}\rangle$ to the final state $|\phi_{Pf}\rangle$, the nanomaterial s is de-excited from the excited state $|s_{ex}\rangle$ to the ground state $|s_g\rangle$, and the exciton-polariton in subsystem M is also de-excited from the state $|1_{(M)}\rangle$. to the vacuum state $|0_{(M)}\rangle$. Therefore, this transition also violates the energy conservation law because the two subsystems decrease their energies simultaneously. On the other hand, Eq. (2.51) also means that Eq. (2.77) originates from the transition from the initial state $|\phi_{Pi}\rangle$ of Eq. (2.34) to the final state $|\phi_{Pf}\rangle$ of Eq. (2.35) via the intermediate state $|\phi_{Q1}\rangle$ of Eq. (2.37a), as schematically illustrated in Fig. 2.4b. Here, in the initial state $|\phi_{Pi}\rangle$, the nanomaterial s is in the excited state $|s_{ex}\rangle$, and the exciton-polariton in subsystem M is in the vacuum state $|0_{(M)}\rangle$. First, by the transition from the initial state $|\phi_{Pi}\rangle$ to the intermediate state $|\phi_{Q1}\rangle$, the nanomaterial s is de-excited to the ground state $|s_g\rangle$. Since the single exciton-polariton is simultaneously generated in subsystem M, i.e., is excited to the state $|1_{(M)}\rangle$, this transition follows the energy conservation law. Next, by the transition from the intermediate state $|\phi_{Q1}\rangle$ to the final state $|\phi_{Pf}\rangle$, the nanomaterial p is excited from the ground state $|p_g\rangle$ to the excited state $|p_{ex}\rangle$, and the exciton-polariton in subsystem M is de-excited from the state $|1_{(M)}\rangle$ to the vacuum state $|0_{(M)}\rangle$. Therefore, this transition also follows the energy conservation law.

Because the two successive transitions in Fig. 2.4a violate the energy conservation law, these transition processes are called virtual processes (refer to the beginning of the previous section). This violation is allowed within a very short duration Δt. In other words, the uncertainty principle $\Delta E \Delta t \geq \hbar/2$ allows a large energy uncertainty ΔE if Δt is small. Therefore, within a sufficiently short duration, the vacuum fluctuations can trigger the successive transitions from the initial state to the intermediate state, and then to the final state, even though they violate the energy conservation law. Since the DP energy is transferred by these transitions, the DP is also called a virtual photon.

The second term in the third row in Eq. (2.55) corresponds to Eq. (2.77) and Fig. 2.4b. It is easily found that it becomes infinity if $E(k) = E_\alpha$ because its denominator is $E(k) - E_\alpha$. The transition process represented by this term is called a resonant process. On the other hand, the first term corresponds to Eq. (2.76) and Fig. 2.4a. The relevant transition process is called a non-resonant process because the denominator of this term is $E(k) + E_\alpha$, which takes a finite value at $E(k) = E_\alpha$.

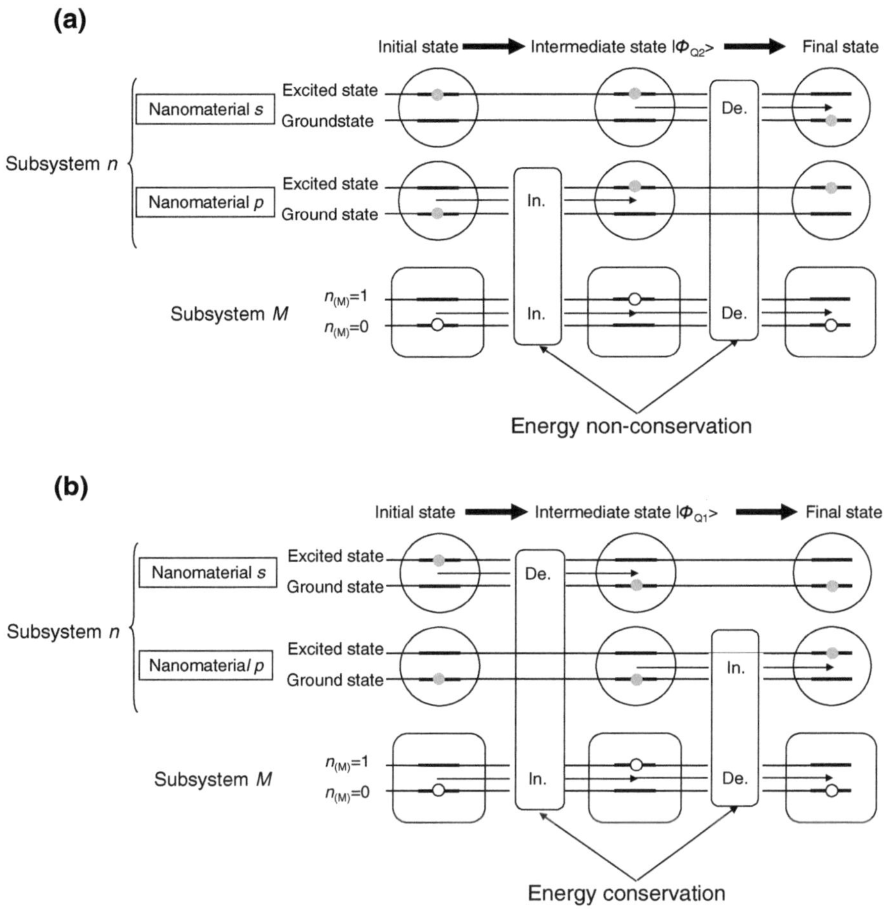

Fig. 2.4 Energy transfer via intermediate states in the Q space. **a** The case of Eq. (2.76). **b** The case of Eq. (2.77). "In." and "De." represent the energy increase and decrease in subsystems n and M, respectively

2.2.2 *Size-Dependent Resonance and Hierarchy*

The terms $Y(\Delta_{s+})$ and $Y\left(\Delta_{p+}\right)$ are picked up from Eq. (2.75) and used in order to discuss the spatial properties of DPs. In the future, it is expected that a more-detailed theoretical model will be able to eliminate the leaking electromagnetic field of Eq. (2.77), and therefore, the terms $Y(\Delta_{s-})$ and $Y\left(\Delta_{p-}\right)$ are excluded in the discussion. The effective interaction energy is thus expressed as

$$V_{eff}(\boldsymbol{r}) = -\frac{p_s p_p}{3(2\pi)\varepsilon_0} W_+ \left\{ \frac{\exp\left(-\frac{r}{a'_s}\right)}{a_s'^2 r} + \frac{\exp\left(-\frac{r}{a'_p}\right)}{a_p'^2 r} \right\}, \tag{2.78a}$$

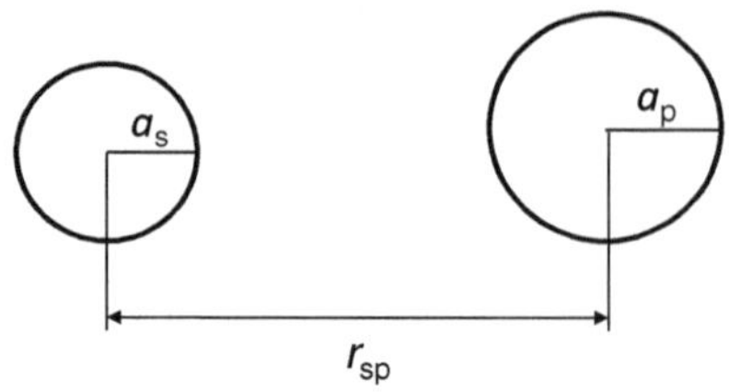

Fig. 2.5 Radii a_s and a_p of the two spherical nanomaterials and their center-to-center separation r_{sp}

where

$$a'_\alpha = \frac{a_\alpha}{2\pi\sqrt{m_{pol}/m_\alpha}} \qquad (\alpha = s, p). \tag{2.78b}$$

Here, $W_{+\alpha}$ in Eq. (2.75) was rewritten as W_+ by removing the subscript α because it does not strongly depend on the size of the nanomaterial a_α. Furthermore, the relation $\Delta_{\alpha+} = 1/a'_\alpha$ was used based on Eqs. (2.67) and (2.72). In the case where the center-to-center separation between the nanomaterials s and p is r_{sp} (refer to Fig. 2.5), the magnitude of the detectable propagating light energy, generated as a result of the interaction, is given by the volume integral of the spatial derivative $\nabla_r V_{eff}\left(\boldsymbol{r}_p - \boldsymbol{r}_s\right)$ of Eq. (2.78a), which is given by

$$\begin{aligned} I\left(r_{sp}\right) &= \left| \int\int \nabla_{r_p} V_{eff}\left(\boldsymbol{r}_p - \boldsymbol{r}_s\right) d^3 r_s d^3 r_p \right|^2 \\ &= \left(\frac{p_s p_p}{3\,(2\pi)\,\varepsilon_0} W_+ \right)^2 \left[8\pi \sum_{\alpha=s}^{p} a'^2_\alpha \left\{ \frac{a_s}{a'_\alpha} \cosh\left(\frac{a_s}{a'_\alpha}\right) - \sinh\left(\frac{a_s}{a'_\alpha}\right) \right\} \right. \\ &\quad \left. \times \left\{ \frac{a_p}{a'_\alpha} \cosh\left(\frac{a_p}{a'_\alpha}\right) - \sinh\left(\frac{a_p}{a'_\alpha}\right) \right\} \left(\frac{a'_\alpha}{r_{sp}} + \frac{a'^2_\alpha}{r^2_{sp}} \right) \exp\left(-\frac{r_{sp}}{a'_\alpha} \right) \right]^2 . \end{aligned} \tag{2.79}$$

(refer to Appendix E for details of the derivation). In order to convert this to the propagating light intensity, the right-hand side is divided by $(a_s^3 + a_p^3)^2$ so that the resultant quantity has dimensions of optical power per unit area and is expressed as

$$\begin{aligned} I_d\left(r_{sp}\right) &= \frac{1}{\left(a_s^3 + a_p^3\right)^2} \left[\sum_{\alpha=s}^{p} a'^2_\alpha \left\{ \frac{a_s}{a'_\alpha} \cosh\left(\frac{a_s}{a'_\alpha}\right) - \sinh\left(\frac{a_s}{a'_\alpha}\right) \right\} \right. \\ &\quad \left. \times \left\{ \frac{a_p}{a'_\alpha} \cosh\left(\frac{a_p}{a'_\alpha}\right) - \sinh\left(\frac{a_p}{a'_\alpha}\right) \right\} \left(\frac{a'_\alpha}{r_{sp}} + \frac{a'^2_\alpha}{r^2_{sp}} \right) \exp\left(-\frac{r_{sp}}{a'_\alpha} \right) \right]^2 \end{aligned} \tag{2.80}$$

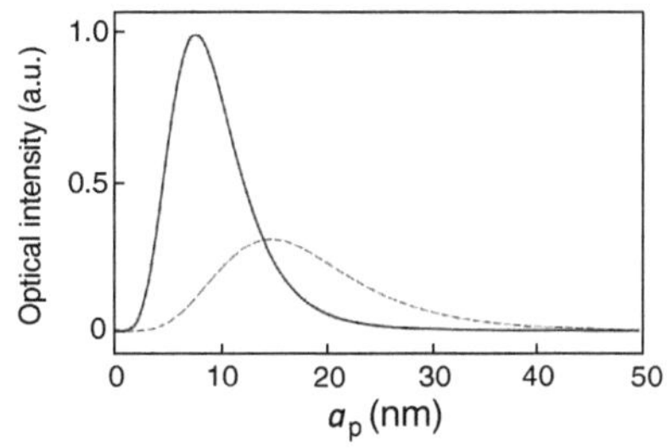

Fig. 2.6 Relation between the radius a_p of the nanomaterial p and the detectable intensity of the propagating light. The solid and broken curves represent the calculated values for $a_s = 10$ and 20 nm, respectively. The surface-to-surface separation between the two nanomaterials is 1 nm

where, for simplicity, all the constants appearing on the left of the symbol $\sum_{\alpha=s}^{p}$ in Eq. (2.79) have been removed.

Now, the value of $2\pi\sqrt{m_{pol}/m_\alpha}$ in the denominator of Eq. (2.78b) must be estimated. The effective mass m_α of the exciton in the nanomaterial α can be assumed to be $0.5\,m_0$ in the case of a semiconductor [12], where m_0 is the mass of the electron in vacuum. On the other hand, the effective mass m_{pol} of the exciton–polariton in subsystem M can be assumed to be $(0.004 \sim 0.03)\,m_0$ based on experimental and theoretical considerations by taking polariton–polariton scattering into account [13, 14]. Using the values of m_α and m_{pol} assumed above, the value of $2\pi\sqrt{m_{pol}/m_\alpha}$ can be estimated to be 0.56–1.54, which is close to unity. Based on this estimation, Eq. (2.80) is numerically evaluated by assuming a_α in Eq. (2.78b) to be equal to a_α, and the result is given in Fig. 2.6 [7], where the surface-to-surface separation between s and p is fixed at 1 nm. The solid and broken curves represent the calculated results for $a_s = 10$ and 20 nm, respectively. This figure shows that the light intensity takes the maximum if a_p is close to a_s, a feature called size-dependent resonance, which means that the interaction energy takes the maximum when the sizes of the nanomaterials s and p are equal.

It is also possible to approximately describe this size-dependent resonance by using a classical electric dipole interaction model in the case where the nanomaterials s and p are placed in vacuum without any surrounding macroscopic materials [15]. In this description, the value of a_p at which the light intensity takes the maximum is slightly different from the value given above, because the contribution of the macroscopic subsystem M is neglected.

Size-dependent resonance means that the magnitude of the energy transfer mediated by the DP takes the maximum when the sizes of the nanomaterials s and p are equal. On the other hand, it has been pointed out that the separation between the nanomaterials s and p must be as short as their sizes because the interaction range mediated by the DP is equivalent to the sizes of the nanomaterials. For these reasons, one finds that the energy transfer mediated by a DP gains a unique feature, named hierarchy.

That is to say, if there is a material s with a complicated shape (its size may not necessarily be nanometric) in the proximity of the nanomaterial p of size a_p, and if their separation is as short as a_p, the energy is preferably transferred from the nanomaterial p to the part of the material s whose size is equal to a_p. On the other hand, if there exists another nanomaterial p' of size a'_p in proximity to the material s, and if their separation is as short as a'_p, the energy is preferably transferred to the other part of the material s whose size is equal to a'_p. Furthermore, these two channels of the preferable energy transfers do not interfere with each other. Thus, it is found that the energy transfer between small materials located in close proximity is independent of the energy transfer between larger materials located farther away. This feature is called the hierarchy, which means that different energy transfer occurs independently for different material sizes and separations.

Some unique phenomena originating from this hierarchy have been observed in the image acquired by a near-field optical microscope (NOM). That is, by bringing a probe with a smaller aperture radius close to a sample and scanning it, one can acquire images of a sample with a size as small as the aperture radius. When using a larger aperture radius, the size of the acquired image is as large as the size of this aperture radius. This means that the NOM probe works as a spatial band-pass filter. By exploiting this hierarchy, multiple DPs with different interaction ranges enable multiple energy transfers, depending on the material sizes and their separations, in order to multiply the signal transmission. Examples of this will be reviewed in Sect. 8.1.

References

1. J.J. Sakurai, *Advanced Quantum Mechanics* (Addison-Wesley, Reading, 1967)
2. T.D. Newton, E.P. Wigner, Rev. Mod. Phys. **21**, 400 (1949)
3. J.P. Sipe, Phys. Rev. A **52**, 1875 (1995)
4. M.O. Scully, M.S. Zubairy, *Quantum Optics* (Cambridge University Press, Cambridge, 1997)
5. M. Ohtsu, K. Kobayashi, T. Kawazoe, T. Yatsui, M. Naruse, *Principles of Nanophotonics* (CRC Press, Bica Raton, 2008)
6. K. Kobayashi, M. Ohtsu, J. Microsc. **194**, 249 (1999)
7. S. Sangu, K. Kobayashi, M. Ohtsu, J. Microscopy **202**, 279 (2001)
8. S. John, T. Quang, Phys. Rev. A **52**, 4083 (1995)
9. H. Suzuura, T. Tsujikawa, T. Tokihiro, Phys. Rev. B **53**, 1294 (1996)
10. A. Shojiguchi, K. Kobayashi, S. Sangu, K. Kitahara, M. Ohtsu, J. Phys. Soc. Jpn. **72**, 2984 (2003)
11. K. Kobayashi, S. Sangu, H. Ito, M. Ohtsu, Phys. Rev. A **63**, 013806 (2001)
12. Y. Liu, T. Morishima, T. Yatsui, T. Kawazoe, M. Ohtsu, Nanotechnology **22**, 215605 (2011)
13. T. Itoh, T. Suzuki, M. Ueta, J. Phys. Soc. Jpn. **42**, 1069 (1977)
14. T. Itoh, T. Suzuki, J. Phys. Soc. Jpn. **45**, 1939 (1978)
15. M. Ohtsu (ed.), *Near-Field Nano/Atom Optics and Technology* (Springer, Berlin, 1998), pp. 16–23

Chapter 3
Energy Transfer and Relaxation by Dressed Photons

Veritas nunquam perit.
Lucius Annaeus Seneca, *Troades*, 614

This chapter describes the energy transfer and subsequent relaxation between closely spaced nanomaterials, mediated by dressed photons. These phenomena are used for realizing nanometer-sized photonic devices (DP devices), to be reviewed in Chap. 5. As an example of the nanomaterial, a semiconductor quantum dot (QD) is adopted in the following discussions.

3.1 Coupled States Originating from Two Energy Levels

This section discusses what kind of coupled states are created through the interaction between two arbitrary energy levels [1]. For simplicity, two energy levels in different QDs are considered, and it is assumed that state $|1\rangle$ in QD1 has the same eigenenergy $\hbar\Omega$ as that of state $|2\rangle$ in QD2. As shown in Fig. 3.1, the interaction energy between the two states is denoted by $\hbar U$, which corresponds to the effective interaction energy in Eq. (2.76). Operators for annihilating and creating an electron-hole pair (exciton) in QD$_i$ are denoted by $\hat{b}_i$ and $\hat{b}_i^\dagger$, respectively. They satisfy the relations[1]

$$\hat{b}_i\,|0\rangle = 0, \quad \hat{b}_i^\dagger\,|0\rangle = |i\rangle \quad (i = 1, 2)\,, \tag{3.1}$$

where $|0\rangle$ is the vacuum state. The Hamiltonian for the two–QD system is given by

$$\hat{H} = \hbar\Omega\left(\hat{b}_1^\dagger\hat{b}_1 + \hat{b}_2^\dagger\hat{b}_2\right) + \hbar U\left(\hat{b}_1^\dagger\hat{b}_2 + \hat{b}_2^\dagger\hat{b}_1\right) \tag{3.2}$$

[1] The equation $\hat{b}_i^\dagger\,|1\rangle = \hat{b}_i^\dagger\,|2\rangle = 0$ also holds because this section considers the states $|0\rangle$, $|1\rangle$, and $|2\rangle$ only.

M. Ohtsu, *Dressed Photons*, Nano-Optics and Nanophotonics,
DOI: 10.1007/978-3-642-39569-7_3,

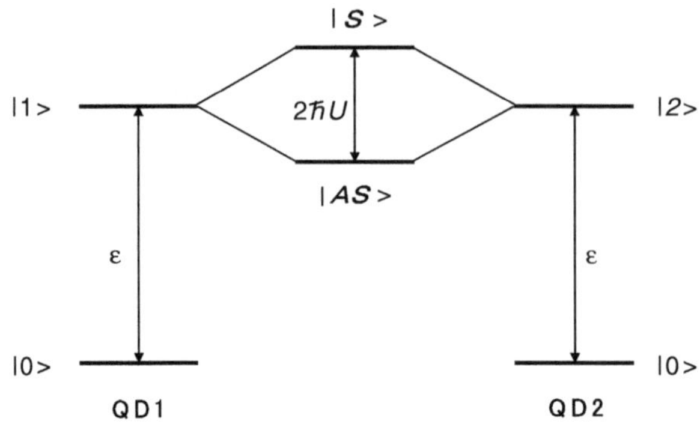

Fig. 3.1 Coupled states originating from the interaction between two quantum dots

Defining the symmetric state $|S\rangle$ and anti-symmetric state $|AS\rangle$ by

$$|S\rangle = \frac{1}{\sqrt{2}} (|1\rangle + |2\rangle) , \tag{3.3a}$$

$$|AS\rangle = \frac{1}{\sqrt{2}} (|1\rangle - |2\rangle) , \tag{3.3b}$$

leads to

$$\hat{H} |S\rangle = \hbar (\Omega + U) |S\rangle , \tag{3.4a}$$

$$\hat{H} |AS\rangle = \hbar (\Omega - U) |AS\rangle . \tag{3.4b}$$

It follows that these states $|S\rangle$ and $|AS\rangle$ are eigenstates of the Hamiltonian $\hat{H}$ with the eigenenergies $\hbar (\Omega + U)$ and $\hbar (\Omega - U)$. Equations (3.4a) and (3.4b) indicate that the symmetric state $|S\rangle$ and the anti-symmetric state $|AS\rangle$ are created after the interaction between the states $|1\rangle$ and $|2\rangle$.

Using the operator of the induced electric dipole moment in QD_i, which is defined by[2]

$$\hat{\boldsymbol{p}}_i = \boldsymbol{p}_i \left(\hat{b}_i + \hat{b}_i^\dagger \right) \qquad (i = 1, 2) , \tag{3.5}$$

the expectation value $\langle \boldsymbol{p}_1 \cdot \boldsymbol{p}_2 \rangle$ of the inner product of the electric dipole moments in the state $|S\rangle$ is [3]

[2] This electric dipole moment is also called a transition dipole moment because it is induced by the transition between the two states, as expressed by the annihilation and creation operators on the right-hand side.

[3] For deriving the third line in this equation, it should be noted that $\langle 1| \hat{b}_1 \hat{b}_2^\dagger |1\rangle$, $\langle 1| \hat{b}_1 \hat{b}_2^\dagger |2\rangle$, $\langle 2| \hat{b}_1 \hat{b}_2^\dagger |2\rangle$, $\langle 1| \hat{b}_1^\dagger \hat{b}_2 |1\rangle$, $\langle 2| \hat{b}_1^\dagger \hat{b}_2 |1\rangle$, and $\langle 2| \hat{b}_1^\dagger \hat{b}_2 |2\rangle$ are all zero because the states $|0\rangle$, $|1\rangle$, and $|2\rangle$ are orthogonal to each other and because other states $|n\rangle$ $(n \neq 0, 1, 2)$ are excluded from the present discussion.

$$
\begin{aligned}
\langle S|\hat{\boldsymbol{p}}_1 \cdot \hat{\boldsymbol{p}}_2 |S\rangle &= \frac{\boldsymbol{p}_1 \cdot \boldsymbol{p}_2}{2} \left(\langle 1| + \langle 2|\right) \left(\hat{b}_1 + \hat{b}_1^\dagger\right) \left(\hat{b}_2 + \hat{b}_2^\dagger\right) \left(|1\rangle + |2\rangle\right) \\
&= \frac{\boldsymbol{p}_1 \cdot \boldsymbol{p}_2}{2} \left(\langle 1| + \langle 2|\right) \left(\hat{b}_1 \hat{b}_2^\dagger + \hat{b}_1^\dagger \hat{b}_2\right) \left(|1\rangle + |2\rangle\right) \\
&= \frac{\boldsymbol{p}_1 \cdot \boldsymbol{p}_2}{2} \left[\langle 2|\, \hat{b}_1 \hat{b}_2^\dagger \,|1\rangle + \langle 1|\, \hat{b}_1^\dagger \hat{b}_2 \,|2\rangle\right] \\
&= \boldsymbol{p}_1 \cdot \boldsymbol{p}_2 .
\end{aligned} \tag{3.6}
$$

This equation indicates that the two dipole moments are parallel in the symmetric state $|S\rangle$. A similar calculation gives

$$
\langle AS|\hat{\boldsymbol{p}}_1 \cdot \hat{\boldsymbol{p}}_2 |AS\rangle = -\boldsymbol{p}_1 \cdot \boldsymbol{p}_2, \tag{3.7}
$$

which shows that they are anti-parallel in the anti-symmetric state $|AS\rangle$.

It follows from these results that the excitation of two QDs with propagating light leads to the symmetric state $|S\rangle$, in which parallel electric dipole moments are created in the two QDs. This is because the two QDs cannot be distinguished spatially owing to diffraction of the propagating light. In other words, since the states $|1\rangle$ and $|2\rangle$ are equally excited, only the symmetric state $|S\rangle$ of Eq. (3.3a) is created. In other words, one can observe, in the far field, only the state that has a large electric dipole moment given by the sum of the two parallel electric dipole moments. In this sense, the symmetric state $|S\rangle$ is called a bright state. In contrast, the anti-symmetric state has a negligibly small electric dipole moment given by the difference of the two anti-parallel electric dipole moments, which cannot be observed in the far field. In this sense, it is called a dark state.

Since the interaction range mediated by a DP is as small as the size of the nanomaterials, as described by Eq. (2.24), the two QDs can be distinguished spatially, and therefore, each QD can be selectively excited. In the case where, for example, the state $|1\rangle$ of QD1 is selectively excited, as well as creating $|S\rangle$, it is found that the anti-symmetric state $|AS\rangle$ can be created too because $|1\rangle$ is expressed as

$$
|1\rangle = \frac{1}{\sqrt{2}} \left(|S\rangle + |AS\rangle\right) \tag{3.8}
$$

from Eqs. (3.3a) and (3.3b). In other words, one can observe the dark state via a DP.

In order to investigate the temporal evolution of the energy transfer between the states $|1\rangle$ and $|2\rangle$ mediated by the DP, it is assumed that the electron, hole, and exciton are in the state $|1\rangle$ of QD1 at time $t = 0$; i.e., the initial state $|\psi(t)\rangle$ of the two QDs is $|\psi(0)\rangle = |1\rangle$. By noting Eq. (3.8) and using the eigenenergies of the right-hand sides of Eqs. (3.4a) and (3.4b), the state vector $|\psi(t)\rangle$ at time t is expressed as

$$|\psi(t)\rangle = \frac{1}{\sqrt{2}}\left[\exp\{-i(\Omega+U)t\}|S\rangle + \exp\{-i(\Omega-U)t\}|AS\rangle\right], \tag{3.9}$$

and this equation can be rewritten as

$$\begin{aligned}|\psi(t)\rangle &= \frac{1}{\sqrt{2}}\{[\cos(Ut) - i\sin(Ut)]\exp(-i\Omega t)|S\rangle \\ &\quad + [\cos(Ut) + i\sin(Ut)]\exp(-i\Omega t)|AS\rangle\} \\ &= \frac{1}{\sqrt{2}}\exp(-i\Omega t)\{\cos(Ut)(|S\rangle + |AS\rangle) - i\sin(Ut)(|S\rangle - |AS\rangle)\} \\ &= \exp(-i\Omega t)\{\cos(Ut)|1\rangle - i\sin(Ut)|2\rangle\}.\end{aligned} \tag{3.10}$$

From this equation, the occupation probability $\rho_{11}(t)$ is given by

$$\rho_{11}(t) = |\langle 1 | \psi(t)\rangle|^2 = \cos^2(Ut), \tag{3.11a}$$

which is the probability that a particle such as an electron, a hole, or an exciton occupies the state $|1\rangle$ of QD1 at time t. The occupation probability $\rho_{22}(t)$ is given by

$$\rho_{22}(t) = |\langle 2 | \psi(t)\rangle|^2 = \sin^2(Ut), \tag{3.11b}$$

which is the probability that a particle occupies the state $|2\rangle$ of QD2 at time t. As shown in Fig. 3.2, these probabilities sinusoidally vary in an anti-correlated manner with period π/U. This means that the eigenenergy $\hbar\Omega$ of the system is periodically transferred between the resonant energy levels of the two QDs, a process known as nutation. It should be noted that nutation is not sustained permanently. It terminates through a relaxation process, such as the energy dissipation caused by the interaction with another system.[4] For example, if the nutation terminates due to energy dissipation at the time when $\rho_{11}(t) = 0$ and $\rho_{22}(t) = 1$ are attained, the energy is transferred from QD1 to QD2.

Next, the characteristics of the transition from the ground state $|0\rangle$ to the excited states $|1\rangle$ and $|2\rangle$ are discussed. In the case of excitation by propagating light, the long-wavelength approximation holds because the value of the electric displacement vector is homogeneous inside the QDs. Therefore, the transition matrix elements between $|0\rangle$ and $|1\rangle$, or $|0\rangle$ and $|2\rangle$, are proportional to the spatial integral of the

[4] If the energy is transferred bi-directionally between the systems under consideration and an external system, this transfer process is called relaxation. On the other hand, if this energy transfer is unidirectional, it is called dissipation.

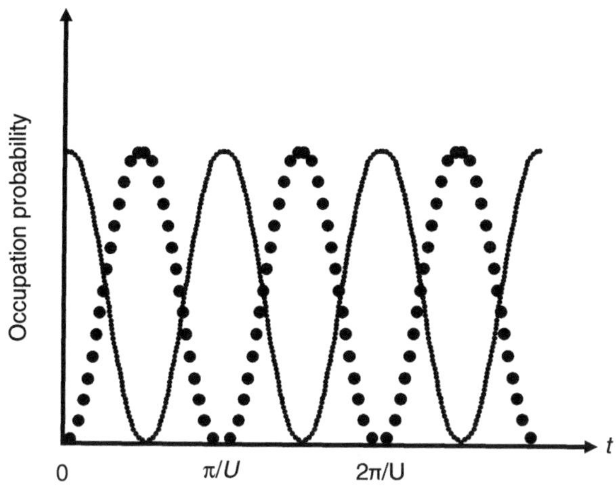

Fig. 3.2 Temporal evolution of occupation probability. The *solid* and *dotted curves* represent $\rho_{11}(t)$ and $\rho_{22}(t)$, respectively

envelope functions representing the motion of the center of gravity of the exciton, as given by Eqs. (F.49) and (F.50) in Appendix F. That is, the electric dipole transition is forbidden if the value of this integral is zero, whereas a nonzero value indicates that the transition is allowed. As an example, in the case of spherical QDs, a transition to a state where both quantum numbers l and m are zero is allowed. In the case of cubic QDs, as another example, the transition is allowed if all of the quantum numbers n_x, n_y, and n_z take odd numbers, whereas it is forbidden if one of them takes an even number.

On the other hand, in the case of excitation by a DP, the long-wavelength approximation does not hold because the interaction range of the DP is as small as the size of the QDs. This suggests that the electric dipole transition that was forbidden in the case of propagating light excitation will be allowed here. It also suggests the possibility of making use of the electric-dipole–forbidden energy levels for DP devices, to be described in the following sections. Moreover, the advantage of using these forbidden energy levels is that the contribution of the propagating light to the DP device operation can be excluded.

To demonstrate the use of the electric-dipole–forbidden transition in the case of excitation by a DP, Fig. 3.3 shows the values of the effective interaction energy V_{eff} between two cubic CuCl QDs embedded in an NaCl crystal, calculated by using Eq. (F.47) in Appendix F. This figure shows the value of V_{eff} as a function of the separation between the QDs. The solid curve is the value for the case where the electric-dipole–allowed levels $(1, 1, 1)$ in two QDs with sizes of 5 nm are used. The dotted curve, on the other hand, is for the case where the energy level $(1, 1, 1)$ in a QD with a size of 5 nm and the energy level $(2, 1, 1)$ in a QD with a size of 7 nm are used. For conventional propagating light, the level $(2, 1, 1)$ is electric-dipole–forbidden. However, the dotted curve takes nonzero values, e.g., 5.05 μeV when the separation

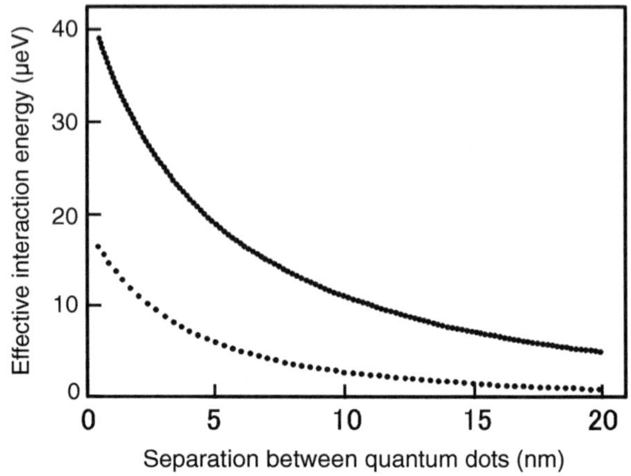

Fig. 3.3 Value of the energy of effective interaction between the two cubic CuCl quantum dots in an NaCl crystal. *Solid curve* is the case where electric-dipole–allowed levels (1, 1, 1) in two quantum dots with the sizes of 5 nm are used. *Dotted curve* is the case where the electric-dipole–allowed level (1, 1, 1) in a quantum dot with a size of 5 nm and the electric-dipole–forbidden level (2, 1, 1) in a quantum dot with a size of 7 nm are used

between the QDs is 6.1 nm, which is as large as one-quarter of the solid curve value. Such a large value means that this level can be regarded as an electric-dipole–allowed level if a DP is used.

3.2 Principles of Dressed-Photon Devices

Use of DPs enables the construction of novel optical devices whose functions cannot be realized by conventional optical devices using propagating light. In a DP device, the energies of the electron, hole, and exciton excited to discrete energy levels in a nanomaterial are transferred to an adjacent nanomaterial, and uni-directional energy transfer is realized by the energy dissipation via interaction with phonons in the heat bath and by destroying quantum coherence [2–8]. As a result, signal transmission becomes possible from one nanomaterial to the other. Various operations (e.g., signal generation, signal control, signal transmission, and input/output interfaces) can be realized by using these energy transfer and signal transmission mechanisms.

In order to describe the dynamic properties of DP devices, quantum master equations of the density matrix operators are solved [9–11, 13] and the characteristics of energy transfer between nanomaterials are analyzed [2–6, 8]. Details of the density matrix, density matrix operator, and quantum master equations are reviewed in Ref. [1]. The following subsections describe temporal evolutions of the populations of electron–hole pairs (excitons) in two or three QDs that couple with a large number of phonons in a heat bath [14].

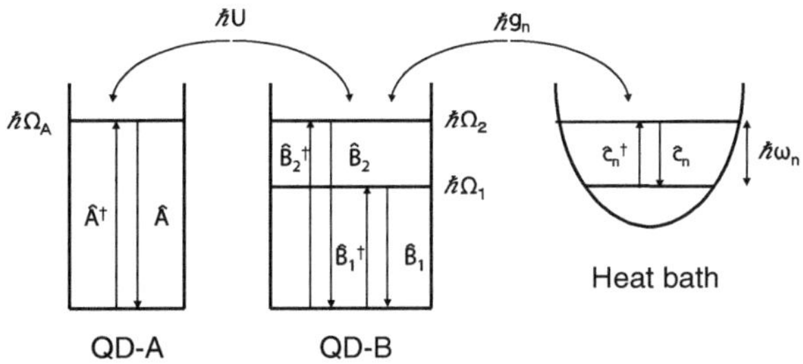

Fig. 3.4 Energy levels of two quantum dots, which interact with the heat bath

3.2.1 Dressed-Photon Devices Using Two Quantum Dots

A DP device composed of two QDs has two terminals, i.e., an input terminal and an output terminal, corresponding to a diode. Figure 3.4 shows the energy levels in two QDs (QD-A and QD-B) interacting with phonons in a heat bath. Here, in QD-A, one energy level of an exciton is assumed, whose eigenenergy is denoted by $\hbar\Omega_A$. On the other hand, in the QD-B, two energy levels are assumed, where the eigenenergy of the upper energy level, denoted by $\hbar\Omega_2$, is assumed to be equal to $\hbar\Omega_A$. The eigenenergy of the lower energy level is denoted by $\hbar\Omega_1$. The following discussion deals with signal transmission using the energy transfer from QD-A to QD-B and subsequent relaxation to the lower energy level in QD-B. Here, the energy transfer by propagating light is neglected because the upper energy level in QD-B is electric-dipole–forbidden.

The Hamiltonian for the system of Fig. 3.4 is expressed as

$$\hat{H} = \hat{H}_0 + \hat{H}_{\text{int}} + \hat{H}_{SR}, \tag{3.12}$$

where the unperturbed Hamiltonian is

$$\hat{H}_0 = \hbar\Omega_A \hat{A}^\dagger \hat{A} + \hbar\Omega_1 \hat{B}_1^\dagger \hat{B}_1 + \hbar\Omega_2 \hat{B}_2^\dagger \hat{B}_2 + \hbar \sum_n \omega_n \hat{c}_n^\dagger \hat{c}_n. \tag{3.13a}$$

The interaction Hamiltonian between QD-A and QD-B is

$$\hat{H}_{\text{int}} = \hbar U \left(\hat{A}^\dagger \hat{B}_2 + \hat{B}_2^\dagger \hat{A} \right), \tag{3.13b}$$

where $\hbar U$ is the energy of interaction mediated by the DP. The interaction Hamiltonian between QD-B and the heat bath, i.e., between the system under consideration and the reservoir, is

$$\hat{H}_{SR} = \hbar \sum_n \left(g_n \hat{c}_n^\dagger \hat{B}_1^\dagger \hat{B}_2 + g_n^* \hat{c}_n \hat{B}_2^\dagger \hat{B}_1 \right), \tag{3.13c}$$

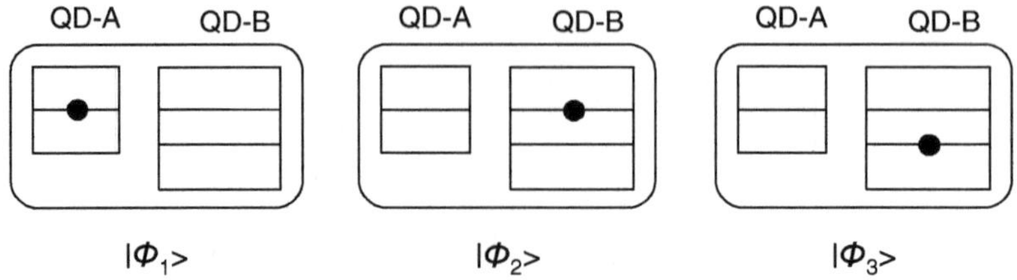

Fig. 3.5 Three bases. *Black circles* represent excitons that occupy the energy levels

which represents the energy dissipation of QD-B after the interaction with QD-A. The interaction energy between the exciton and phonon is denoted by $\hbar g_n$. In Eqs. (3.13a)–(3.13c), $\hat{A}$ and $\hat{A}^\dagger$ are the annihilation and creation operators of the exciton in QD-A with energy $\hbar\Omega_A$ ($= \hbar\Omega_2$), $\hat{B}_1$ and $\hat{B}_1^\dagger$ are those of the exciton in QD-B with energy $\hbar\Omega_1$, $\hat{B}_2$ and $\hat{B}_2^\dagger$ are those of the exciton in QD-B with energy $\hbar\Omega_2$, and $\hat{c}_n$ and $\hat{c}_n^\dagger$ are those of the phonon in the heat bath with energy $\hbar\omega_n$.

In order to analyze the dynamic behavior of the excitons, three bases $|\phi_i\rangle$ $(i = 1 - 3)$ are used, as schematically illustrated in Fig. 3.5. They are

$$\begin{aligned} |\phi_1\rangle &= \left|A^*, B_1, B_2\right\rangle \\ |\phi_2\rangle &= \left|A, B_1, B^*{}_2\right\rangle \\ |\phi_3\rangle &= \left|A, B^*{}_1, B_2\right\rangle \end{aligned} \tag{3.14}$$

in which the symbol $*$ on the right-hand side indicates that the relevant energy level is occupied by an exciton. The quantum master equation for the density matrix, within the Born–Markov approximation, is given by

$$\begin{aligned} \frac{\partial \hat{\rho}}{\partial t} = &-\frac{i}{\hbar}\left[\hat{H}_0 + \hat{H}_{\text{int}}, \hat{\rho}\right] + \frac{\gamma}{2}\left(\left[\hat{B}_1^\dagger \hat{B}_2, \hat{\rho}\hat{B}_2^\dagger \hat{B}_1\right] + \left[\hat{B}_1^\dagger \hat{B}_2\hat{\rho}, \hat{B}_2^\dagger \hat{B}_1\right]\right) \\ &+ \gamma n\left(\left[\hat{B}_1^\dagger \hat{B}_2\hat{\rho}, \hat{B}_2^\dagger \hat{B}_1\right] - \left[\hat{B}_2^\dagger \hat{B}_1, \hat{\rho}\hat{B}_1^\dagger \hat{B}_2\right]\right), \end{aligned} \tag{3.15}$$

which corresponds to Eq. (2.162) of Ref. [1].[5] Here, n represents the number of phonons in the heat bath. The non-radiative relaxation rate is denoted by γ, which represents the rate of relaxation of the exciton from the upper energy level to the lower one in QD-B as a result of phonon scattering, i.e., the interaction with phonons in the heat bath. This rate is proportional to the square of g_n in Eq. (3.13c) and is given by

[5] This equation is derived by replacing the operators $\hat{a}$ and $\hat{a}^\dagger$ in Eq. (2.162) of Ref. [1] with $\hat{B}_1^\dagger \hat{B}_2$ and $\hat{B}_2^\dagger \hat{B}_1$, respectively.

$$\gamma = 2\pi \left| \frac{E_{ex-ph}}{\hbar} \right|^2 D(\omega), \tag{3.16}$$

where E_{ex-ph} is the interaction energy between the exciton and phonon, $D(\omega)$ is the phonon density of states, and ω is equal to $\Omega_2 - \Omega_1$ [12]. The radiative relaxation rate γ_{rad} can be neglected when analyzing transitory phenomena because $\gamma \gg \gamma_{rad}$, where γ_{rad} represents the decay rate of electromagnetic radiation emitted from the lower energy level in QD-B as a result of electron–hole recombination.

By using U in Eq. (3.13b) and Eq. (3.15) can be written for the elements

$$\rho_{mn}(t) \equiv \langle \phi_m | \hat{\rho}(t) | \phi_n \rangle \tag{3.17}$$

of the matrix $\hat{\rho}$ as

$$\frac{d\rho_{11}(t)}{dt} = iU(r)\left[\rho_{12}(t) - \rho_{21}(t)\right], \tag{3.18a}$$

$$\frac{d\rho_{12}(t)}{dt} - \frac{d\rho_{21}(t)}{dt} = 2iU(r)\left[\rho_{11}(t) - \rho_{22}(t)\right] - (n+1)\gamma\left[\rho_{12}(t) - \rho_{21}(t)\right], \tag{3.18b}$$

$$\frac{d\rho_{22}(t)}{dt} = -iU(r)\left[\rho_{12}(t) - \rho_{21}(t)\right] - 2(n+1)\gamma\rho_{22}(t) + 2n\gamma\rho_{33}(t), \tag{3.18c}$$

$$\frac{d\rho_{33}(t)}{dt} = 2(n+1)\gamma\rho_{22}(t) - 2n\gamma\rho_{33}(t), \tag{3.18d}$$

where the diagonal elements $\rho_{mm}(t)$ represent the occupation probabilities of the state $|\phi_m\rangle$ in QD-A and QD-B, and the off-diagonal elements $\rho_{mn}(t)$ represent quantum coherence between the states $|\phi_m\rangle$ and $|\phi_n\rangle$ [1].[6]

In the case where the absolute temperature T is 0 K, i.e., the phonon is in the vacuum state ($n = 0$), these simultaneous equations can be analytically solved. To solve them, the initial conditions $\rho_{11}(0) = 1$ and $\rho_{mn}(0) = 0$ are assumed, which mean that an exciton is created in QD-A by applying an optical input signal. With this assumption, the application of the input signal is represented by the creation of an exciton in QD-A. The solution under these initial conditions is written as

$$\rho_{11}(t) = \frac{1}{Z^2} e^{-\gamma t} \left[\frac{\gamma}{2} \sinh(Zt) + Z\cosh(Zt) \right]^2, \tag{3.19a}$$

$$\rho_{22}(t) = \frac{U^2}{Z^2} e^{-\gamma t} \sinh^2(Zt), \tag{3.19b}$$

[6] Quantum coherence represents the degree of correlation between the two states $|\phi_m\rangle$ and $|\phi_n\rangle$, which interact with each other, as given by Eq. (3.17) This corresponds to the degree of interference of the wave functions representing these states and is quantitatively evaluated by the cross-correlation coefficient between these wave functions.

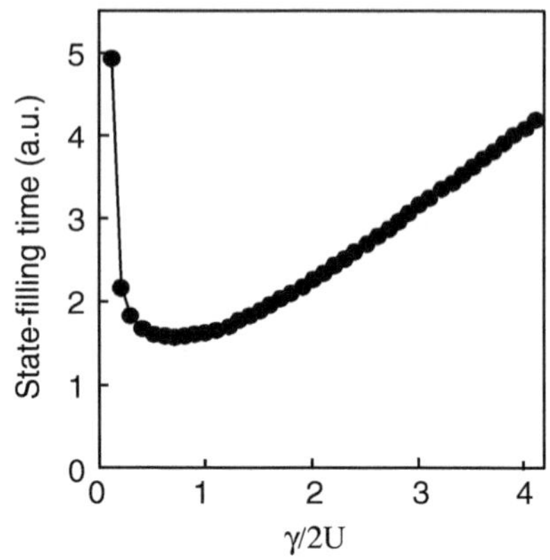

Fig. 3.6 Relation between $\gamma/2U$ and state-filling time τ_s

$$\rho_{12}(t) - \rho_{21}(t) = 2i\frac{U}{Z^2}e^{-\gamma t}\sinh(Zt)\left[\frac{\gamma}{2}\sinh(Zt) + Z\cosh(Zt)\right], \quad (3.19c)$$

where

$$Z \equiv \sqrt{(\gamma/2)^2 - U^2}, \quad (3.19d)$$

(refer to Sect. G.1 of Appendix G). Since the sum of the diagonal elements is unity, one obtains

$$\rho_{33}(t) = 1 - [\rho_{11}(t) + \rho_{22}(t)]. \quad (3.19e)$$

This equation represents the probability of creating an exciton in the lower energy level in QD-B, which corresponds to the probability of generating an output signal. In other words, the output signal corresponds to the light emitted by annihilating this exciton. It should be noted that the emitted light intensity decays with the radiative relaxation rate γ_{rad}.

It follows from Eqs. (3.19a)–(3.19d) that the temporal evolution of the population is quite different at $\gamma < 2U$ (Z; imaginary number) and $\gamma > 2U$ (Z; real number). Although these equations seem to be undefined at $\gamma = 2U$ ($Z = 0$), taking a limit value, there is a definite solution regardless of whether $Z \to +0$ or -0 is taken. In Fig. 3.6, the state-filling time τ_S is plotted as a function of the ratio $\gamma/2U$, where τ_S is defined by $\rho_{33}(\tau_S) = 1 - e^{-1}$. This figure shows that τ_S decreases with increasing γ, which is because the energy transfer time decreases. However, τ_S increases when γ is larger than $2U$. This is because the upper energy level in QD-B is broadened with increasing γ, which resultantly decreases the magnitude of the energy transferred between the QDs. It follows from the figure that the fastest energy transfer is obtained when $\gamma = 2U$ is satisfied.

The term $2n\gamma\rho_{33}(t)$ on the right-hand side of Eq. (3.18c) indicates that the finite temperature effect caused by the finite number of phonons ($n \neq 0$) induces back transfer of the energy from the heat bath to the two QDs. Within the Born–Markov approximation adopted here, this term increases the population $\rho_{22}(t)$. Since the

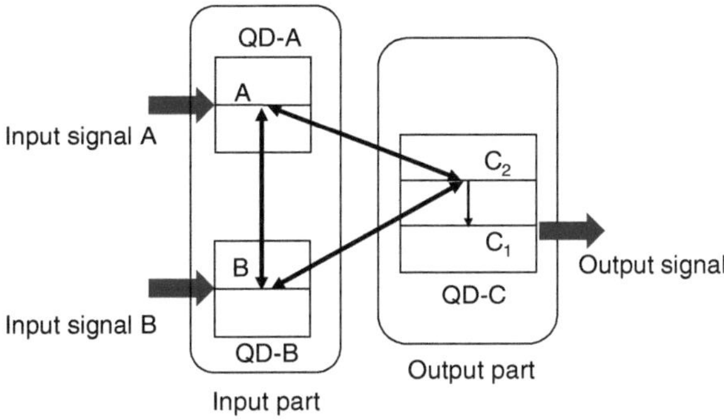

Fig. 3.7 Logic gates by using the energy transfer between three symmetrically arranged quantum dots. *Double-headed arrow* and *downward arrow* represent the energy transfer and subsequent non-radiative relaxation, respectively

population $\rho_{33}(t)$ is proportional to $\rho_{22}(t)$, the back transfer becomes large due to the increase of $\rho_{33}(t)$, giving residual populations $\rho_{11}(t)$ and $\rho_{22}(t)$ in the energy level of QD-A and in the upper energy level of QD-B, respectively.

So far, the theoretical modeling of the population dynamics in the two-QD system has assumed the exact resonance condition ($\hbar\Omega_A = \hbar\Omega_2$). The following discussion deals with the slightly off-resonance condition. That is, by introducing a deviation $\hbar\Delta\Omega$ between the energies $\hbar\Omega_2$ and $\hbar\Omega_A$, the factor on the right-hand side of Eq. (3.19a) is modified. As a result, it is found that the ratio of the factors between off-resonance and on-resonance conditions is approximately proportional to $\gamma^2/\left(\gamma^2 + \Delta\Omega^2\right)$. Therefore, an energy transfer efficiency of more than 50 % of that under exact resonance is achieved if $|\Delta\Omega| < \gamma$. When the QD size and γ are set to 7.1 nm and $4.1 \times 10^{12}\text{s}^{-1}$, respectively, a deviation of approximately 10 % in the QD size is allowed while maintaining $|\Delta\Omega| < \gamma$. With recent nanofabrication techniques, it is feasible to make QDs within this size deviation. In fact, experimental results show consistent population dynamics, as discussed in Refs. [4] and [5].

3.2.2 Dressed-Photon Devices Using Three Quantum Dots

A DP device composed of three QDs has three terminals, i.e., two input terminals and one output terminal, corresponding to a triode. Dynamics of the energy transfer, mediated by DPs, are analyzed here by assuming that three QDs (QD-A, QD-B, and QD-C) are symmetrically placed at the vertices of an isosceles triangle, as shown in Fig. 3.7. In this system, two identical QDs (QD-A and QD-B) are resonantly coupled with each other via a DP, which form the input part, whereas the third QD (QD-C) that is larger than the other two corresponds to an output part. It will be demonstrated in the following discussion that these QDs work as a novel logic gate device.

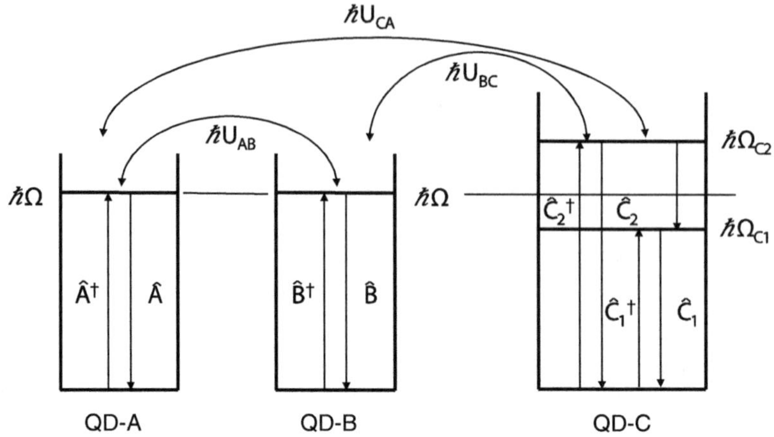

Fig. 3.8 Energy levels of excitons, their interactions, and non-radiative relaxation

As a result of the energy transfer from the input to the output parts, an exciton is created in the upper energy level of QD-C, which relaxes to the lower energy level with the non-radiative relaxation rate γ of Eq. (3.16) due to the interaction with phonons in the heat bath. Since the value of the non-radiative relaxation rate γ is much larger than that of γ_{rad}, only the contribution of γ is considered for analyzing transient phenomena, similarly to the previous subsection. Furthermore, the QDs are spatially arranged in order to realize an interaction energy $\hbar U$ between QD-A and QD-B that is larger than the interaction energy $\hbar U'$ between QD-A and QD-C (and also larger than that between QD-B and QD-C).

The Hamiltonian $\hat{H}_S$ for the system in Fig. 3.7 is

$$\begin{aligned}
\hat{H}_S &= \hat{H}_0 + \hat{H}_{\text{int}}, \\
\hat{H}_0 &= \hbar\Omega\left(\hat{A}^\dagger\hat{A} + \hat{B}^\dagger\hat{B}\right) + \hbar\sum_{i=1}^{2}\Omega_{C_i}\hat{C}_i^\dagger\hat{C}_i, \\
\hat{H}_{\text{int}} &= \hbar U\left(\hat{A}^\dagger\hat{B} + \hat{B}^\dagger\hat{A}\right) + \hbar U'\left(\hat{B}^\dagger\hat{C}_2 + \hat{C}_2^\dagger\hat{B} + \hat{C}_2^\dagger\hat{A} + \hat{A}^\dagger\hat{C}_2\right),
\end{aligned} \tag{3.20}$$

where the annihilation and creation operators of an exciton with eigenenergy $\hbar\Omega$ in QD-A are denoted by $\hat{A}$ and $\hat{A}^\dagger$, those of an exciton with eigenenergy $\hbar\Omega$ in QD-B are denoted by $\hat{B}$ and $\hat{B}^\dagger$, and those for an exciton with eigenenergy $\hbar\Omega_{C_i}$ in energy level i in QD-C are denoted by $\hat{C}_i$ and $\hat{C}_i^\dagger$. As shown in Fig. 3.8, $\hbar U_{AB} = \hbar U$ is the interaction energy between QD-A and QD-B, whereas $\hbar U_{BC} = \hbar U_{CA} = \hbar U'$ is that between QD-B and QD-C, and also between QD-C and QD-A.

Similar to the Subsect. 3.2.2 the dynamics of the three-QD system are analyzed by using the density matrix under the initial condition that one exciton or two excitons

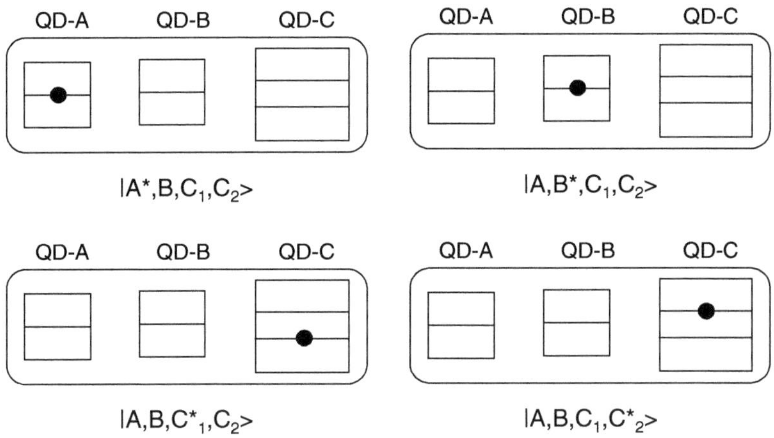

Fig. 3.9 Bases of the one-exciton state. *Black circles* represent excitons that occupy the energy levels

are created by a DP. This corresponds to one or two optical input signals. The quantum master equation for the density matrix operator $\hat{\rho}$ is [2]

$$\frac{\partial}{\partial t}\hat{\rho}(t) = -\frac{i}{\hbar}\left[\hat{H}_0 + \hat{H}_{\text{int}}, \hat{\rho}(t)\right] \\ + \frac{\gamma}{2}\left\{2\hat{C}_1^{\dagger}\hat{C}_2\hat{\rho}(t)\,\hat{C}_2^{\dagger}\hat{C}_1 - \hat{C}_2^{\dagger}\hat{C}_1\hat{C}_1^{\dagger}\hat{C}_2\hat{\rho}(t) - \hat{\rho}(t)\,\hat{C}_2^{\dagger}\hat{C}_1\hat{C}_1^{\dagger}\hat{C}_2\right\}. \tag{3.21}$$

This equation represents a system with phonon number $n = 0$, i.e., an absolute temperature of 0 K, as was the case in Eqs. (3.19a)–(3.19d), and thus, the effect of the heat bath is represented only by the non-radiative relaxation rate γ.

(a) XOR logic gate

It is possible to use the three QDs in Fig. 3.7 as an XOR logic gate. In order to find the necessary condition for this gate, a one-exciton state is considered, which is the state where any one of the three QDs is occupied by an exciton by applying an input signal, as shown in Fig. 3.9. It is most advantageous to choose the basis so that the number of density matrix elements is minimized in the quantum master equation [12]. It is minimized if the following four states are adopted:

$$|S_1\rangle = \frac{1}{\sqrt{2}}\left(|A^*, B, C_1, C_2\rangle + |A, B^*, C_1, C_2\rangle\right),$$

$$|AS_1\rangle = \frac{1}{\sqrt{2}}\left(|A^*, B, C_1, C_2\rangle - |A, B^*, C_1, C_2\rangle\right),$$

$$
\begin{aligned}
|P_1\rangle &= \left|A, B, C_1^*, C_2\right\rangle, \\
\left|P_1'\right\rangle &= \left|A, B, C_1, C_2^*\right\rangle,
\end{aligned}
\tag{3.22}
$$

where the symmetric and anti-symmetric states (refer to Eqs. (3.3a) and (3.3b)) are abbreviated to S and AS with the subscript 1 on the left–hand side. The symbol $*$ on the right-hand side indicates that the relevant energy level is occupied by an exciton.

The quantum master equations for the matrix elements are written by using Eqs. (3.21) and (3.22) as

$$
\frac{\partial}{\partial t}\rho_{S_1,S_1}(t) = i\sqrt{2}U'\left\{\rho_{S_1,P_1'}(t) - \rho_{P_1',S_1}(t)\right\}, \tag{3.23a}
$$

$$
\begin{aligned}
\frac{\partial}{\partial t}\rho_{S_1,P_1'}(t) = &\left\{i(\Delta\Omega - U) - \frac{\gamma}{2}\right\}\rho_{S_1,P_1'}(t) \\
&+ i\sqrt{2}U'\left\{\rho_{S_1,S_1}(t) - \rho_{P_1',P_1'}(t)\right\},
\end{aligned}
\tag{3.23b}
$$

$$
\begin{aligned}
\frac{\partial}{\partial t}\rho_{P_1',S_1}(t) = &-\left\{i(\Delta\Omega - U) + \frac{\gamma}{2}\right\}\rho_{P_1',S_1}(t) \\
&- i\sqrt{2}U'\left\{\rho_{S_1,S_1}(t) - \rho_{P_1',P_1'}(t)\right\},
\end{aligned}
\tag{3.23c}
$$

$$
\frac{\partial}{\partial t}\rho_{P_1',P_1'}(t) = -\gamma\rho_{P_1',P_1'}(t) - i\sqrt{2}U'\left\{\rho_{S_1,P_1'}(t) - \rho_{P_1',S_1}(t)\right\}, \tag{3.23d}
$$

$$
\frac{\partial}{\partial t}\rho_{P_1,P_1}(t) = \gamma\rho_{P_1',P_1'}(t), \tag{3.23e}
$$

where

$$
\Delta\Omega \equiv \Omega_{C_2} - \Omega. \tag{3.24}
$$

It should be noted that the matrix elements for the anti-symmetric state $|AS_1\rangle$ never appear in these equations when the QDs are arranged in a spatially symmetric way, as shown in Fig. 3.7, because the interaction energy $\hbar U'$ between QD-A and QD-C is equal to that between QD-B and QD-C. As a result, the number of equations and the number of terms are minimized.

These differential equations can be solved analytically with the help of a Laplace transform (refer to Sect. G.2 of Appendix G). When QD-A is initially occupied by an exciton, that is, $\rho_{S_1,S_1}(0) = \rho_{AS_1,AS_1}(0) = \rho_{S_1,AS_1}(0) = \rho_{AS_1,S_1}(0) = 1/2$, the probability $\rho_{P_1,P_1}(t)$ that the exciton is transferred to the lower energy level C1 in QD-C is given by

$$
\begin{aligned}
\rho_{P_1,P_1}(t) &= \gamma\int_0^t \rho_{P_1',P_1'}(t')dt' \\
&= \frac{1}{2} + \frac{4U'^2}{\omega_+^2 - \omega_-^2}\left\{\cos\phi_+\cos(\omega_+ t + \phi_+) - \cos\phi_-\cos(\omega_- t + \phi_-)\right\}e^{-\left(\frac{\gamma}{2}\right)t}
\end{aligned}
\tag{3.25}
$$

which corresponds to the probability of generating an output signal. Here,

$$\begin{aligned}
\omega_{\pm} &= \frac{1}{\sqrt{2}}\Big[(\Delta\Omega - U)^2 + W_+ W_- \\
&\quad \pm \sqrt{\{(\Delta\Omega - U)^2 + W_+^2\}\{(\Delta\Omega - U)^2 + W_-^2\}}\Big]^{\frac{1}{2}}, \\
\phi_{\pm} &= \tan^{-1}\left(\frac{2\omega_{\pm}}{\gamma}\right), \\
W_{\pm} &= 2\sqrt{2}U' \pm \frac{\gamma}{2}.
\end{aligned} \tag{3.26}$$

The second term in the second row of Eq. (3.25) represents the sinusoidally oscillating nutation. Its denominator

$$\omega_+^2 - \omega_-^2 = \sqrt{\{(\Delta\Omega - U)^2 + W_+^2\}\{(\Delta\Omega - U)^2 + W_-^2\}} \tag{3.27}$$

takes the minimum at

$$\Delta\Omega = U. \tag{3.28}$$

This means, from Eqs. (3.24) and (3.28), that the efficiency of the energy transfer from the input to the output parts takes the maximum at

$$\Omega + U = \Omega_{C2}. \tag{3.29}$$

The left-hand side of this equation corresponds to the eigenenergy of the symmetric state

$$|S\rangle = \frac{1}{\sqrt{2}}\left(|1\rangle_A |0\rangle_B + |0\rangle_A |1\rangle_B\right), \tag{3.30}$$

which is generated as a result of the interaction ($\hbar U$) between QD-A and QD-B, as was given by Eq. (3.4a). Therefore, Eq. (3.29) means that the symmetric state of Eq. (3.30) is resonant with the upper energy level C_2 in QD-C. Under this resonant condition, the energy can be transferred most efficiently from the symmetric state $|S\rangle$ of the input part to the upper energy level C_2 in QD-C.

In the case where Eq. (3.29) holds, the output signal is generated from QD-C if one input signal is applied to either QD-A or QD-B. If the two input signals are applied to QD-A and QD-B simultaneously, an output signal is not generated from QD-C because Eq. (3.29) does not hold. These relations between the input and output signals represent XOR logic gate operation.

(b) AND logic gate

It is also possible to use the three QDs in Fig. 3.7 as an AND logic gate. In order to find the necessary condition for this gate, a two-exciton state is considered, which

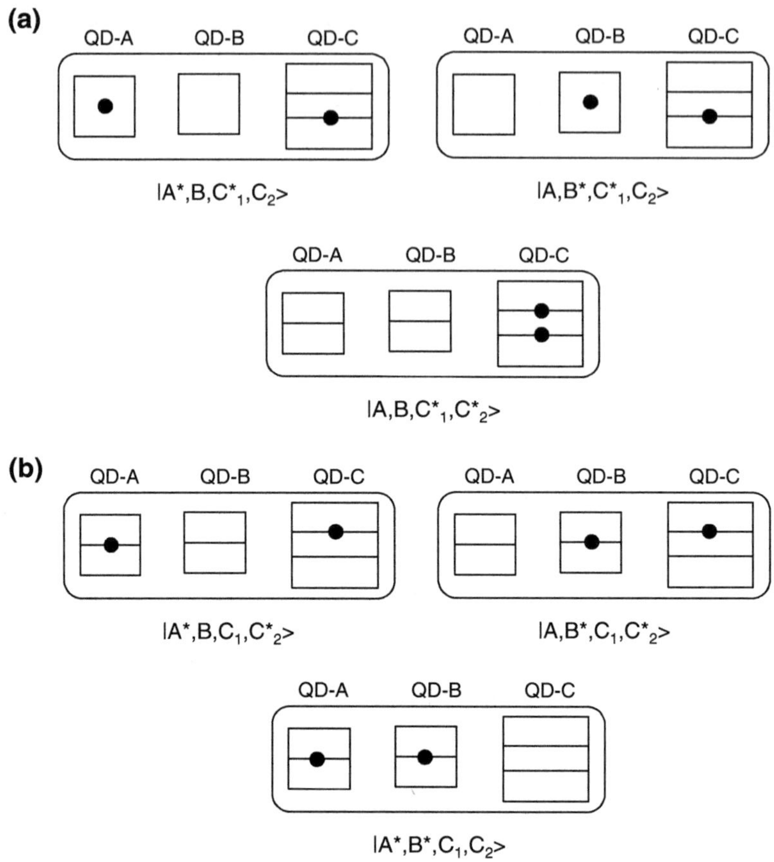

Fig. 3.10 Bases of the two-exciton state. *Black circles*represent excitons that occupy the energy levels. **a** and **b** represent the states with and without C_1 occupation, respectively

is the state where any two of the three QDs are occupied by excitons by applying two input signals, as shown in Fig. 3.10. The following six states were chosen for minimizing the number of density matrix elements in the quantum master equation. The two-exciton states are classified by whether the lower energy level C_1 in QD-C is occupied or not. For the states with C_1 occupation, the following three states

$$
\begin{aligned}
|S_2\rangle &= \frac{1}{\sqrt{2}}\left(\left|A^*, B, C_1^*, C_2\right\rangle + \left|A, B^*, C_1^*, C_2\right\rangle\right), \\
|AS_2\rangle &= \frac{1}{\sqrt{2}}\left(\left|A^*, B, C_1^*, C_2\right\rangle - \left|A, B^*, C_1^*, C_2\right\rangle\right), \\
|P_2\rangle &= \left|A, B, C_1^*, C_2^*\right\rangle
\end{aligned}
\tag{3.31}
$$

are adopted, as shown in Fig. 3.10a. For the states without C_1 occupation, the states

$$\begin{aligned}
\left|S_2'\right\rangle &= \frac{1}{\sqrt{2}}\left(\left|A^*, B, C_1, C_2^*\right\rangle + \left|A, B^*, C_1, C_2^*\right\rangle\right), \\
\left|AS_2'\right\rangle &= \frac{1}{\sqrt{2}}\left(\left|A^*, B, C_1, C_2^*\right\rangle - \left|A, B^*, C_1, C_2^*\right\rangle\right), \\
\left|P_2'\right\rangle &= \left|A^*, B^*, C_1, C_2\right\rangle,
\end{aligned} \tag{3.32}$$

are adopted, as shown in Fig. 3.10b. The quantum master equations for the matrix elements are written by using Eqs. (3.21), (3.31), and (3.32) as

$$\frac{\partial}{\partial t}\rho_{S_2',S_2'}(t) = i\sqrt{2}U'\left\{\rho_{S_2',P_2'}(t) - \rho_{P_2',S_2'}(t)\right\} - \gamma\rho_{S_2',S_2'}(t), \tag{3.33a}$$

$$\begin{aligned}
\frac{\partial}{\partial t}\rho_{S_2',P_2'}(t) = &-\left\{i\left(\Delta\Omega + U\right) + \frac{\gamma}{2}\right\}\rho_{S_2',P_2'}(t) \\
&+ i\sqrt{2}U'\left\{\rho_{S_2',S_2'}(t) - \rho_{P_2',P_2'}(t)\right\},
\end{aligned} \tag{3.33b}$$

$$\begin{aligned}
\frac{\partial}{\partial t}\rho_{P_2',S_2'}(t) = &\left\{i\left(\Delta\Omega + U\right) - \frac{\gamma}{2}\right\}\rho_{P_2',S_2'}(t) \\
&- i\sqrt{2}U'\left\{\rho_{S_2',S_2'}(t) - \rho_{P_2',P_2'}(t)\right\},
\end{aligned} \tag{3.33c}$$

$$\frac{\partial}{\partial t}\rho_{P_2',P_2'}(t) = -i\sqrt{2}U'\left\{\rho_{S_2',P_2'}(t) - \rho_{P_2',S_2'}(t)\right\} \tag{3.33d}$$

Because the QDs are arranged in a spatially symmetric way, as shown in Fig. 3.7, the matrix elements for the anti-symmetric states ($|AS_2\rangle$, $|AS_2'\rangle$) never appear in these equations, similar to the case of Eqs. (3.23a)–(3.23e).

Under the initial condition $\rho_{P_2',P_2'}(0) = 1$, that is, when both QD-A and QD-B are initially occupied by one exciton, the probability of an exciton transferring to the lower energy level C_1 in QD-C is derived with the help of a Laplace transform (refer to Sect. G.3 of Appendix G) and is expressed as

$$\begin{aligned}
\rho_{S_2,S_2}(t) + \rho_{P_2,P_2}(t) &= \gamma\int_0^t \rho_{S_2',S_2'}(t')dt' \\
&= 1 + \frac{8U'^2}{\omega_+'^2 - \omega_-'^2}\left\{\cos\phi_+'\cos\left(\omega_+'t + \phi_+'\right)\right. \\
&\quad \left. - \cos\phi_-'\cos\left(\omega_-'t + \phi_-'\right)\right\}e^{-\left(\frac{\gamma}{2}\right)t}
\end{aligned} \tag{3.34}$$

which corresponds to the probability of generating an output signal. Here,

$$\omega'_{\pm} = \frac{1}{\sqrt{2}} \Big[(\Delta\Omega + U)^2 + W_+ W_- \pm \sqrt{\{(\Delta\Omega + U)^2 + W_+^2\}\{(\Delta\Omega + U)^2 + W_-^2\}} \Big]^{\frac{1}{2}},$$

$$\phi'_{\pm} = \tan^{-1}\left(\frac{2\omega'_{\pm}}{\gamma}\right). \tag{3.35}$$

The denominator

$$\omega'^2_{+} - \omega'^2_{-} = \sqrt{\{(\Delta\Omega + U)^2 + W_+^2\}\{(\Delta\Omega + U)^2 + W_-^2\}} \tag{3.36}$$

of the second term on the right-hand side of Eq. (3.34) takes the minimum at

$$\Delta\Omega = -U. \tag{3.37}$$

This means, from Eqs. (3.26) and (3.37), that the efficiency of the energy transfer from the input part to the output part takes the maximum at

$$\Omega - U = \Omega_{C2} \tag{3.38}$$

which means, from Eq. (3.4b), that the anti-symmetric state

$$|AS\rangle = \frac{1}{\sqrt{2}} \left(|1\rangle_A |0\rangle_B - |0\rangle_A |1\rangle_B \right) \tag{3.39}$$

generated as a result of the interaction ($\hbar U$) between QD-A and QD-B is resonant with the upper energy level C_2 in QD-C. Under this resonant condition, the energy can be transferred most efficiently from the anti-symmetric state $|AS\rangle$ of the input part to the upper energy level C_2 of QD-C.

In the case where Eq. (3.38) holds, an output signal is generated from QD-C if the input signals are applied to QD-A and QD-B simultaneously. If only one input signal is applied to either QD-A or QD-B, an output signal is not generated because Eq. (3.38) does not hold. These relations between the input and output signals represent AND logic gate operation.

(c) Example of numerical calculations

Cubic CuCl QDs in an NaCl crystal are taken as an example. The sizes of QD-A and QD-B are fixed at 10 nm, and that of QD-C is $\sqrt{2}$ times larger, i.e., 14.1 nm. By adjusting the separations between these QDs, the interaction energy between QD-A and QD-B is assumed to be $\hbar U = 89\,\mu\text{eV}$, and that between QD-A and QD-C (or, QD-B and QD-C) is $\hbar U' = 14\,\mu\text{eV}$. Figures. 3.11 and 3.12 show the calculated

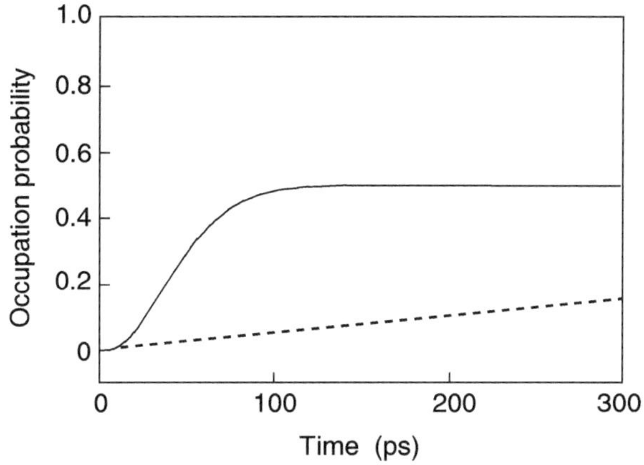

Fig. 3.11 Temporal evolutions of the occupation probabilities of the exciton in QD-C in the case of $\Delta\Omega = U$, which corresponds to the XOR logic gate operation. *Solid* and *broken curves* represent the case of one input signal and two input signals, respectively. $\hbar U = 89\,\mu\text{eV}$, $\hbar U' = 14\,\mu\text{eV}$

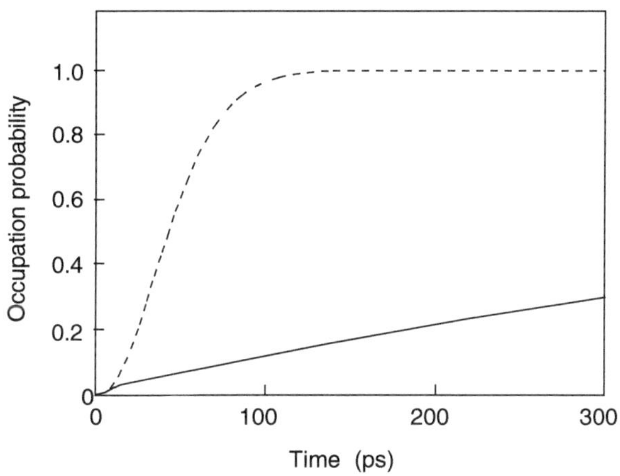

Fig. 3.12 Temporal evolutions of the occupation probabilities of the exciton in QD-C in the case of $\Delta\Omega = -U$, which corresponds to the AND logic gate operation. *Solid* and *broken curves* represent the case of one input signal and two input signals, respectively. signals, respectively. $\hbar U = 89\,\mu\text{eV}$, $\hbar U' = 14\,\mu\text{ev}$

results of the temporal evolution of the probabilities that the lower energy level C1 in QD-C is occupied by an exciton for $\Delta\Omega = U$ and $\Delta\Omega = -U$, respectively, based on the formalisms reviewed in Sects. (a) and (b).

Table 3.1 Relation between the input and output signals. Dependence on the energy difference $\Delta\Omega$, and corresponding logic operations are also shown

Input signal		Output signal: C	
A	B	XOR logic gate ($\Delta\Omega = U$)	AND logic gate ($\Delta\Omega = -U$)
0	0	0	0
1	0	0.5	0
0	1	0.5	0
1	1	0	1

The solid curves in these figures represent the values for the case of one input signal; i.e., the initial condition is the one-exciton state. The broken curves are for the case of two input signals; i.e., the initial condition is the two-exciton state. Figure 3.11 represents the XOR gate operation because the value of the solid curve is larger.[7] Figure 3.12, on the other hand, represents the AND gate operation because the value of the broken curve is larger. These figures demonstrate that the logic gates are realized by using three QDs. Table 3.1 summarizes the operations of these gates.

References

1. M. Ohtsu, K. Kobayashi, T. Kawazoe, T. Yatsui, M. Naruse, *Principles of Nanophotonics* (CRC Press, Bica Raton, 2008)
2. S. Sangu, K. Kobayashi, A. Shojiguchi, M. Ohtsu, Phys. Rev. B **69**, 115334 (2004)
3. A. Shojiguchi, K. Kobayashi, S. Sangu, K. Kitahara, M. Ohtsu, J. Phys. Soc. Jpn. **72**, 2984 (2003)
4. K. Kobayashi, S. Sangu, T. Kawazoe, M. Ohtsu, J. Lumin. **112**, 117 (2005)
5. K. Kobayashi, S. Sangu, T. Kawazoe, M. Ohtsu, Errattum to. J. Lumin. **114**, 315 (2005)
6. K. Kobayashi, S. Sangu, A. Shojiguchi, T. Kawazoe, K. Kitahara, M. Ohtsu, J. Microsc. **210**, 247 (2003)
7. M. Ohtsu, K. Kobayashai, T. Kawazoe, S. Sangu, T. Yatsui, IEEE J. Sel. Top. Quant. Electron. **8**, 839 (2002)
8. S. Sangu, K. Kobayashi, A. Shojiguchi, T. Kawazoe, M. Ohtsu, J. Appl. Phys. **93**, 2937 (2003)
9. U. Weiss, *Quantum Dissipative Systems*, 2nd edn. (World Scientific Publishing, Singapore, 1999)
10. K. Blum, *Density Matrix Theory and Applications*, 2nd edn. (Plenum, New York, 1996)

[7] The value of the solid curve in Fig. 3.11 asymptotically approaches not 1.0 but 0.5 with increasing time, which means that the probability of generating an output signal is 0.5. This originates from the fact that the initial state is $(|S\rangle + |AS\rangle)/\sqrt{2}$; i.e., the symmetric and anti-symmetric states are equally excited. Therefore, for reliable operation of the XOR logic gate, the average of the final state should be taken by repeating the device operation.

11. H.-P. Breuer, F. Petruccione, *The Theory of Open Quantum Systems* (Oxford University Press, New York, 2002)
12. H.J. Carmichael, *Statistical Methods of Quantum Optics 1* (Springer, Berlin, 1999)
13. H. Haken, *Light 1* (North-Holland, Amsterdam, 1986)
14. S. Sangu, K. Kobayashi, A. Shojiguchi, T. Kawazoe, M. Ohtsu, in *Progress, in Nano-Electro-Optics V*, ed. by M. Ohtsu, Theory and Principles of Operation of Nanophotonic Functional Devices (Springer, Berlin, 2006), pp. 1–62

Chapter 4
Coupling Dressed Photons and Phonons

Mihi contuenti semper suasit rerum natura nihil incredibile existimare de ea. Caius Plinius Secundus Major, *Naturalis Historya,* XI, 2

This chapter presents theoretical formulations of the coupling between dressed photons (DPs) and phonons based on the physical picture of DPs reviewed in Chap. 2. After showing some novel phenomena involving photo-dissociation of molecules, a theoretical model for the interaction between DPs and phonons is described.

4.1 Novel Molecular Dissociation and the Need for a Theoretical Model

4.1.1 Unique Phenomena of Molecular Dissociation by Dressed Photons

First, conventional molecular dissociation is discussed by taking a diatomic molecule as the simplest example. Comparing a nucleus and an electron in an atom, the nucleus moves more slowly than the electron because of its larger mass. In other words, the electron shifts its position instantaneously by following the movement of the nucleus, whereas the nucleus is almost independent of the electron movement. Therefore, one can use the approximation that the inter-nuclear distance R of the diatomic molecule is maintained constant, while only the electron moves, which is called the Born–Oppenheimer approximation [1]. This is also called the adiabatic approximation because the state of the nucleus remains unchanged. The inter-nuclear force is repulsive or attractive depending on R. As is shown in Fig. 4.1, the coupling length between the two atoms corresponds to the inter-nuclear distance R_0 at which the potential energy of the inter-atomic interaction takes the minimum. In the limit

M. Ohtsu, *Dressed Photons*, Nano-Optics and Nanophotonics,
DOI: 10.1007/978-3-642-39569-7_4, © Springer-Verlag Berlin Heidelberg 2014

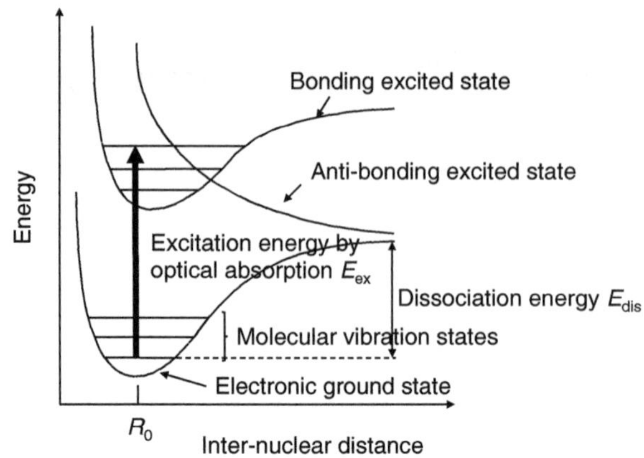

Fig. 4.1 Relation between the inter-nuclear distance in a molecule and the energies of the electron states and the molecular vibration states

of infinite distance ($R \to \infty$), the two atoms do not interact any more, which corresponds to dissociation. The difference between the interaction energies at $R = \infty$ and at R_0 corresponds to the dissociation energy R_{dis}.

The adiabatic approximation is effective also in the case of molecular dissociation induced by propagating light because the nucleus does not respond to the propagating light. Therefore, the molecule does not dissociate even if it is illuminated with propagating light whose photon energy is as high as R_{dis}. However, if propagating light with much higher photon energy is applied to the molecule, it is absorbed by the electron, and the electron is excited to the binding excited state shown in Fig. 4.1. This transition is represented by an upward arrow in Fig. 4.1. The energy of this binding excited state takes the minimum at a value of R that is larger than R_0 of the ground state because the inter-nuclear binding force is weaker in the electronic excited state. Here, since R still remains at R_0 due to adiabaticity, the nuclei start vibrating. Quantization of this vibrational motion creates discrete energy levels, which are called the vibrational energy levels and are represented by several horizontal solid lines in Fig. 4.1. This intra-molecular vibration relaxes without emitting photons. Through this relaxation, the electron transits from the binding excited state to the anti-binding excited state in Fig. 4.1, whose energy does not have any local minima. Since the energy of the anti-binding state takes the minimum at $R = \infty$, R increases to infinity, and finally, the molecule dissociates.

As described above, photo-dissociation requires the electronic transition from the ground to the excited state. The energy required for this transition is called the excitation energy E_{ex}, which is larger than E_{dis}. The principle governing this transition is called the Frank–Condon principle [2], in which only the electron is excited by photo-absorption, whereas the molecular vibration remains un-excited.

Second, it should be pointed out that a novel molecular dissociation scheme is possible when the molecules are excited by DPs. As an example, experiments on dissociating gaseous $Zn(C_2H_5)_2$ molecules (DEZn) have been demonstrated by cutting the bonds between the zinc (Zn) and the ethyl groups. Since the values of E_{ex}

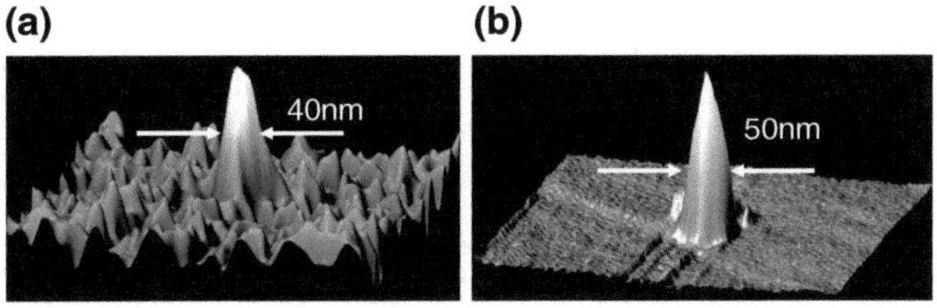

Fig. 4.2 Atomic force microscopic images of a Zn particle deposited by means of dressed photons using light sources with photon energies of **a** 1.81 eV (red light with wavelength of 684 nm) and **b** 2.54 eV (blue light with wavelength of 488 nm)

and E_{dis} of DEZn are 4.59 eV and 2.26 eV, respectively, the molecule is not dissociated by propagating light with a photon energy lower than these values (the optical wavelengths corresponding to these energies are 270 nm and 549 nm, respectively). However, when DPs are generated at the apex of a tapered probe by injecting red light with a photon energy of 1.81 eV (wavelength of 684 nm) from the tail of the probe (refer also to Sect. 1.1 of Chap. 1), any DEZn molecules jumping into the DP field are dissociated even though the photon energy is lower than both E_{ex} and E_{dis}. As a result, Zn atoms are deposited on a sapphire substrate, as shown in Fig. 4.2(a) [3]. Figure 4.2(b) shows the result obtained by using blue light with a photon energy of 2.45 eV (wavelength of 488 nm). Although the photon energy is larger than E_{dis}, it is still lower than E_{ex}. Here too, the molecules are dissociated, creating a small Zn particle on the substrate, like the case shown in Fig. 4.2(a).

Figures 4.2(a) and (b) indicate that some sort of molecular excitations are involved in the dissociation. Nonlinear excitation processes, such as multi-photon absorption, can be neglected because the intensity of the light injected into the probe is too low. Also, since these dissociation phenomena never occur when using propagating light, the adiabatic approximation is not effective for describing these phenomena. Therefore, the photo-chemical process involved in the novel dissociation demonstrated by Fig. 4.2 is called a non-adiabatic process. The rest of this chapter will present a theoretical model for describing this process.

Although the molecule is not dissociated by propagating light with a photon energy lower than E_{dis}, it will be found from the theoretical model described below that the molecule, when jumping into the DP field at the probe apex, absorbs not only the DP energy but also phonon energy [4, 5]. That is, the molecule can receive energy from phonons, exciting molecular vibrations, while the electrons remain in the ground state. As a result, the molecule gains vibrational energy by phonon absorption, and the excitation exceeds the potential barrier even though low-photon-energy light is injected into the tail of the probe.

Another dissociation process can be assumed by using only the phonon energy; i.e., the molecule absorbs multiple phonons, while the electrons stay in the ground state, which is represented by the upward broken arrow in Fig. 4.3. In this process, the state of the nucleus changes, whereas the electronic states remain unchanged; it is by virtue of this that the process is called a non-adiabatic process. However,

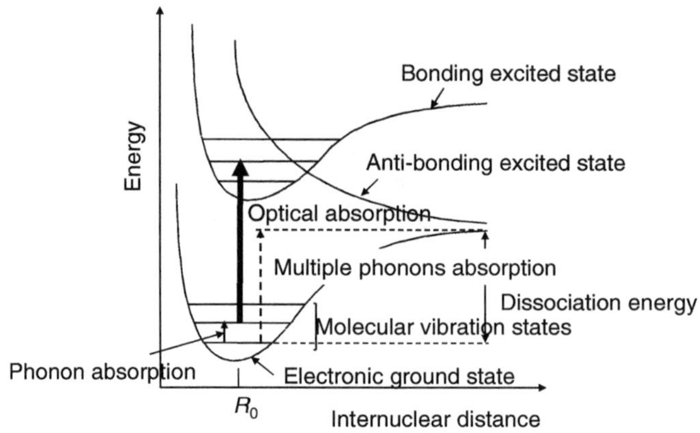

Fig. 4.3 Dissociation process involving multiple phonons

it should be noted that the phonon energy is several tens of meV, which is about 1/100-th of E_{dis}. Accordingly, multiple phonons need to be absorbed for dissociating the molecule. Here, if only one phonon is exchanged per interaction between the molecule and the probe apex, the probability of absorbing multiple phonons is very low because it is a multi-step process. Therefore, for theoretically explaining the unique experimental results in Fig. 4.2 based on the absorption of multiple phonons, it should be assumed that multiple phonons cohere with each other, and thus, multiple phonons are exchanged per interaction between the molecule and the probe apex.

Based on the assumptions above, it can be considered that the molecule interacts not only with the DP but also with phonons. This means that a quasi-particle, that is, a coupled state of the DP and phonons, is created at the probe apex, and the energy of this quasi-particle is exchanged between the molecule and the probe apex. The following sections give theoretical formulations for describing this quasi-particle.

4.1.2 Lattice Vibrations in the Probe

In order to study the interaction between the molecule and the probe apex, the probe apex is approximated as a one-dimensional material, and its one-dimensional lattice vibration is analyzed. Here, it should be noted that the size of this material is finite because this approximation is applied only to the probe apex. Since the representative size of the probe is the radius of curvature a of the apex (corresponding to the constant a in the Yukawa function of eq. (2.24) in Chap. 2), the probe is coarse-grained with this size.

Since the size of the probe apex is finite and nanometric, translational symmetry is broken, and thus, the momentum (or wave-number) has a large uncertainty [6]. For analyzing the vibrational (phonon) modes, therefore, it is more advantageous to

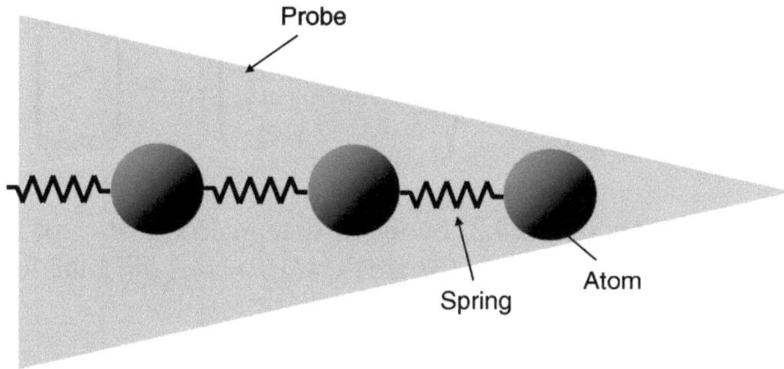

Fig. 4.4 Spring model for representing the constituent components of the probe apex

use a system Hamiltonian than the conventional method of simultaneous equations based on the symmetry in the crystal structure of the material [7].

Constituent elements of the probe apex, assumed as a result of coarse-graining, are called "atoms" for convenience, and they are connected by springs, as shown in Fig. 4.4. The number of atoms is finite and is denoted by N. The mechanical motions of these atoms correspond to the lattice vibrations, whose system Hamiltonian is expressed as

$$H = \sum_{i=1}^{N} \frac{\boldsymbol{p}_i^2}{2m_i} + \sum_{i=1}^{N-1} \frac{k}{2}(\boldsymbol{x}_{i+1} - \boldsymbol{x}_i)^2 + \sum_{i=1,N} \frac{k}{2}\boldsymbol{x}_i^2, \tag{4.1}$$

where $\boldsymbol{x}_i$, $\boldsymbol{p}_i$, and m_i are the displacement from an equilibrium point, its conjugate momentum, and the mass of an atom at site i, respectively, and k is the spring constant. Both ends ($i = 1$ and $i = N$) of the spring are assumed to be fixed, and one-dimensional longitudinal motions are considered in the following [1].

Based on eq.(4.1), the equations of motion, as given by the Hamilton equation, are

$$\frac{d}{dt}\boldsymbol{x}_i = \frac{\partial H}{\partial \boldsymbol{p}_i}, \tag{4.2a}$$

$$\frac{d}{dt}\boldsymbol{p}_i = -\frac{\partial H}{\partial \boldsymbol{x}_i}. \tag{4.2b}$$

Using a matrix form, from eqs. (4.2a) and (4.2b) one can derive

$$M\frac{d^2}{dt^2}\boldsymbol{x} = -k\Gamma\boldsymbol{x}, \tag{4.3}$$

where

[1] Therefore, these lattice vibrations correspond to longitudinal acoustic and optical phonons. In the case of a three-dimensional material, two transverse acoustic phonon and two transverse optical phonon have to be also taken into account.

$$M = \begin{pmatrix} m_1 & 0 & \cdot & 0 \\ 0 & m_2 & \cdot & \cdot \\ \cdot & \cdot & \cdot & \cdot \\ 0 & \cdot & \cdot & m_N \end{pmatrix}, \Gamma = \begin{pmatrix} 2 & -1 & & & \\ -1 & 2 & \cdot & & \\ & \cdot & \cdot & -1 & \\ & & -1 & 2 \end{pmatrix}, \boldsymbol{x} = \begin{pmatrix} \boldsymbol{x}_1 \\ \boldsymbol{x}_2 \\ \cdot \\ \boldsymbol{x}_N \end{pmatrix}. \tag{4.4}$$

Multiplying both sides of eq.(4.3) by the inverse matrix $\sqrt{M}^{-1}$ of the diagonal matrix $\sqrt{M}$ (the matrix elements of $\sqrt{M}$ are $(\sqrt{M})_{ij} = \delta_{ij}\sqrt{m_i}$), the left-hand side of eq.(4.3) is rewritten as

$$\sqrt{M}\frac{d^2}{dt^2}\boldsymbol{x}, \tag{4.5a}$$

and the right-hand side is

$$-k\sqrt{M}^{-1}\Gamma\sqrt{M}^{-1}\sqrt{M}\boldsymbol{x}. \tag{4.5b}$$

By using the notation

$$\boldsymbol{x}' = \sqrt{M}\boldsymbol{x}, \quad A = \sqrt{M}^{-1}\Gamma\sqrt{M}^{-1}, \tag{4.6}$$

the symmetric matrix A can be diagonalized by an orthonormal matrix P whose elements will be given by eq.(4.20). As a result, the diagonalized matrix Λ is

$$\Lambda = P^{-1}AP \tag{4.7}$$

with the matrix elements

$$(\Lambda)_{pq} = \delta_{pq}\frac{\Omega_p^2}{k}, \tag{4.8}$$

where Ω_p is the angular frequency of vibration. This leads us to the equation of motion for a set of harmonic oscillators:

$$\frac{d^2}{dt^2}\boldsymbol{x}' = -kA\boldsymbol{x}' = -kP\Lambda P^{-1}\boldsymbol{x}'. \tag{4.9}$$

By multiplying both sides of eq.(4.9) by the matrix P^{-1} from the left and defining a normal coordinate $\boldsymbol{y}$ as

$$\boldsymbol{y} = P^{-1}\boldsymbol{x}', \tag{4.10}$$

one derives

$$\frac{d^2}{dt^2}\boldsymbol{y} = -k\Lambda\boldsymbol{y} \tag{4.11a}$$

and

$$\frac{d^2}{dt^2}y_p = -\Omega_p^2 y_p. \tag{4.11b}$$

These two equations represent the motions of mutually independent harmonic oscillators. The number of normal coordinates for describing these harmonic oscillators is equal to that of the atoms N, and each coordinate is labeled by a mode number p, which does not depend on the wave-number or momentum. By using the orthonormal matrix P, the relation between $\boldsymbol{x}$ and $\boldsymbol{y}$ is expressed as

$$\boldsymbol{x} = \sqrt{M}^{-1} P\boldsymbol{y} \tag{4.12}$$

and

$$\boldsymbol{x}_i = \frac{1}{\sqrt{m_i}} \sum_{p=1}^{N} P_{ip} \boldsymbol{y}_p. \tag{4.13}$$

By rewriting the Hamiltonian in terms of the normal coordinate $\boldsymbol{y}_p$ and the conjugate momentum $\pi_p = (d/dt)\boldsymbol{y}_p$, and moreover, by replacing them with the corresponding operators $\hat{\boldsymbol{y}}_p$ and $\hat{\pi}_p$ for quantization, one has

$$\hat{H}\left(\hat{y}, \hat{\pi}\right) = \sum_{p=1}^{N} \frac{\hat{\pi}_p^2}{2} + \sum_{p=1}^{N} \Omega_p^2 \frac{\hat{y}_p^2}{2}. \tag{4.14}$$

Here, the commutation relation

$$\left[\hat{y}_p, \hat{\pi}_q\right] = \hat{y}_p \hat{\pi}_q - \hat{\pi}_q \hat{y}_p = i\hbar\delta_{pq} \tag{4.15}$$

is imposed. When the operators $\hat{c_p}$ and $\hat{c_p}^\dagger$ are defined as

$$\hat{c}_p = \frac{1}{\sqrt{2\hbar\Omega_p}} \left(\Omega_p \hat{y}_p + i\hat{\pi}_p\right), \tag{4.16}$$

$$\hat{c}_p^\dagger = \frac{1}{\sqrt{2\hbar\Omega_p}} \left(\Omega_p \hat{y}_p - i\hat{\pi}_p\right), \tag{4.17}$$

they satisfy the boson commutation relation

$$\left[\hat{c}_p, \hat{c}_q^\dagger\right] \equiv \hat{c}_p \hat{c}_q^\dagger - \hat{c}_q^\dagger \hat{c}_p = \delta_{pq}. \tag{4.18}$$

The operators $\hat{c_p}$ and $\hat{c_p}^\dagger$ respectively represent annihilation and creation operators for a phonon with eigenenergy $\hbar\Omega_p$. By using them, eq.(4.14) is rewritten as

$$\hat{H}_{phonon} = \sum_{p=1}^{N} \hbar\Omega_p \left(\hat{c}_p^\dagger \hat{c}_p + \frac{1}{2}\right). \tag{4.19}$$

When all the atoms are identical, that is, $m_i = m$, the matrix A in eq.(4.7) is diagonalized by the orthonormal matrix P with elements

$$P_{ip} = \sqrt{\frac{2}{N+1}} \sin\left(\frac{ip}{N+1}\pi\right) \quad (1 \le i, p \le N), \tag{4.20}$$

yielding the eigen angular frequency

$$\Omega_p = 2\sqrt{\frac{k}{m}} \sin\left[\frac{p}{2(N+1)}\pi\right] \tag{4.21}$$

(refer to eqs. (H.63) and (H.64) in Appendix H). In this case, all the vibration modes are delocalized; that is, they spread over the whole probe apex. In contrast, if there are some doped impurity atoms or defects in the probe apex, the modes cannot be simply expressed by using the sinusoidal functions of eqs. (4.20) and (4.21). If the masses of the impurity atoms are assumed to be different from the other atoms, and the spring constant remains unchanged, the behaviors of the vibration modes strongly depend on the positions of the impurity atoms and their masses.

In particular, if the masses of the impurity atoms are smaller than those of other atoms, special vibration modes, called localized modes, manifest themselves [8–11]. Figure 4.5(a) shows the vibration amplitude as a function of the atom site number, where the total number of sites is 30. Squares and circles represent two localized modes with the highest and the next-highest phonon energies, respectively, whereas the triangles represent the delocalized mode with the lowest energy. In the localized modes, the vibration amplitudes are large around the sites of impurity atoms, whereas those of the delocalized modes spread over the whole probe apex. The relation between the mode number and the phonon energy is plotted in Fig. 4.5(b).

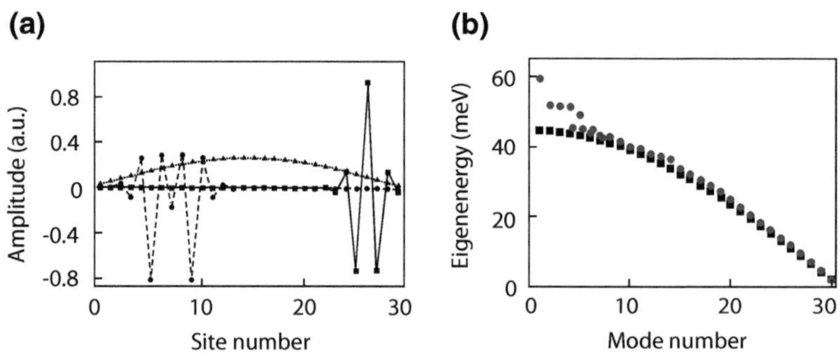

Fig. 4.5 Amplitude and eigenenergy of vibration modes. The number of modes, N, is 30. impurity atoms are at sites 5, 9, 18, 25, 26, and 27. Their masses are 0.5-times that of the other atoms. $\hbar\sqrt{k/m} = 22.4$ meV. **a** The vibration amplitudes. Squares and circles represent the first and second localized modes, respectively. Triangles represent the delocalized mode. **b** Eigenenergy. Circles and squares represent the results with and without impurity atoms, respectively

Squares represent the eigenenergies of phonons in the case of no impurity atoms, and the circles show those in the case of six impurity atoms, located at sites 5, 9, 18, 25, 26, and 27. The mass of the impurity atom is assumed to be 0.5-times that of the other atoms. It follows from this figure that the phonon energies of the localized modes are higher than those of the delocalized modes, which can be understood by the law of inertia, because the lighter atoms are more nimble in their motions than the heavier atoms.

Since the glass fiber typically used for an actual probe is not a perfect crystal but an amorphous material, it always contains lattice defects, as well as doped impurity atoms. In addition, since the probe apex has a tapered profile, the effective masses of the atoms depend on the position in the probe apex. Therefore, high-energy localized modes whose vibration amplitudes are spatially inhomogeneous can be created in the probe.

4.2 Transformation of the Hamiltonian

4.2.1 Diagonalization by Unitary Transformation

This subsection discusses the interaction between the DP and phonons, suggested by the experimental results of molecular dissociation in Subsect. 4.1.1. That is, as a result of the interaction with the DP, many phonons can cohere with each other, thus enabling a unique interaction that is different from the interaction between photons and phonons in a macroscopic material. This originates from the fact that the probe apex is a finite system and nanometric in size.

On the other hand, the DP localizes at one site of the lattice in the probe apex, where the extent of localization corresponds to the size of the atoms [2]. The Hamiltonian for this model is given by

$$\hat{H} = \sum_{i=1}^{N} \hbar\omega \tilde{a}_i^\dagger \tilde{a}_i + \left\{ \sum_{i=1}^{N} \frac{\hat{p}_i^2}{2m_i} + \sum_{i=1}^{N-1} \frac{k}{2} (\hat{x}_{i+1} - \hat{x}_i)^2 + \sum_{i=1,N} \frac{k}{2} \hat{\boldsymbol{x}}_i^2 \right\}$$
$$+ \sum_{i=1}^{N} \hbar\chi \tilde{a}_i^\dagger \tilde{a}_i \hat{\boldsymbol{x}}_i + \sum_{i=1}^{N-1} \hbar J \left(\tilde{a}_i^\dagger \tilde{a}_{i+1} + \tilde{a}_{i+1}^\dagger \tilde{a}_i \right), \quad (4.22)$$

[2] This size is represented by the material size a in the Yukawa function of eq. (2.24) in Chap. 2. The position in the one-dimensional space, coarse-grained by the unit of the material size a, is what we call the *site* here.

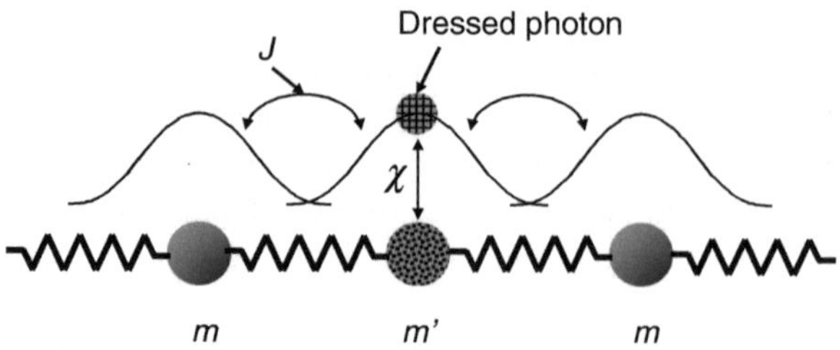

Fig. 4.6 Schematic explanation of the interaction between the dressed photon and phonons, and hopping of the dressed photon

where $\tilde{a}_i$ and $\tilde{a}_i^\dagger$ respectively denote the annihilation and creation operators of a DP with energy $\hbar\omega$ at site i in the lattice[3], and $\hat{\boldsymbol{x}}_i$ and $\hat{\boldsymbol{p}}_i$ respectively represent the displacement and conjugate momentum operators of the vibration. The mass of the atom at the site i is designated by m_i, and the atoms are assumed to be connected by springs with a spring constant k. The third and fourth terms stand for the DP–vibration interaction with the interaction energy $\hbar\chi$ and DP hopping with hopping energy $\hbar J$, respectively, as shown in Fig. 4.6. The quantities χ and J are called the coupling constant and hopping constant, respectively.

After the vibration field is quantized in terms of phonon operators of mode p and eigenenergy ($\hat{c}_i^\dagger$, $\hat{c}_i$, $\hbar\Omega_p$), eq.(4.22) can be rewritten as

$$\hat{H} = \sum_{i=1}^{N} \hbar\omega\tilde{a}_i^\dagger\tilde{a}_i + \sum_{p=1}^{N} \hbar\Omega_p\hat{c}_p^\dagger\hat{c}_p + \sum_{i=1}^{N}\sum_{p=1}^{N} \hbar\chi_{ip}\tilde{a}_i^\dagger\tilde{a}_i\left(\hat{c}_p^\dagger + \hat{c}_p\right)$$
$$+ \sum_{i=1}^{N-1} \hbar J\left(\tilde{a}_i^\dagger\tilde{a}_{i+1} + \tilde{a}_{i+1}^\dagger\tilde{a}_i\right), \tag{4.23}$$

with the coupling constant χ_{ip} at the site i for the phonon of mode p. By replacing $\hat{\boldsymbol{x}}_i$ in the third term of eq.(4.22) with eq.(4.13), and by transforming it by using eqs. (4.16) and (4.17), the third term of eq.(4.23) represents that the site-dependent coupling constant χ_{ip} is expressed by the original coupling constant χ of eq.(4.22):

$$\chi_{ip} = \chi P_{ip}\sqrt{\frac{\hbar}{2m_i\Omega_p}}. \tag{4.24}$$

In addition to eq.(4.18), annihilation and creation operators for a DP and a phonon satisfy the boson commutation relations as follows:

[3] As has been described in Sect. 2.1 in Chap. 2, annihilation and creation operators of the DP are given by $\sum_{\boldsymbol{k}\lambda}\tilde{a}_{\boldsymbol{k}\lambda}$ and $\sum_{\boldsymbol{k}\lambda}\tilde{a}_{\boldsymbol{k}\lambda}^\dagger$, respectively, which contain an infinite number of modulation sidebands. However, for simplicity, only one sideband is considered here, which is resonant with and interacts with the phonons. The eigenenergy and the annihilation and creation operators of this sideband are denoted by $\hbar\omega$, $\tilde{a}_i$, and $\tilde{a}_i^\dagger$, respectively, in the first term.

$$\left[\tilde{a}_i, \tilde{a}_j^\dagger\right] = \delta_{ij},$$
$$\left[\tilde{a}_i, \hat{c}_p\right] = \left[\tilde{a}_i, \hat{c}_p^\dagger\right] = \left[\tilde{a}_i^\dagger, \hat{c}_p\right] = \left[\tilde{a}_i^\dagger, \hat{c}_q^\dagger\right] = 0,$$
$$\left[\tilde{a}_i, \tilde{a}_j\right] = \left[\tilde{a}_i^\dagger, \tilde{a}_j^\dagger\right] = \left[\hat{c}_p, \hat{c}_q\right] = \left[\hat{c}_p^\dagger, \hat{c}_q^\dagger\right] = 0. \tag{4.25}$$

Based on the discussions in Subsect. 4.1.2, in the probe apex, there are localized phonons, which govern the spatial features of the DP. However, eq.(4.23) is not easily handled because of the third power of the operators in the third term. To overcome this difficulty, this term will be eliminated by diagonalizing a part of the Hamiltonian using a unitary transformation (refer to Section H.1 in Appendix H) [6, 12, 13].

For this diagonalization, an anti-Hermitian operator $\hat{S}$ is used which is defined as

$$\hat{S} = \sum_{i=1}^{N}\sum_{p=1}^{N} \frac{\chi_{ip}}{\Omega_p}\tilde{a}_i^\dagger \tilde{a}_i \left(\hat{c}_p^\dagger - \hat{c}_p\right) \tag{4.26}$$

(refer to eq. (H.14) in Appendix H). This equation yields

$$\hat{S}^\dagger = \sum_{i=1}^{N}\sum_{p=1}^{N} \frac{\chi_{ip}}{\Omega_p}\tilde{a}_i^\dagger \tilde{a}_i \left(\hat{c}_p - \hat{c}_p^\dagger\right) = -\hat{S}, \tag{4.27}$$

by which $\hat{S}$ is confirmed to be an anti-Hermitian operator. Unitary operators $\hat{U}$ and $\hat{U}^\dagger$ for the unitary transformation are given by

$$\hat{U} = e^{\hat{S}}, \tag{4.28}$$

$$\hat{U}^\dagger = \mathrm{e}^{-\hat{S}} = \hat{U}^{-1}, \tag{4.29}$$

These operators lead us to the exact transformation of the annihilation and creation operators for the DP and phonons:

$$\hat{\alpha}_i^\dagger \equiv \hat{U}^\dagger \tilde{a}_i^\dagger \hat{U} = \tilde{a}_i^\dagger \exp\left\{-\sum_{p=1}^{N} \frac{\chi_{ip}}{\Omega_p}\left(\hat{c}_p^\dagger - \hat{c}_p\right)\right\}, \tag{4.30a}$$

$$\hat{\alpha}_i \equiv \hat{U}^\dagger \tilde{a}_i \hat{U} = \tilde{a}_i \exp\left\{\sum_{p=1}^{N} \frac{\chi_{ip}}{\Omega_p}\left(\hat{c}_p^\dagger - \hat{c}_p\right)\right\}, \tag{4.30b}$$

$$\hat{\beta}_p^\dagger \equiv \hat{U}^\dagger \hat{c}_p^\dagger \hat{U} = \hat{c}_p^\dagger + \sum_{p=1}^{N} \frac{\chi_{ip}}{\Omega_p}\tilde{a}_i^\dagger \tilde{a}_i, \tag{4.31a}$$

$$\hat{\beta}_p \equiv \hat{U}^\dagger \hat{c}_p \hat{U} = \hat{c}_p + \sum_{p=1}^{N} \frac{\chi_{ip}}{\Omega_p} \tilde{a}_i^\dagger \tilde{a}_i. \tag{4.31b}$$

These transformed operators can be regarded as the annihilation and creation operators of a new quasi-particle, which represents the coupled state of the DP and phonons[4]. Since they are derived by a unitary transformation, they satisfy the same boson commutation relations as those of the DP and phonons:

$$\left[\hat{\alpha}_i, \hat{\alpha}_j^\dagger\right] = \hat{U}^\dagger \tilde{a}_i \hat{U} \hat{U}^\dagger \tilde{a}_j^\dagger \hat{U} - \hat{U}^\dagger \tilde{a}_j^\dagger \hat{U} \hat{U}^\dagger \tilde{a}_i \hat{U} = \hat{U}^\dagger \left[\tilde{a}_i, \tilde{a}_j^\dagger\right] \hat{U} = \delta_{ij}, \tag{4.32}$$

$$\left[\hat{\beta}_p, \hat{\beta}_q^\dagger\right] = \delta_{pq}, \tag{4.33}$$

$$\left[\tilde{\alpha}_i, \tilde{\beta}_p\right] = \left[\tilde{\alpha}_i, \tilde{\beta}_p^\dagger\right] = \left[\tilde{\alpha}_i^\dagger, \tilde{\beta}_p\right] = \left[\tilde{\alpha}_i^\dagger, \tilde{\beta}_p^\dagger\right] = 0, \tag{4.34a}$$

$$\left[\tilde{\alpha}_i, \tilde{\alpha}_j\right] = \left[\tilde{\alpha}_i^\dagger, \tilde{\alpha}_j^\dagger\right] = \left[\tilde{\beta}_p, \tilde{\beta}_q\right] = \left[\tilde{\beta}_p^\dagger, \tilde{\beta}_q^\dagger\right] = 0. \tag{4.34b}$$

Using these annihilation and creation operators, the Hamiltonian of eq.(4.23) can be rewritten as

$$\hat{H} = \sum_{i=1}^{N} \hbar\omega \hat{\alpha}_i^\dagger \hat{\alpha}_i + \sum_{p=1}^{N} \hbar\Omega_p \hat{\beta}_p^\dagger \hat{\beta}_p - \sum_{i=1}^{N}\sum_{j=1}^{N}\sum_{p=1}^{N} \frac{\hbar\chi_{ip}\chi_{jp}}{\Omega_p} \hat{\alpha}_i^\dagger \hat{\alpha}_i \hat{\alpha}_j^\dagger \hat{\alpha}_j$$
$$+ \sum_{i=1}^{N-1} \hbar \left(\hat{J}_i \hat{\alpha}_i^\dagger \hat{\alpha}_{i+1} + \hat{J}_i^\dagger \hat{\alpha}_{i+1}^\dagger \hat{\alpha}_i\right), \tag{4.35}$$

with

$$\hat{J}_i = J \exp\left\{\sum_{p=1}^{N} \frac{\left(\chi_{ip} - \chi_{i+1p}\right)}{\Omega_p} \left(\hat{\beta}_p^\dagger - \hat{\beta}_p\right)\right\}. \tag{4.36}$$

The third term of eq.(4.35) is proportional to the fourth power of the operator $\hat{\alpha}_i^\dagger \hat{\alpha}_i \hat{\alpha}_j^\dagger \hat{\alpha}_j$. However, since this term is represented as $\hat{N}_i \hat{N}_j$ by using the number operator of the quasi-particle $\hat{N}_i = \hat{\alpha}_i^\dagger \hat{\alpha}_i$, it is found that this term does not represent any interactions involving annihilation or creation of quasi-particles. Furthermore, since the eigenstates of the first three terms are number states of the quasi-particle, this Hamiltonian has been diagonalized, except for the fourth term. In the fourth term, the hopping constant J has been replaced by a site-dependent operator $\hat{J}_i$, which means

[4] The quati-particle represented by eqs. (4.30a) and (4.30b) is the DP which is dressed by the phonon energies. Therefore, its eigenenergy is modulated, giving an infinite number of sidebands, as was described at the end of Sect. 2.1 in Chap. 2.

that the phonons interact with the DP indirectly through the hopping operator $\hat{J}_i$. As will be described in Sect. 4.3, this interaction can be expressed as a site-dependent hopping of DP if the mean field approximation is employed. Therefore, it suffices to consider only the operators $\hat{\alpha}_i$ and $\hat{\alpha}_i^\dagger$ when analyzing the behavior of the DP.

4.2.2 Physical Picture of the Quasi-Particle

In order to create a physical picture of the quasi-particle introduced in the previous subsection, the creation operator $\hat{\alpha}_i^\dagger$ is applied to the vacuum state $|0\rangle$. Then, from eq.(4.30a) one derives

$$\begin{aligned}\hat{\alpha}_i^\dagger |0\rangle &= \tilde{a}_i^\dagger \exp\left\{-\sum_{p=1}^{N} \frac{\chi_{ip}}{\Omega_p}\left(\hat{c}_p^\dagger - \hat{c}_p\right)\right\}|0\rangle \\ &= \tilde{a}_i^\dagger \prod_{p=1}^{N} \exp\left\{-\frac{\chi_{ip}}{\Omega_p}\left(\hat{c}_p^\dagger - \hat{c}_p\right)\right\}|0\rangle \\ &= \tilde{a}_i^\dagger \prod_{p=1}^{N} \exp\left\{-\frac{1}{2}\left(\frac{\chi_{ip}}{\Omega_p}\right)^2\right\} \exp\left(-\frac{\chi_{ip}}{\Omega_p}\hat{c}_p^\dagger\right)|0\rangle,\end{aligned} \tag{4.37}$$

where the relation $\hat{c}_p|0\rangle = 0$ was used in the second row in order to derive the third row. The third row of this equation represents that the state of the DP at site i is associated with multimode coherent phonons[5]. In other words, it means that the DP is dressed by the energies of an infinite number of phonons.

When $\hat{\beta}_p^\dagger$ is applied to the vacuum state $|0\rangle$, one has

---

[5] Since the multimode phonons are confined in the nanometric space in the probe apex, their state functions easily cohere, and thus, these multimode phonons can stay in a coherent state. In other words, the lattice vibrations of these modes are excited in an in-phase manner. Therefore, the coherent state of the phonons in the nanometric space is independent of heat generation. In contrast, the phases of the lattice vibrations in a macroscopic material are random, causing heating of the material.

A laser can generate a nearly coherent state of light above the oscillation threshold based on building up the optical energy by repeated photon emission and absorption. If the annihilation and creation operators of the phonons in eq.(4.30a) are replaced with those of photons, it is readily understood that the exponential function on the right-hand side of eq.(4.30a), called a displacement operator function, represents the infinitely repeated photon emission and absorption. The coherent state of photons represents a state in which the width of the probability density function of the optical electric field amplitude is kept at a minimum; i.e., the probability density is cohered.

In the case of the phonons described above, for a short time after the phonons are excited by the incident light, they stay in the coherent state (refer to Fig. 4.7(a)). Afterwards, the coherent state is destroyed by the relaxation due to phonon–phonon scattering. In the case of a laser, the magnitude of this scattering corresponds to the cavity loss, governing the oscillation threshold. Therefore, again in the case of phonons, if light with sufficiently high energy to overcome the scattering loss is injected into the probe apex, the coherent state of the phonons is sustained. The minimum optical energy for this sustainment corresponds to the above-mentioned oscillation threshold of the laser.

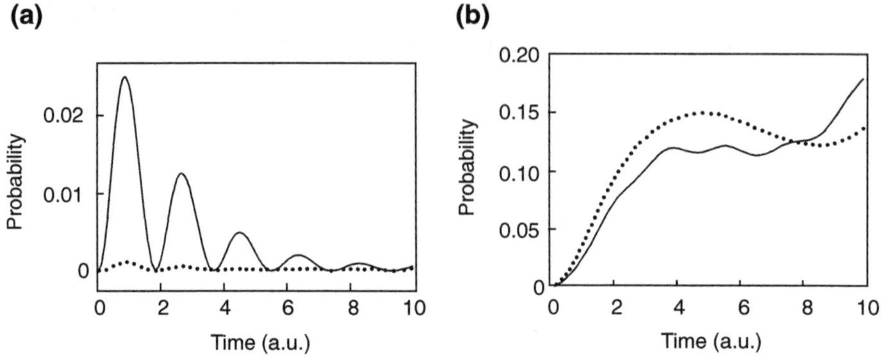

Fig. 4.7 Temporal evolution of the excitation probability of phonons. The number of modes, N, is 20. The impurity atoms are at sites 4, 6, 13, and 19 . Their masses are 0.2-times that of the other atoms. $\chi = 10.0\text{fs}^{-1}\text{nm}^{-1}$. **a** and **b** show the results for the localized mode and de-localized mode, respectively. Solid curve represents the result when the dressed photon is initially generated at impurity atom site 4. Dotted curve is the result when the dressed photon is initially generated at site 5

$$\hat{\beta}_p^{\dagger} |0\rangle = \hat{c}_p^{\dagger} |0\rangle , \tag{4.38}$$

which is expressed only by the bare phonon operator $\hat{c}_p^{\dagger}$ (before the transformation) of the same mode p. That is, the phonon is not affected by the DP. Therefore, in the following discussions, it is possible to focus on the quasi-particle expressed by $\hat{\alpha}_i^{\dagger}$ and $\hat{\alpha}_i$. Note that this is valid only if the average number of DPs, i.e., the expectation value of $\hat{\alpha}_i^{\dagger}\hat{\alpha}_i$ in the second term of eq.(4.31a), is not so large that the number fluctuation is more important than the number itself. That is, the Hamiltonian of eq.(4.35) is effective for describing the state of the probe with a small number of DPs, where the quantum nature is clearly observed.

The coherent state means that an infinite number of quasi-particles cohere with each other (refer to section H.2 in Appendix H). However, since it is not an eigenstate of the Hamiltonian, the number of quasi-particles fluctuates. When the light is injected into the probe apex, the phonons are excited by this fluctuation. If the phonons are in the vacuum state and the DP is generated at site i in the one-dimensional lattice by the injected light, the initial condition of the system at time $t = 0$ is expressed as $|\psi\rangle = \tilde{a}_i^{\dagger} |0\rangle \equiv \hat{\alpha}_i^{\dagger} |\gamma\rangle$. Under this condition, the probability that the phonons still stay in the vacuum state at time t is given by

$$P'(t) = \left| \langle\psi| \exp\left(-\frac{iH't}{\hbar}\right) |\psi\rangle \right|^2 , \tag{4.39a}$$

from which the probability of exciting the phonons of the mode p is given by $P(t) = 1 - P'(t)$, i.e.,

$$P(t) = 1 - \exp\left\{2\left(\frac{\chi_{ip}}{\Omega_p}\right)^2 \left[\cos\left(\Omega_p t\right) - 1\right]\right\} \tag{4.39b}$$

(refer to eq. (H.46) in Appendix H). Here, the hopping term was neglected for simplicity. This means that the $\hat{H}'$ in eq.(4.39a) corresponds to the Hamiltonian of eq.(4.35), from which the fourth term was excluded.

The probability expressed by eq.(4.39b) oscillates with the period and takes the maximum at time $2\pi/\Omega_p$. Since the frequencies of the localized modes are larger than those of the delocalized modes, the delocalized modes are excited after the excitation probability of the localized modes reaches the maximum. Figures 4.7(a) and (b) show the temporal evolution of the excitation probability $P_{p_0}(t)$ for the localized and delocalized modes, respectively, calculated from

$$\begin{aligned} P_{p_0}(t) &= P(t : p = p_0)\, P'(t : p \neq p_0) \\ &= \left[1 - \exp\left\{2\left(\frac{\chi_{ip_0}}{\Omega_{p_0}}\right)^2 \left[\cos\left(\Omega_{p_0} t\right) - 1\right]\right\}\right] \exp\left\{\sum_{p \neq p_0} 2\left(\frac{\chi_{ip}}{\Omega_p}\right)^2 \left[\cos\left(\Omega_p t\right) - 1\right]\right\} \end{aligned} \tag{4.40}$$

(refer to eq. (H.47) in Appendix H), where a specific phonon mode p_0 is excited, and other modes remain in the vacuum state. Comparing the solid curves in both figures, it is found that the localized mode is excited soon after the light is injected, and then the excitation probability of the delocalized mode gradually increases. Furthermore, the broken curve in Fig. 4.7(a) shows that the localized mode is not excited if the DP is not generated at the localized site, i.e., at the impurity site. The solid curve in Fig. 4.7(a) shows that the excitation probability of the localized mode decreases sinusoidally, because the excitation probabilities of other modes gradually increase triggered by the fluctuations. In summary, the localized modes can be excited by the fluctuations in the number of phonons if the phonons are in the coherent state.

It should be noted that the present theoretical model does not consider the temperature dependence of the system. If the phonon energy follows the Boltzmann distribution, the localized modes are hardly excited because of their high energies. Also, it should be noted that temperature is an important parameter for the incoherent phonons, whereas the phonons in the coherent state are independent of the temperature [14]. However, since the probe temperature used for molecular dissociation has not yet been evaluated in detail, further studies are required to judge whether the system is in thermal equilibrium or not.

4.2.3 The Equilibrium Positions of Atoms

Since the equilibrium positions of the atoms connected by the spring are displaced as a result of the DP–phonon interaction, this subsection derives the value of this displacement in order to use it in the next section.

The expectation value $\langle \hat{\boldsymbol{x}}_j \rangle_i$ of the displacement $\hat{\boldsymbol{x}}_j$ of site j is derived when the DP is localized at site j (state $\hat{\alpha}_i^\dagger |0\rangle$). Here, the state $\hat{\alpha}_i^\dagger |0\rangle$ is expressed, using eq.(4.37), as

$$\hat{\alpha}_i^\dagger |0\rangle = \tilde{a}_i^\dagger A \exp\left(\sum_{p=1}^{N} \gamma_{ip} \hat{c}_p^\dagger\right) |0\rangle \tag{4.41}$$

with

$$\gamma_{ip} = -\frac{\chi_{ip}}{\Omega_p}, \tag{4.42a}$$

$$A = \exp\left\{-\frac{1}{2}\sum_{p=1}^{N} \gamma_{ip}^2\right\}. \tag{4.42b}$$

By noting that the coherent state is the eigenstate of the annihilation operator and is normalized (refer to eqs. (H.31) and (H.32) in Appendix H), the expectation value $\langle \hat{\boldsymbol{x}}_j \rangle_i$ is

$$\begin{aligned}
\langle \hat{\boldsymbol{x}}_j \rangle_i &= \langle 0| \hat{\alpha}_i \hat{\boldsymbol{x}}_j \hat{\alpha}_i^\dagger |0\rangle \\
&= \langle 0| \tilde{a}_i A \exp\left(\sum_{p'=1}^{N} \gamma_{ip'} \hat{c}_{p'}\right) \sum_{p=1}^{N} P_{jp} \frac{\hbar}{2m_j \Omega_p} \\
&\quad \times \left(\hat{c}_p^\dagger + \hat{c}_p\right) A \exp\left(\sum_{p''=1}^{N} \gamma_{ip''} \hat{c}_{p''}^\dagger\right) \tilde{a}_i^\dagger |0\rangle \\
&= \sum_{p=1}^{N} 2\gamma_{ip} P_{ip} \frac{\hbar}{2m_j \Omega_p} = -\sum_{p=1}^{N} \frac{\hbar \chi P_{ip} P_{jp}}{\sqrt{m_i m_j} \Omega_p^2} \\
&= -\frac{2}{\chi} \sum_{p=1}^{N} \frac{\chi_{ip} \chi_{jp}}{\Omega_p},
\end{aligned} \tag{4.43}$$

where eq.(4.24) was used to derive the last row.

By inserting this into the third term of eq.(4.35), the Hamiltonian is rewritten as

$$\begin{aligned}
\hat{H} &= \sum_{i=1}^{N} \hbar\omega \hat{\alpha}_i^\dagger \hat{\alpha}_i + \sum_{p=1}^{N} \hbar\Omega_p \hat{\beta}_p^\dagger \hat{\beta}_p + \sum_{i=1}^{N} \sum_{j=1}^{N} \frac{\hbar\chi}{2} \langle \boldsymbol{x}_j \rangle_i \hat{\alpha}_i^\dagger \hat{\alpha}_i \hat{\alpha}_j^\dagger \hat{\alpha}_j \\
&\quad + \sum_{i=1}^{N-1} \hbar \left(\hat{J}_i \hat{\alpha}_i^\dagger \hat{\alpha}_{i+1} + \hat{J}_i^\dagger \hat{\alpha}_{i+1}^\dagger \hat{\alpha}_i\right).
\end{aligned} \tag{4.44}$$

The third term of this equation represents the DP–phonon interaction even though it does not explicitly contain any operators for phonons.

4.3 Localization Mechanism of Dressed Photons

4.3.1 Conditions for Localization

The DP does not localize at any specific site in the absence of DP–phonon interactions because the values of the physical quantities, such as the energy and hopping constant J, for the DP are independent of the site number. However, by the DP–phonon interaction, the spatial behavior of the DP is drastically modified due to the localized modes of phonons. This behavior can be analyzed by diagonalizing the first, third, and fourth terms of eq.(4.44), which contain the operators for the DP. For this diagonalization, the mean field approximation is employed for the third term. That is, the average value of the number of DPs is inserted into the operator $\hat{N}_i(=\hat{\alpha}_i^\dagger\hat{\alpha}_i)$ or $\hat{N}_j(=\hat{\alpha}_j^\dagger\hat{\alpha}_j)$ in this term. This average value can be assumed to be $1/N$ in the case when the number of DPs is unity and its field is homogeneously distributed over the whole one-dimensional lattice. Thus, the third term is approximated as

$$\sum_{i=1}^{N}\sum_{j=1}^{N}\frac{\hbar\chi}{2}\left\langle \boldsymbol{x}_j\right\rangle_i\hat{\alpha}_i^\dagger\hat{\alpha}_i\hat{\alpha}_j^\dagger\hat{\alpha}_j \simeq \sum_{i=1}^{N}\sum_{j=1}^{N}\frac{\hbar\chi}{2}\left\langle \boldsymbol{x}_j\right\rangle_i\frac{1}{N}\hat{\alpha}_i^\dagger\hat{\alpha}_i \equiv -\sum_{i=1}^{N}\hbar\omega_i\hat{\alpha}_i^\dagger\hat{\alpha}_i. \quad (4.45)$$

The angular frequency ω_i on the right-hand side is expressed as

$$\omega_i = -\sum_{j=1}^{N}\frac{\chi\left\langle \boldsymbol{x}_j\right\rangle_i}{2N} \equiv \sum_{j=1}^{N}\sum_{p=1}^{N}\frac{\hbar\chi^2 P_{ip}P_{jp}}{2N\sqrt{m_i m_j}\Omega_p^2}, \quad (4.46)$$

where eq.(4.43) was used to replace the middle part with the right-hand side.

Furthermore, for simplicity, by replacing the operator $\hat{J}_i$ of eq.(4.36) with the hopping constant J,

$$\hat{J}_i = J, \quad (4.47)$$

i.e., neglecting the site-dependence of the hopping, the first, third, and fourth terms of eq.(4.44) are expressed in a quadratic form

$$\hat{H}_{DP} = \sum_{i=1}^{N}\hbar(\omega-\omega_i)\hat{\alpha}_i^\dagger\hat{\alpha}_i + \sum_{i=1}^{N-1}\hbar J\left(\hat{\alpha}_i^\dagger\hat{\alpha}_{i+1}+\hat{\alpha}_{i+1}^\dagger\hat{\alpha}_i\right), \quad (4.48)$$

or in the matrix form

$$\hat{H}_{DP} = \hbar\hat{\boldsymbol{\alpha}}^{\dagger}\begin{pmatrix} \omega-\omega_1 & J & \cdot & 0 \\ J & \omega-\omega_2 & \cdot & \cdot \\ \cdot & \cdot & \cdot & J \\ 0 & \cdot & J & \omega-\omega_N \end{pmatrix}\hat{\boldsymbol{\alpha}}, \quad \hat{\boldsymbol{\alpha}} = \begin{pmatrix} \hat{\alpha}_1 \\ \hat{\alpha}_2 \\ \cdot \\ \hat{\alpha}_N \end{pmatrix}. \tag{4.49}$$

The effect from the phonons is involved in ω_i in the diagonal elements of this matrix.

Denoting an orthonormal matrix to diagonalize the Hamiltonian of eq.(4.49) by Q, one has

$$\hat{H}_{DP} = \sum_{r=1}^{N} \hbar\Omega_r \hat{A}_r^{\dagger}\hat{A}_r, \tag{4.50}$$

where $\hbar\Omega_r$ is the r-th eigenvalue, and the relations

$$\hat{A}_r = \sum_{i=1}^{N}\left(Q^{-1}\right)_{ri}\hat{\alpha}_i = \sum_{i=1}^{N} Q_{ir}\hat{\alpha}_i, \tag{4.51}$$

$$\left[\hat{A}_r, \hat{A}_s^{\dagger}\right] \equiv \hat{A}_r\hat{A}_s^{\dagger} - \hat{A}_s^{\dagger}\hat{A}_r = \delta_{rs} \tag{4.52}$$

hold. Using these relations, the time evolution of the number operator for the DP at site i is derived by using

$$\hat{N}_i = \hat{\alpha}_i^{\dagger}\hat{\alpha}_i = \left(\sum_{r=1}^{N} Q_{ir}\hat{A}_r^{\dagger}\right)\left(\sum_{s=1}^{N} Q_{is}\hat{A}_s\right) \tag{4.53}$$

and eq.(4.52). The result is expressed in the Heisenberg representation as

$$\begin{aligned} \hat{N}_i(t) &= \exp\left(i\frac{\hat{H}_{DP}}{\hbar}t\right)\hat{N}_i \exp\left(-i\frac{\hat{H}_{DP}}{\hbar}t\right) \\ &= \sum_{r=1}^{N}\sum_{s=1}^{N} Q_{ir}Q_{is}\hat{A}_r^{\dagger}\hat{A}_s \exp\{i(\Omega_r - \Omega_s)t\} \\ &= \sum_{r=1}^{N}\sum_{s=1}^{N} Q_{ir}Q_{is}\hat{A}_r^{\dagger}\hat{A}_s \cos\{(\Omega_r - \Omega_s)t\}. \end{aligned} \tag{4.54}$$

Since $\hat{N}_i(t)$ is a Hermitian operator, the exponential function in the second row was replaced with the cosinusoidal function in the third row. Since one can express the state $|\psi_j\rangle$ of the DP localized at site j at time $t = 0$ as

$$|\psi_j\rangle = \hat{\alpha}_j^{\dagger}|0\rangle = \sum_{r=1}^{N} Q_{jr}\hat{A}_r^{\dagger}|0\rangle, \tag{4.55}$$

the expectation value of the number of DPs at site i at time t under the initial condition of eq.(4.55) is given by

$$\langle N_i(t)\rangle_j = \langle \psi_j | \hat{N}_i(t) | \psi_j \rangle = \sum_{r=1}^{N}\sum_{s=1}^{N} Q_{ir}Q_{jr}Q_{is}Q_{js}\cos\{(\Omega_r - \Omega_s)t\}. \quad (4.56)$$

The value of $\langle N_i(t)\rangle_j$ given by this equation can be regarded as the observation probability of a DP at an arbitrary site i and time t, initially located at site j. In the case of no DP–phonon interaction ($\omega_i = 0$) and an infinite number of sites ($N \to \infty$), this function is analytically expressed in terms of a Bessel function $J_n(x)$ of the first kind as

$$\langle N_i(t)\rangle_j = \left\{ J_{j-i}(2Jt) - (-1)^i J_{j+i}(2Jt) \right\}^2 \quad (4.57)$$

(refer to eq. (H.71) in Appendix H). Here, the argument J in the Bessel function is the hopping constant. Equation (4.57) means that the expectation value is small at sites distant from the site j at which the DP is generated at time $t = 0$, and furthermore, it decreases with time. This decrease suggests diffusion of the DP from the initial site j.

Although the quantity ω_i in the diagonal elements of eq.(4.49) reflects the inhomogeneity of the phonon field, this inhomogeneity is eliminated if all the phonon modes in the right-hand side of eq.(4.46) are summed up, and, as a result, this equation is expressed in a symmetric form

$$\omega_i = \frac{\hbar\chi^2}{2Nk}\sum_{j=1}^{N}\frac{1}{N+1}\sum_{n=1}^{N}\frac{\sin\left(\frac{in}{N+1}\pi\right)\sin\left(\frac{jn}{N+1}\pi\right)}{1-\cos\left(\frac{n}{N+1}\pi\right)} \quad (4.58)$$

(refer to eq. (H.76b) in Appendix H). Therefore, to maintain the inhomogeneity, one has to sum up only the localized modes. This summation is allowed because the DP can interact only with the localized modes and does not interact with the delocalized modes which spread over the whole lattice. Furthermore, since the localized modes are selectively excited soon after the light injection, as was described in Subsect. 4.2.2, summation of only the localized modes corresponds to analyzing the behavior of the DP and phonons in the time domain immediately after the phonon excitation.

Based on the considerations described above, the temporal evolution of the observation probability $\langle N_i(t)\rangle_j$ of a DP at each site was derived by summing up only the localized modes, and the results are presented in Figs. 4.8(a) and (b) for the DP–phonon coupling constants $\chi = 0$ and $\chi = 1.4 \times 10^3 \text{fs}^{-1}\text{nm}^{-1}$, respectively. Without the DP–phonon coupling, the DP spreads over the whole lattice as a result of hopping, as shown in Fig. 4.8(a), which can be approximated by eq.(4.57). That is, the DP is reflected at the end of the finite lattice, and freely hops, meaning that the DP is not localized at any site. On the other hand, Fig. 4.8(b) shows the result with DP–phonon coupling, which shows that the DP slowly moves from one impurity

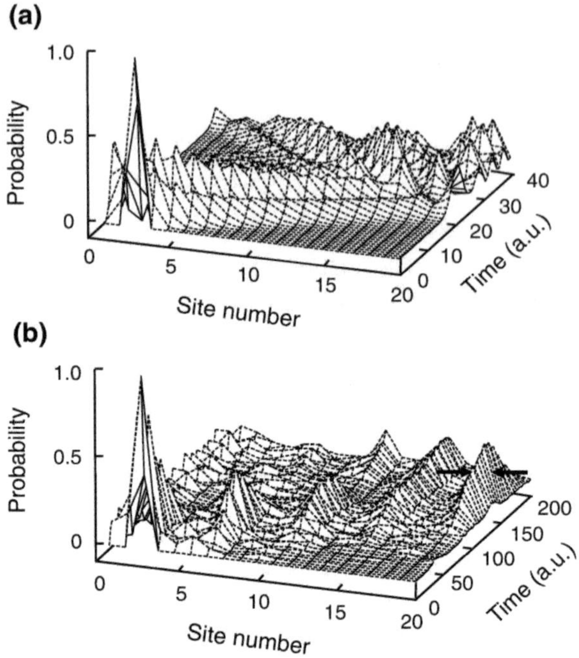

Fig. 4.8 Temporal evolution of the observation probability of the dressed photons at each site. The number of modes, N, is 20. The impurity atoms are at sites 3, 7, 11, 15, and 19. Their masses are 0.2-times that of the other atoms. As an initial condition, the dressed photon is assumed to be generated at impurity atom site 3. $\hbar\omega = 1.81$ eV, $\hbar J = 0.5$ eV, $\chi/\sqrt{kJ/\hbar} = 15.3$. **a** and **b** show the results without ($\chi = 0$) and with ($\chi = 1.4 \times 10^3 \mathrm{fs}^{-1}\mathrm{nm}^{-1}$) coupling dressed-photons and phonons, respectively

site to the other instead of freely hopping (note that the values on the time-axis of Fig. 4.8(b) are five times those of Fig. 4.8(a)).

Since the effect of localization represented by χ is contained in the diagonal elements in the Hamiltonian, whereas the hopping constant J is contained in the off-diagonal elements, localization or hopping can be judged by comparing the values of these constants. Since the DP is localized when $\omega_i > J$, one can derive

$$\chi > N\sqrt{\frac{kJ}{\hbar}}, \tag{4.59}$$

which is the criterion for localization. Here, the relations $\omega_i \sim \hbar\chi^2 P_{ip}^2/Nm_i\Omega_p^2$ (obtained from eq.(4.46)) and $P_{ip}/\Omega_p \sim \sqrt{m_i/Nk}$ (obtained from eqs. (4.20) and (4.21)) were inserted into $\omega_i > J$.

It is found that the width of the curve in Fig. 4.8(b) (identified by the gap between the two facing arrows) is narrow, and that this curve takes a large value at impurity sites. That is, the DP localizes only at the impurity sites and its extent of localization is narrow.

4.3.2 Position of Localization

The previous subsection obtained the criterion for DP localization (eq.(4.59)) at a specific site by analyzing the diagonal elements of the Hamiltonian. This subsection analyzes the off-diagonal elements, which are represented by the fourth term of eq.(4.35) with the site-dependent hopping operator as a result of the unitary transformation. Since the site-dependence, represented by eq.(4.36), contains the phonon operators and thus cannot be readily treated, the mean field approximation is employed, similarly to the previous subsection. That is, since the phonons are in the coherent state $|\gamma\rangle$, which is represented by the third row of eq.(4.37), the expectation value $J_i \left(= \langle\gamma|\, \hat{J}_i \,|\gamma\rangle\right)$ of the hopping operator $\hat{J}_i$ of eq.(4.36) is derived and inserted into the fourth-term of eq.(4.35).

Since the coherent state $|\gamma\rangle$ of the phonons is the eigenstate of the annihilation operator $\hat{c}_p$ (refer to eq. (H.31) in Appendix H), it satisfies the relation

$$\hat{c}_p\,|\gamma\rangle = \gamma_p\,|\gamma\rangle\,, \tag{4.60}$$

where γ_p is an eigenvalue. Thus, the relation

$$\exp\left(-\sum_p \kappa_p \hat{c}_p\right)|\gamma\rangle = \exp\left(-\sum_p \kappa_p \gamma_p\right)|\gamma\rangle \tag{4.61}$$

holds, where is a real number constant. Since eqs. (4.31a) and (4.31b) suggest that the difference between the annihilation and creation operators for the phonons remains unchanged even after the unitary transformation

$$\hat{\beta}_p^\dagger - \hat{\beta}_p = \hat{c}_p^\dagger - \hat{c}_p, \tag{4.62}$$

eqs. (4.60)–(4.62) lead us to derive the expectation value J_i, and the result is expressed as

$$\begin{aligned}
J_i &= \langle\gamma|\, \hat{J}_i \,|\gamma\rangle \\
&= J\,\langle\gamma|\exp\left(\sum_{p=1}^{N} C_{ip}\left(\hat{c}_p^\dagger - \hat{c}_p\right)\right)|\gamma\rangle \\
&= J\exp\left(-\tfrac{1}{2}\sum_{p=1}^{N} C_{ip}^2\right)\langle\gamma|\exp\left(\sum_{p'=1}^{N} C_{ip'}\hat{c}_{p'}^\dagger\right)\exp\left(-\sum_{p''=1}^{N} C_{ip''}\hat{c}_{p''}\right)|\gamma\rangle \\
&= J\exp\left(-\tfrac{1}{2}\sum_{p=1}^{N} C_{ip}^2\right)\exp\left(\sum_{p'=1}^{N} C_{ip'}\gamma_{p'}\right)\exp\left(-\sum_{p''=1}^{N} C_{ip''}\gamma_{p''}\right)\langle\gamma\mid\gamma\rangle \\
&= J\exp\left(-\tfrac{1}{2}\sum_{p=1}^{N} C_{ip}^2\right).
\end{aligned} \tag{4.63}$$

Here, eq. (H.28) in Appendix H was used for the transformation from the second row to the third row. The constant C_{ip} is defined by

$$C_{ip} \equiv \frac{\chi_{ip} - \chi_{i+1p}}{\Omega_p}. \tag{4.64}$$

Since the last row of eq.(4.63) does not contain the eigenvalue γ_p, it is found that the result of the mean field approximation is independent of as long as the phonons are in the coherent state. On the other hand, the argument in the exponential function represents the summation of C_{ip}^2 over all the modes ($p = 1 - N$) of the phonons, including the localized modes. Therefore, unlike the case of Fig. 4.8, all of the phonon modes will be summed up in the following calculations. In this case, the expectation value J_i of eq.(4.63) corresponds to the off-diagonal elements in the Hamiltonian of eq.(4.49) and represents the effect of the localized modes, i.e., the inhomogeneity of the phonon field. On the other hand, in the diagonal elements is independent of the localized modes because eq.(4.58) is inserted into . Figure 4.9 shows the calculated result of the site-dependence of J_i for $\chi = 40.0\text{fs}^{-1}\text{nm}^{-1}$, where the impurity site numbers are 4, 6, 13, and 19. It follows from this figure that the hopping constants are highly modified around the impurity sites.

It is not straightforward to grasp the possibility of localization at each site from the time dependence of the spatial profile of the DP shown in Fig. 4.8, and therefore, eigenstates of the DP energy have to be investigated. The eigenstate $|r\rangle$ of the Hamiltonian with the eigenvalue $\hbar\Omega_r$ is given by the superposition of the states $\hat{\alpha}_i^\dagger |0\rangle$ of the DP localized at all sites i and expressed as

$$|r\rangle = \sum_{i=1}^{N} Q_{ir}\hat{\alpha}_i^\dagger |0\rangle. \tag{4.65}$$

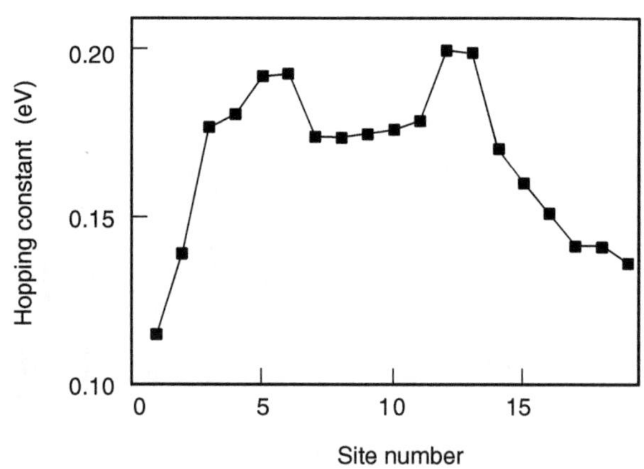

Fig. 4.9 Values of the site-dependent hopping constant. The number of modes, N, is 20. The impurity atoms are at sites 4, 6, 13, and 19. Their masses are 0.2-times that of the other atoms. $\hbar\omega = 1.81$ eV, $\hbar J = 0.5$ eV, $\chi = 40.0\text{fs}^{-1}\text{nm}^{-1}$

Here, the coefficient Q_{ir} is the (i, r)-th element of the orthonormal matrix Q used for diagonalizing the Hamiltonian of eq.(4.49); i.e., this coefficient can be regarded as the spatial coordinate representation of the eigenstate of the DP energy. Thus, by considering the column vector of the matrix Q for representing the eigenstate of the maximum energy, as an example, the values of the squares of this vector's elements, $|Q_{ir}|^2$, are calculated and displayed in Fig. 4.10. These values represent the occupation probabilities of the DP at each site. In the absence of DP–phonon coupling (curve A: $\chi = 0$), the DP hops and, thus, its field is distributed over the whole probe apex. In the case of the DP–phonon coupling (curve B: $\chi = 40.0\mathrm{fs}^{-1}\mathrm{nm}^{-1}$), the DP can localize at an impurity site. Although this figure represents the results for one mode only, there exist other modes in which the DP localizes at the other impurity sites. By increasing the coupling constant (curve C: $\chi = 54.0\mathrm{fs}^{-1}\mathrm{nm}^{-1}$), the DP can localize at the end of the lattice (see the right end of the curve C). This originates from the finite size of the lattice and is called the "finite-size effect" [6, 15]. In this case, besides the modes shown by the curve C, there exist several other modes that localize at the left end or at the impurity sites. Further increases in the coupling constant decrease the value of J_i, as is understood from eq.(4.63), which suppresses the DP hopping. However, since the angular frequency $\omega - \omega_i$ of the DP becomes negative if the value of χ becomes larger than a certain value, the present theoretical model becomes invalid.

The above discussions enable us to find the site of the DP localization by analyzing the off-diagonal elements. Comparing with the curves in Fig. 4.8(b), it is found that the curves B and C in Fig. 4.10 have peaks at the impurity sites, and their widths (shown by the gap between the two facing arrows) are broader than that of the curve in Fig. 4.8(b), and furthermore, the tails of the curves extend to the sites that are adjacent to the relevant impurity sites. That is, unlike the DP localization represented by the diagonal elements, the extent of the this localization is broader. Since the extent of localization is determined by the competition between the effects of localization (χ) and hopping (J), a larger value of χ decreases this extent.

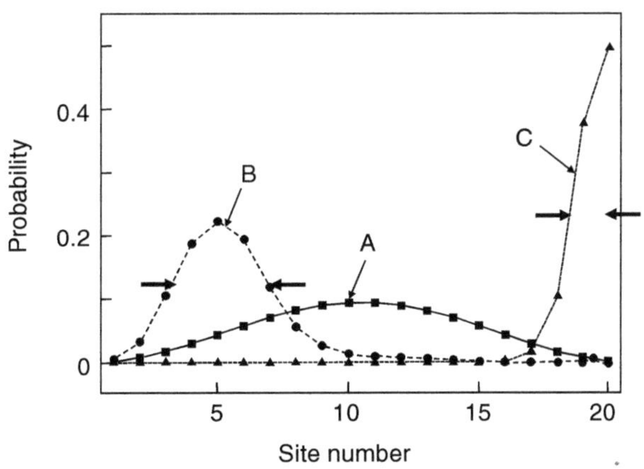

Fig. 4.10 Occupation probability of the dressed photon at each site. Curves A, B, and C represent the results for $\chi = 0$, 40.0, and $54.0\mathrm{fs}^{-1}\mathrm{nm}^{-1}$, respectively. Other numerical values are the same as those in Fig. 4.9

The "atoms" in the theoretical model described above correspond to nanomaterials whose sizes are equivalent to the radius of curvature of the probe apex, and Fig. 4.10 shows that the DP field is as broad as several atomic sizes due to the coupling with the localized modes of the phonons. The quasi-particle created by this coupling is called a dressed-photon–phonon (DPP). When the DP is localized at the end of the lattice, the DPP field penetrates the probe surface, and the penetration length is equivalent to the radius of curvature at the top of the probe apex. If a gas molecule comes flying into this penetration area, the DPP energy is transferred to the molecule and, as a result, the molecule is excited to a vibrational excited state by multiple phonons in the DPP and, successively, to a higher electronic state. By these successive excitations, the molecule can be dissociated even though the photon energy of the light injected into the probe is lower than the dissociation energy of the molecule. These energy transfer and excitations are the origin of the novel dissociation phenomenon shown in Fig. 4.2. A detailed comparison of this theoretical model and experimental results will be given in Sect. 6.1 of Chap. 6.

The reason that we are able to successfully derive the DP–phonon coupling and the resultant localization features is that the Hamiltonian of eq.(4.1) retained the site number i even after it was transformed to eq.(4.51). It should be noted that the two subscripts i and p in the above discussions represent different quantities from each other: the former is the site number, and the latter is the mode number of the phonon.

4.4 Light Absorption and Emission via Dressed-Photon–Phonons

Since the DP is a photon that is dressed by the energy of the electron–hole pair, its eigenenergy has a large number of modulation sidebands, as was described at the end of Sect. 2.1 in Chap. 2. Among them, the eigenenergy $\hbar\omega'_k$ of the upper sideband is larger than the photon energy $\hbar\omega_o$ of the incident light. Furthermore, since the DPP described in the previous section is a photon that is dressed not only by the energy of the electron–hole pair but also by the energies of the multiple coherent phonons, it also has modulation sidebands whose number is larger than that of the DP. Among these sidebands, the eigenenergy of the upper sideband is larger than $\hbar\omega_o$. Therefore, as is schematically explained by Fig. 4.11, if the DPP energy is transferred from nanomaterial 1 to nanomaterial 2 and if the electron–hole pair in nanomaterial 2 is resonant with one of the upper sidebands of the DPP, the electron–hole pair is excited by absorbing the eigenenergy $\hbar\omega''_k$ of this sideband. Since $\hbar\omega''_k$ is larger than the photon energy $\hbar\omega_o$ of the incident light, this excitation process can be regarded as energy up-conversion. Details of novel technologies exploiting this conversion will be reviewed in Chap. 7. This section describes the fundamental processes of light absorption and emission for this conversion. Since not only the electronic states but also the phonon states are involved in the energy states of the nanomaterial, for simplicity, only one specific sideband component that is resonant

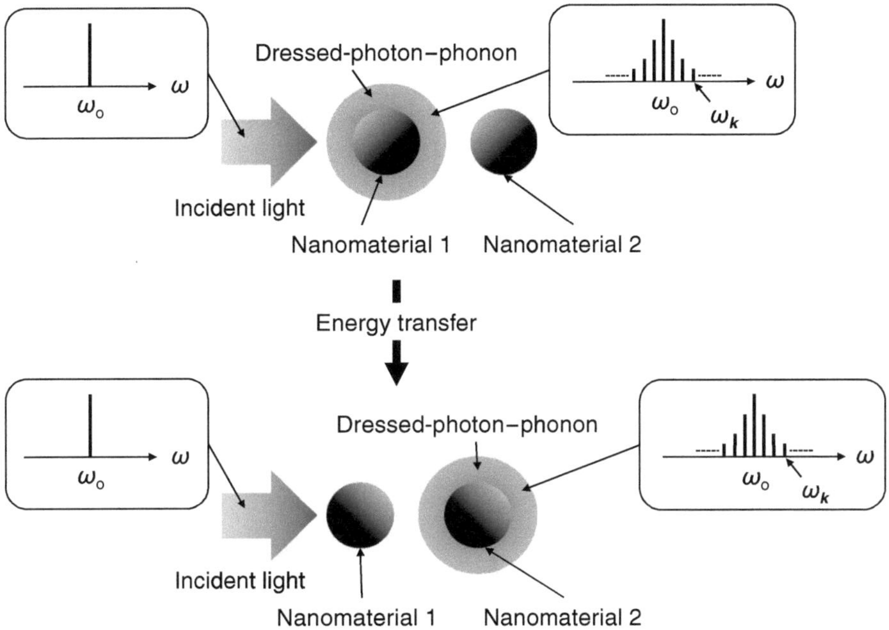

Fig. 4.11 Schematic explanation of the dressed-photon–phonon. Four insets represent the spectral profiles of the incident light and the modulated dressed-photon–phonon. The latter has modulation sidebands

with the phonon states is considered from among the large number of sidebands in the following discussions[6].

In the operators for the DPP given by eqs. (4.30a) and (4.30b), the annihilation ($\tilde{a}_i$) and creation ($\tilde{a}_i^\dagger$) operators for the DP are involved in the transition of the electron between the ground state $|E_g; el\rangle$ and the excited state $|E_{ex}; el\rangle$. Furthermore, the phonon operators ($\hat{c}_p$, $\hat{c}_p^\dagger$) in the exponential functions of eqs. (4.30a) and (4.30b) are involved in the transition of the phonons between the thermal equilibrium state (ground state) $|E_{thermal}; phonon\rangle$ and the excited state $|E_{ex}; phonon\rangle$. Therefore, in order to analyze the DPP-mediated interaction between nanomaterials, one has to consider the states represented by the direct product of the electronic state and phonon state of the nanomaterials, e.g., $|E_g; el\rangle \otimes |E_{thermal}; phonon\rangle$, $|E_g; el\rangle \otimes |E_{ex}; phonon\rangle$, $|E_{ex}; el\rangle \otimes |E_{thermal}; phonon\rangle$, and $|E_{ex}; el\rangle \otimes |E_{ex}; phonon\rangle$[7]. The origins of the energy up-conversion can be analyzed in terms of these states.

[6] Upward and downward arrows in Figs. 4.12(a)–(c) represent the absorption and emission by one of these sidebands, respectively.

[7] The direct product of the electronic state and the phonon state suggests that the eigenenergy of the electron-hole pair in the nanomaterials is also modulated and has sidebands, which is the dual relation with the modulation of the DPP eigenenergies. These modulation sidebands correspond to an infinite number of phonon states. The large number of horizontal lines in Figs. 4.12(a)–(c) represent these phonon states.

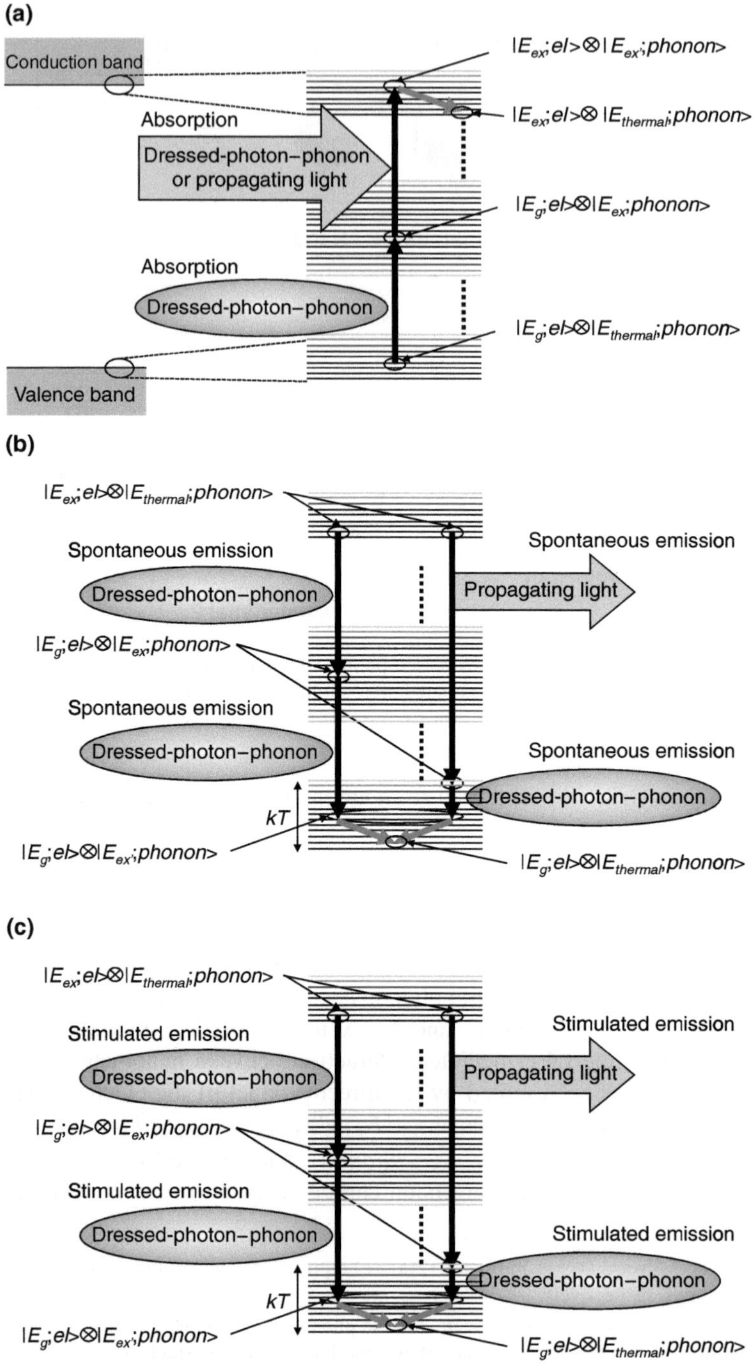

Fig. 4.12 Optical absorption and emission processes by dressed-photon–phonon. **a** absorption. **b** Spontaneous emission. The downward arrows at the left and right correspond to routes 1 and 2 in Table 4.3, respectively. **c** Stimulated emission. The downward arrows at the left and right correspond to routes 1 and 2 in Table 4.4, respectively

Table 4.1 Excitation and de-excitation processes for energy conversion by DPP

Class of energy conversion	excitation	De-excitation
From light to light (7.1*)	Absorption of light	Spontaneous emission of light
From light to electricity (7.2*)	Absorption of light	Stimulated emission of light, Extracting electric current
From electricity to light (7.3*)	Injecting electric current (Absorption of light**)	Spontaneous emission of light (Stimulated emission of light**)

*Section numbers in Chap. 7, in which the details of conversion technologies are reviewed
** Used for device fabrication

It should be noted that this conversion corresponds to the non-adiabatic process described in Subsect. 4.1.1 because the phonons are involved. For comparison, since only an electric dipole-allowed transition is involved in the conventional interactions between propagating light and matter, in those interactions, it is sufficient to consider electronic states $|E_g; el\rangle$ and $|E_{ex}; el\rangle$. Such interactions correspond to the adiabatic process described in Subsect. 4.1.1.

For energy conversion, it is essential to excite or de-excite electrons or electron-hole pairs. If the DPP is involved in this excitation or de-excitation, energy up-conversion becomes possible. Table 4.1 classifies the types of energy conversions and their excitation or de-excitation processes in which the DPP is involved.

The rest of this section discusses the relation between the DPP and absorption, spontaneous emission, and stimulated emission [16]. In the case of a semiconductor, for example, absorption or emission of the propagating light is not possible if its photon energy is lower than the bandgap energy E_g of the semiconductor material, i.e., if its wavelength is longer than the cut-off wavelength $\lambda_c = E_g/hc$. However, absorption or emission becomes possible if a DPP is involved, and as a result, energy up-conversion is realized. In this case, since the light incident on the material has a lower photon energy than E_g, excitation or de-excitation of the electrons or electron–hole pairs takes place in multiple steps. Here, a two-step process is considered for simplicity.

First, the light absorption process is described. Since the energies of the incident light and DP are lower than the bandgap energy E_g, the two-step process is required for exciting an electron from the valence band to the conduction band, as is shown in Fig. 4.12(a). The steps are:

First step The initial state of the electron is the ground state $|E_g; el\rangle$, which corresponds to the valence band in the semiconductor. On the other hand, the phonon is in the thermal equilibrium state $|E_{thermal}; phonon\rangle$, which depends on the crystal lattice temperature[8]. Therefore, the initial state is expressed by the direct product of these two states: $|E_g; el\rangle \otimes |E_{thermal}; phonon\rangle$. In the excitation by absorbing the

[8] In the case when the crystal lattice temperature is 0 K, this state is the vacuum state $|0; phonon\rangle$.

Table 4.2 Two-step absorption

First step	Initial state: $\|E_g; el\rangle \otimes \|E_{thermal}; phonon\rangle^{(1)}$ ⇓Electric dipole-forbidden transition⇓ Intermediate state: $\|E_g; el\rangle \otimes \|E_{ex}; phonon\rangle^{(2)}$	
Second step	Route 1 (by DPP)	Route 2 (by propagating light)
	⇓Electric dipole-allowed transition⇓ $\|E_{ex}; el\rangle \otimes \|E_{ex'}; phonon\rangle^{(3)}$	
	⇓Relaxation Final state: $\|E_{ex}; el\rangle \otimes \|E_{thermal}; phonon\rangle$	

(1) $|E_g; el\rangle$ and $|E_{thermal}; phonon\rangle$ represent the ground state of the electron and the thermal equilibrium state of the phonon, respectively
(2) $|E_{ex}; phonon\rangle$ represents the excited state of the phonon
(3) $|E_{ex}; el\rangle$ and $|E_{ex'}; phonon\rangle$ represent the excited states of the electron and the phonon, respectively

DPP, the electron is not excited to the conduction band but remains in the ground state $|E_g; el\rangle$ because the energies of the incident light and the DP are lower than the bandgap energy E_g of the material. However, the phonon is excited to one of the excited states $|E_{ex}; phonon\rangle$ depending on the DP energy, and thus, the final state of the transition is expressed as $|E_g; el\rangle \otimes |E_{ex}; phonon\rangle$. It should be noted that this transition is electric dipole-forbidden because the electron stays in the ground state even after the transition. This state, $|E_g; el\rangle \otimes |E_{ex}; phonon\rangle$, is the intermediate state of the two-step excitation.

Second step In the excitation from the above intermediate state to the final state, the electron is excited to the excited state $|E_{ex}; el\rangle$, i.e., the conduction band[9]. This transition is electric dipole-allowed because it is a transition from the ground state $|E_g; el\rangle$ to the excited state $|E_{ex}; el\rangle$ of the electron. Therefore, this transition is possible not only due to the DPP but also the propagating light. As a result of this transition, the system reaches the state $|E_{ex}; el\rangle \otimes |E_{ex'}; phonon\rangle$, which is represented by the direct product of the excited state $|E_{ex}; el\rangle$ of the electron and the excited state $|E_{ex'}; phonon\rangle$ of the phonon. Since the phonon promptly relaxes to the thermal equilibrium state $|E_{thermal}; phonon\rangle$ after this excitation, the final state of this two-step excitation is expressed by the direct product of the excited state of the electron and the thermal equilibrium state of the phonon: $|E_{ex}; el\rangle \otimes |E_{thermal}; phonon\rangle$. Table 4.2 summarizes the two-step absorption process.

Second, the light emission process is described. The two-step process is also required in this case for the same reason as described above. Spontaneous emission occurs by the following two steps, as is schematically explained by Fig. 4.12(b) and summarized in Table 4.3.

[9] It is also possible to excite the phonon to a higher excitation state while the electron is still in the ground state, as in the case of the first step. As an example, optical frequency up-conversion by dye grains involves this excitation process. Refer to Sect. 7.1 in Chap. 7 for the details of this up-conversion.

Table 4.3 Two-step spontaneous emission

First step	Initial state: $\vert E_{ex}; el\rangle \otimes \vert E_{thermal}; phonon\rangle$ ⇓Electric dipole-allowed transition⇓	
	Route 1 (Spontaneous emission: DPP generation) Intermediate state: $\vert E_g; el\rangle \otimes \vert E_{ex}; phonon\rangle$	Route 2 (Spontaneous emission: Propagating light generation) Intermediate state: $\vert E_g; el\rangle \otimes \vert E_{ex}; phonon\rangle$
Second step	⇓Electric dipole-forbidden transition⇓ (Spontaneous emission: DPP generation) $\vert E_g; el\rangle \otimes \vert E_{ex'}; phonon\rangle$	
	⇓Relaxation Final state: $\vert E_g; el\rangle \otimes \vert E_{thermal}; phonon\rangle$	

Table 4.4 Two-step stimulated emission

First step	DP incident on the electron in the conduction band ⇓ Initial state: $\vert E_{ex}; el\rangle \otimes \vert E_{thermal}; phonon\rangle$ ⇓Electric dipole-allowed transition⇓	
	Route 1(Stimulated emission: DPP generation) Intermediate state: $\vert E_g; el\rangle \otimes \vert E_{ex}; phonon\rangle$	Route 2 (Stimulated emission: Propagating light generation) Intermediate state: $\vert E_g; el\rangle \otimes \vert E_{ex}; phonon\rangle$
Second step	⇓Electric dipole-forbidden transition⇓ (Stimulated emission: DPP generation) $\vert E_g; el\rangle \otimes \vert E_{ex'}; phonon\rangle$	
	⇓Relaxation Final state: $\vert E_g; el\rangle \otimes \vert E_{thermal}; phonon\rangle$	

First step The initial state is expressed by the direct product of the excited state of the electron in the conduction band and the thermal equilibrium state of the phonon: $|E_{ex}; el\rangle \otimes |E_{thermal}; phonon\rangle$. De-excitation to the ground state $|E_g; el\rangle$ of the electron, i.e., to the valence band, is an electric dipole-allowed transition because it corresponds to the opposite process of the second step of absorption described above. Therefore, this emission process generates not only a DPP but also propagating light. As a result, the system reaches the intermediate state $|E_g; el\rangle \otimes |E_{ex}; phonon\rangle$. Here, the excited state $|E_{ex}; phonon\rangle$ of the phonon after DPP emission (route 1 in Table 4.3) has a much higher eigenenergy than that of the thermal equilibrium state $|E_{thermal} : phonon\rangle$. This is because the DP couples with the phonon, resulting in phonon excitation. On the other hand, the excited state $|E_{ex}; phonon\rangle$ of the phonon after the propagating light emission (route 2 in Table 4.3) has an eigenenergy as low as that of $|E_{thermal} : phonon\rangle$. This is because the propagating light does not couple with the phonon.

Second step This step is an electric dipole-forbidden transition because it corresponds to the opposite process of the first step of absorption described above. Thus, only the DPP is generated by this emission process. As a result, the electron is de-excited to the ground state $|E_g; el\rangle$, i.e., to the valence band, and the system is expressed as $|E_g; el\rangle \otimes |E_{ex'}; phonon\rangle$. After this transition, the phonon promptly relaxes to the thermal equilibrium state, and thus, the final state is expressed as $|E_g; el\rangle \otimes |E_{thermal}; phonon\rangle$.

Finally, the stimulated emission process is explained by Fig. 4.12(c) and summarized in Table 4.4, which are similar to Fig. 4.12(b) and Table 4.3, respectively. The only difference is that the DPP is incident on the electron in the conduction band to trigger the stimulated emission for the transition from the initial state to the intermediate state in the first step.

References

1. P. Atkins, J. De Paula, *Physical Chemistry, the*, 9th edn. (Oxford Univ. Press, Oxford, 2010), p. 372
2. P. Atkins, J. De Paula, *Physical Chemistry, the*, 9th edn. (Oxford Univ. Press, Oxford, 2010), pp. 495–497
3. T. Kawazoe, Y. Yamamoto, M. Ohtsu, Appl. Phys. Lett. **79**, 1184 (2004)
4. Y. Tanaka, K. Kobayashi, Physica E **40**, 297 (2007)
5. Y. Tanaka, K. Kobayashi, J. Microscopy **229**, 228 (2008)
6. C. Falvo, V. Pouthier, J. Chem. Phys. **122**, 014701 (2005)
7. M.E. Striefler, G.R. Barsch, Phys. Rev. B **12**, 4553 (1975)
8. D.N. Payton, W.M. Visscher, Phys. Rev. **154**, 802 (1967)
9. A.J. Sievers, A.A. Maradudin, S.S. Jaswal, Phys. Rev. **138**, A272 (1965)
10. S. Mizuno, Phys. Rev. B **65**, 193302 (2002)
11. T. Yamamoto, K. Watanabe, Phys. Rev. Lett. **96**, 255503 (2006)
12. A.S. Davydov, G.M. Pestryakov, Phys. Stat. Sol. (b) **49**, 505 (1972)
13. L. Jacak, P. Machnikowski, J. Krasynj, P. Zoller, Eur. Phys. J **D22**, 319 (2003)
14. K. Mizoguchi, T. Furuichi, O. Kojima, M. Nakayama, S. Saito, A. Syouji, K. Sakai, Appl. Phys. Lett. **87**, 093102 (2005)
15. V. Pouthier, C. Girardet, J. Chem. Phys. **112**, 5100 (2000)
16. T. Kawazoe, M.A. Mueed, M. Ohtsu, Appl. Phys. B **104**, 747 (2011)

Chapter 5
Devices Using Dressed Photons

Natura semina nobis scientiae dedit, scientiam non dedit.
Lucius Annaeus Seneca,*Epistulaea,* CXX, 4

This chapter reviews novel dressed-photon (DP) devices based on the operating principles described in Chap. 3 [1]. These principles can be summarized as follows:

(1) Using the near field optical interaction between two closely spaced quantum dots (QDs), the QDs are coupled to create symmetric and anti-symmetric states.
(2) Uni-directional energy transfer from the first QD to the second QD is realized by means of relaxation in the second QD.

Practical DP devices utilize the electric dipole-forbidden energy levels of excitons in QDs. This makes it impossible for excitons to be exrefd by propagating light, and as a result, device malfunction is avoided.

5.1 Structure and Function of Dressed-Photon Devices

Table 5.1 summarizes the functions and names of some typical DP devices that will be reviewed in this section. A prototype integrated circuit formed of DP devices is schematically illustrated in Fig. 5.1. Although Table 5.1 does not contain an output interface, it can be readily realized by using a metal nanoparticle, as will be demonstrated in Part (b) of Sect. 5.1.1.

5.1.1 Devices Utilizing Energy Dissipation

In DP devices, the input signal energy is transferred from one QD to another other, i.e., from an input terminal to an output terminal. After this energy transfer, the exciton

M. Ohtsu, *Dressed Photons*, Nano-Optics and Nanophotonics,
DOI: 10.1007/978-3-642-39569-7_5, © Springer-Verlag Berlin Heidelberg 2014

Table 5.1 Functions and names of DP devices reviewed in this section

Function	Name
Signal generation	Delayed-feedback optical pulse generator (Sect. 5.1.1(g)), Super-radiant optical pulse generator (Sect. 5.1.2(b))
Signal control	Optical switch (AND logic gate) (Sect. 5.1.1(a)), NOT logic gate (Sect. 5.1.1(b)), DA converter (Sect. 5.1.1(d)), Frequency up-convertor (5.1.1(f)), Buffer memory (Sect. 5.1.2(a))
Signal transmission	Energy transmitter (Sect. 5.1.1(e))
Input interface	Nano-optical condenser (Sect. 5.1.1(c))

created in the QD of the output terminal subsequently relaxes from the higher energy level to a lower energy level, and a part of its energy is dissipated, resulting in selective and uni-directional energy transfer. Thus, energy dissipation plays an essential role in the device function. This subsection reviews some examples of these functions.

(a) Optical switch

An optical switch controls signal transmission from the input terminal to the output terminal by using a control signal. Since a variety of fundamental optical functions, such as light emission and modulation, are involved in device operation, the optical switch is regarded as the most basic representative example of an optical device. Based on a slight modification of the operating principles in Chap. 3, an optical switch can be constructed by using three different-sized QDs, as schematically explained in Fig. 5.2. Since the output signal is generated by applying two input signals to the two input terminals, this device can also be regarded as an AND logic gate. Three cubic semiconductor QDs, as an example, are used for the input (QD-I), output (QD-O), and control (QD-C) terminals. These QDs are placed in close proximity to each other, and their separations are as short as their sizes. Their side lengths are assumed to be L, $\sqrt{2}L$, and $2L$, respectively.

By replacing the L in Eq. (F.23) of Appendix F with $\sqrt{2}L$ and $2L$, one can readily find that the exciton energy of level (1, 1, 1) in QD-I and that of level (2, 1, 1) in QD-O are equal to that of level (2, 2, 2) in QD-C. That is, these three energy levels are resonant with each other. Moreover, it is confirmed that the energy of level (1, 1, 1) in QD-O and that of level (2, 1, 1) in QD-C are also resonant with each other.

By referring to Appendix F, Eq. (F.50) takes a nonzero value when the envelope function of the exciton is an even function. Since it takes a value of zero when the envelope function is an odd function, it is found that the energy levels (2, 1, 1) in QD-O and QD-I are both electric dipole-forbidden. However, by utilizing the near-field optical interaction, energy transfers from the energy level (1, 1, 1) in QD-I to the forbidden levels (2, 1, 1) in QD-O and (2,2,2) in QD-C become possible, as represented by wavy arrows in Fig. 5.2. After these transfers, relaxation to the lower energy levels occurs promptly, as shown by downward arrows. Based on this transfer and subsequent relaxation, the operation of the optical switch can be explained as follows:

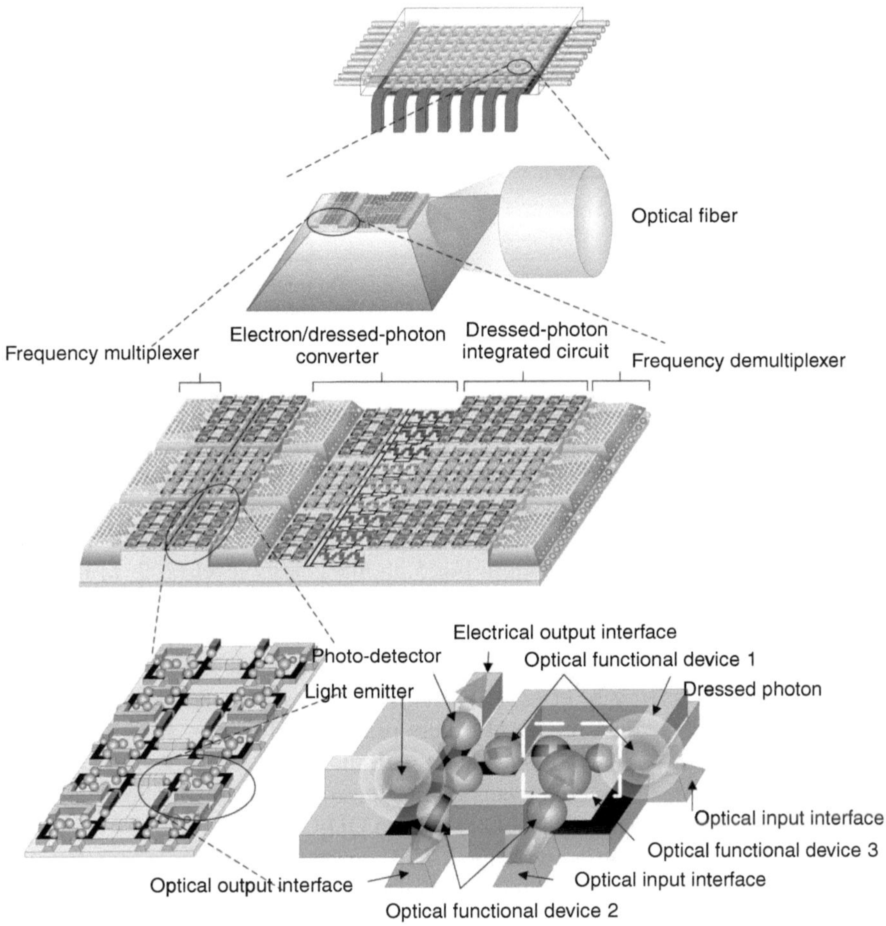

Fig. 5.1 Prototype of integrated circuit composed of dressed-photon (DP) devices

Switch off: After an exciton is created in the energy level (1, 1, 1) in QD-I by applying an input signal, the energy is transferred to the upper energy levels in QD-O and QD-C, and the exciton relaxes to the lower energy levels (the relaxation rate is γ in Eq. (3.16)), as shown by the wavy and downward arrows in Fig. 5.2a, eventually occupying the lower energy level (1, 1, 1) in QD-C. Subsequently, the exciton is annihilated with the time constant γ'^{-1} by interactions with phonons in the heat bath. Here, γ' is the non-radiative relaxation rate, which is equivalent to γ in Eq. (3.16); however, the value of ω in $D(\omega)$ on the right-hand side has a width $\Delta\omega$ depending on the crystal lattice temperature. The exciton is annihilated also by emitting a photon, where the decay time constant of the emitted light intensity is the inverse of the radiative relaxation rate γ_{rad}. Since these annihilations are both energy dissipation

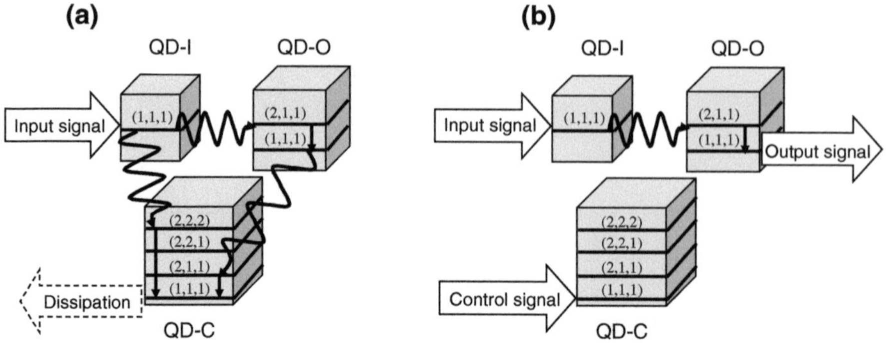

Fig. 5.2 Structure of an optical switch. **a** and **b** represent off- and on-states, respectively

phenomena, QD-O is not occupied by the exciton, and thus, an output signal is not generated, which means that the optical switch is off.

Switch on: When the exciton is created in the energy level (1, 1, 1) in QD-C by applying a control signal, as shown in Fig. 5.2b, the energy level (2, 2, 2) in QD-C becomes off-resonant with the energy level (1, 1, 1) in QD-I (the state-filling effect), and thus, the exciton in the energy level (1, 1, 1) in QD-I cannot be transferred to the energy level (2, 2, 2) in QD-C. As a result, energy transfer is allowed only from QD-I to the upper energy level (2, 1, 1) in QD-O. Subsequently, relaxation to the lower energy level (1, 1, 1) in QD-O occurs promptly. However, since the energy level (2, 1, 1) in QD-C is also off-resonant with the lower energy level (1, 1, 1) in QD-O due to the state filling effect, energy transfer to QD-C is not possible, and thus, the exciton in the lower energy level (1, 1, 1) in QD-O is annihilated by emitting a photon, which is used as an output signal. This means that the optical switch is on.

The operation of the optical switch based on the principle reviewed above has been confirmed by using three cubic CuCl QDs embedded in a NaCl host crystal [2]. CuCl QDs with side lengths of 3.5, 4.6, and 6.3 nm were used as QD-I, QD-O, and QD-C, respectively. Figure 5.3 shows the experimental results, which represents the spatial intensity distribution of 383 nm-wavelength light emitted from the lower energy level (1, 1, 1) in QD-O at a temperature of 15 K. In Fig. 5.3a, the image is dark at and around QD-O when the optical switch is off. In contrast, Fig. 5.3b shows a bright image at and around QD-O when the optical switch is on. Moreover, these figures show that the device size is smaller than 20 nm. Figure 5.4a shows the temporal evolution of the output signal, i.e., the light intensity from QD-O, emitted when a control signal pulse (pulse width: 10 ps) is applied to QD-C. The optical intensity increases rapidly with a rise time[1] of 90 ps, which depends on the magnitude of the interaction energy given by Eq. (2.76). After the control signal pulse decays, the output signal also decays with a small amplitude oscillation. This oscillation originates from the nutation between

[1] Curves for the intensities of rising and falling signals are approximated by exponential functions $\exp(t/\tau_r)$ and $\exp\left(-t/\tau_f\right)$, respectively. Their time constants τ_r and τ_f are defined as the rise and fall times, respectively.

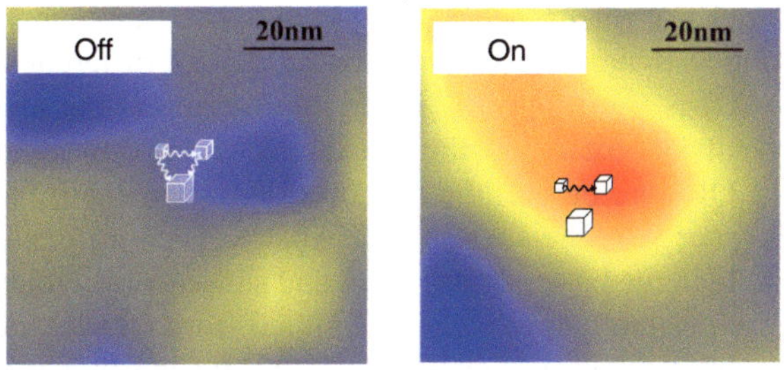

Fig. 5.3 Measured spatial distributions of the intensity of light emitted from the output terminal QD-O in an optical switch composed of CuCl quantum dots. **a** and **b** represent off- and on-states, respectively

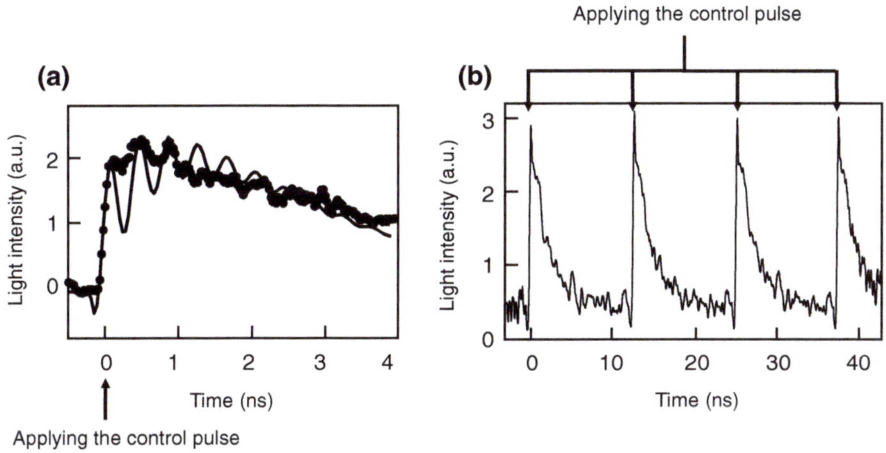

Fig. 5.4 Measured temporal evolution of the intensity of light emitted from the output terminal QD-O in an optical switch composed of CuCl quantum dots. **a** *Solid curve* represents the calculated results, which are fitted to the measured results represented by *closed circles*. **b** Example of repetitive operation

QD-I and QD-O. The period of oscillation is found to be about 400 ps from this figure. The decay time (the fall time[1]) is about 4 ns, which depends on the value of the radiative relaxation rate γ_{rad}. The solid curve in Fig. 5.4a represents the results calculated by using Eq. (2.76) [3], which agrees well with the experimental results. Figure 5.4b shows repetitive operation, in which on and off operations are repeated by applying a control signal pulse train.

The switching frequency can be increased by decreasing the rise and fall times in Fig. 5.4a. The rise time can be decreased by decreasing the separation between QDs. To decrease the fall time, several methods have been experimentally demonstrated, e.g., decreasing the emission lifetime by bringing a metal nanoparticle close to QD-O, as will be reviewed in Part (b).

In contrast to the use of three QDs described above, another possible scheme for realizing an optical switch is to use two cubic QDs of different sizes, where the smaller QD (QD-IO) is used not only as the input terminal but also as the output terminal. As shown in Fig. 5.5a, after the input signal creates an exciton in the barrier layer in which the QDs are buried, the exciton reaches the energy level (1, 1, 1) in QD-IO. Then, the energy of the exciton is dissipated when it transfers to the upper energy level (2, 1, 1) in the larger QD (QD-C) and subsequently relaxes to the lower energy level. Therefore, no output signal is emitted from QD-IO, meaning that the switch is off. In contrast, when an exciton is created in the lower energy level in QD-C by applying the control signal, as shown in Fig. 5.5b, the energy transfer is blocked by the state filling effect, and as a result, the exciton in the energy level (1, 1, 1) in QD-IO is annihilated, generating an output signal, which means that the switch is on. The first experimental demonstration of an optical switch based on this principle was carried out not by using two QDs but by using two closely spaced ZnO/ZnMgO quantum wells in a ZnO nanorod [4].

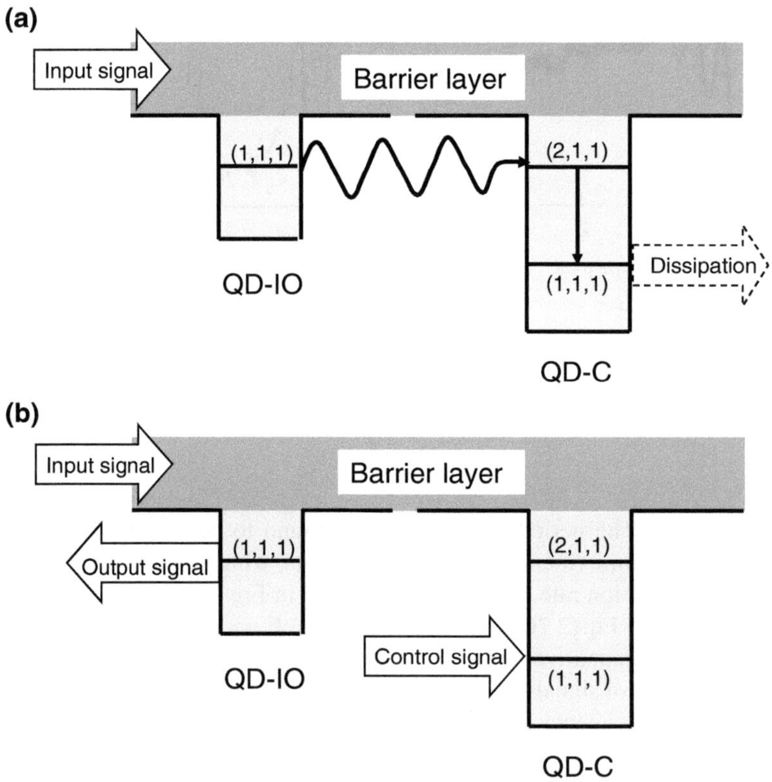

Fig. 5.5 Optical switch using two quantum dots. **a** and **b** represent off- and on-states, respectively

Table 5.2 Figures of merit for several optical switches

Types of devices	Volume V	Switching time T_{sw}	Energy E of the control signal	Contrast C	Figure of merit FOM[c]
Using dressed photons	$(\alpha\lambda/10)^3$	100 ps	$h\nu$	10~25	1
Optical MEMS	$(\alpha\lambda)^{3}$[a]	1 μs	10^{-18} J	10^4	10^{-5}
Mach-Zehnder interferometer-type	$(\alpha\lambda)^3$	1 ps	10^{-18} J	10^2	10^{-2}
Using non-resonant optical nonlinearity of the third-order	$(\alpha\lambda)^3$	10 fs	$10^6 h\nu$[b]	10^3	10^{-3}
Using resonant optical nonlinearity	$(\alpha\lambda)^3$	1 ns	$10^3 h\nu$	10^4	10^{-4}
Using quantized energy levels in quantum wells	$(\alpha\lambda)^3$	100 fs	$10^3 h\nu$	10^3	10^{-1}

[a] α: a real number, which is larger than unity. λ: optical wavelength
[b] $h\nu$: photon energy
[c] The figures of merit for conventional optical switches are normalized to that for the DP optical switch

For applications such as optical information transmission and processing, important physical quantities of a practical optical switch are its volume V, switching time T_{sw}, and control signal energy E. Moreover, the contrast C is also important because low intensity light may be emitted from the output terminal even when the control signal is not applied to the switch. The contrast C is defined as the ratio of the output signal intensities emitted with and without applying the control signal. Using these quantities, a figure of merit can be defined by

$$FOM = \frac{C}{V T_{SW} E}. \tag{5.1}$$

Table 5.2 shows a comparison of the figures of merit of the above-mentioned DP optical switch using CuCl QDs (Figs. 5.3, 5.4) and conventional optical switches. Here, it should be noted that the conventional optical switches do not simultaneously have the values of the quantities shown in the table, and therefore, the figures of merit in this table represent the theoretically allowed maximum values. Comparing these values, it is found that the figure of merit for the DP optical switch is about 10–100 times larger than those of the conventional optical switches, confirming its superior performance.

(b) NOT logic gate

Figure 5.6 shows the structure of a NOT logic gate using two cubic QDs of different sizes [5], where the larger and smaller QDs are used as an input terminal QD-I and an output terminal QD-O, respectively. Their side lengths are $\sqrt{2}\,(L + \delta L)$ and L, respectively; i.e., their ratio is slightly detuned from $\sqrt{2} : 1$ by an amount $\delta L/L$, and as a result, the energy level (1, 1, 1) of the exciton in QD-O is slightly off-resonant with the energy level (2, 1, 1) in QD-I.

As shown in Fig. 5.6a, continuous-wave (CW) light, whose power is temporally invariant, is applied to QD-O, creating an exciton in energy level (1, 1, 1). However,

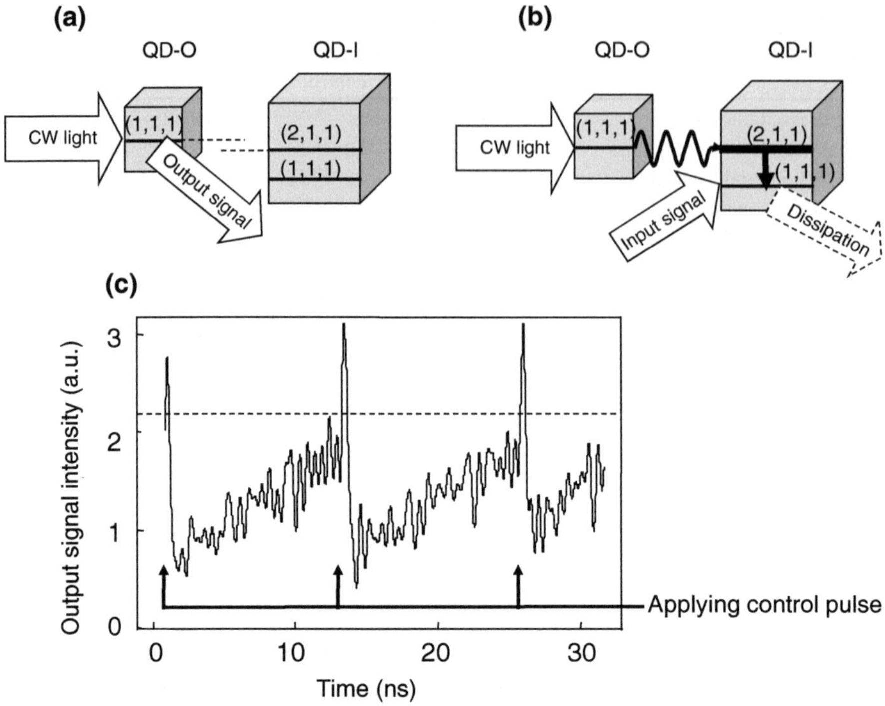

Fig. 5.6 NOT logic gate. **a** and **b** represent the cases without and with the input signal, respectively. **c** Measured temporal evolution of the output signal intensity of a NOT logic gate composed of cubic CuCl quantum dots. *Horizontal broken line* represents the value of the output signal intensity in the absence of the input signal

the energy is not transferred to the non-resonant upper energy level (2, 1, 1) in QD-I, and a photon is created by annihilation of the exciton in QD-O. The generated photon is used as an output signal, which means that the output signal is generated without applying an input signal to this device.

Next, as shown in Fig. 5.6b, an input signal is applied to QD-I in order to create an exciton in energy level (1, 1, 1), which induces the state-filling effect, similarly to the case of the optical switch described in Part (a) above. This state-filling also induces broadening of the width of the upper energy level (2, 1, 1) in QD-I, which is called collision broadening, mainly due to many-body effects [6]. As a result, this level becomes resonant with the energy level (1, 1, 1) in QD-O. Due to this induced resonance, the energy is transferred from the energy level (1, 1, 1) in QD-O to the upper energy level (2, 1, 1) in QD-I after the CW light applied to QD-O creates an exciton. Due to subsequent relaxation to the lower energy level (1, 1, 1) in QD-I, the energy is dissipated, and therefore, an output signal is not generated.

In summary, Fig. 5.6 shows that an optical signal is generated in the absence of an input signal, and on the contrary, an output signal is not generated when an input signal is applied. This behavior represents the operation of a NOT logic gate. In this figure, the CW light applied to QD-O corresponds to the power supply for an electronic circuit and works as the energy source to generate the output signal in the absence of the input signal. Figure 5.6c shows the experimental results obtained by using two cubic CuCl QDs. In this figure, the optical intensity of the output signal decreases the instant the input optical pulse is applied, confirming NOT logic gate operation.

Practical implementation of DP devices requires size- and position- controlled QDs, high-reproducibility fabrication technologies, and room temperature operation. Two-dimensional arrays of room-temperature operation NOT logic gates have been successfully fabricated by using InAs QDs, meeting the above requirements [7]. After the QDs were grown by molecular beam epitaxy, a two-dimensional array of mesa-shaped DP devices was formed, as shown in Fig. 5.7a, by using electron-beam lithography and Ar-ion milling. Figure 5.7b is a scanning transmission electron microscope image of the cross-sectional profile of the mesa-shaped device, which has a base area of 300 nm × 300 nm and a height of 85 nm. Inside the mesa, two oblate hemispherical QDs were grown and aligned vertically, separated by a 24 nm-thick barrier layer. The average diameter and height of the large QD were 42 and 11 nm, respectively. Those of the small QD, grown below the large QD, were 38 and 10 nm, respectively. This image shows that a hemispherical gold nanoparticle with a base diameter of 50 nm and a height of 30 nm was fixed on top of the mesa. This gold nanoparticle increases the extraction efficiency of the output signal generated in the mesa. Since the refractive index of the GaAs barrier layer is large, the output signal of the present DP device would normally be back-scattered from the large QD to the substrate [8]. With this gold nanoparticle, however, most of the scattered light can be forward-scattered outside of the device, where it can be detected. Figure 5.7c is an optical microscopic image of the fabricated two-dimensional array of DP devices. The DP devices were arranged two-dimensionally in a 20 × 20 array with 1 μm separation.

Figure 5.8 shows the measured spatial distribution of the optical intensity of the output signal from the two-dimensional array of mesa-shaped devices, each of which has a base area of 200 nm × 200 nm and a height of 85 nm. In this figure, the spots at the positions of the NOT logic gates are dark, because output signals are not generated from these gates when applying both the CW light and the input signals. It should be pointed out that this array contains several AND logic gates composed of two QDs, whose operating principle was described above with reference to Fig. 5.5. The spots at these devices are bright because output signals are generated when applying both the CW light and the input signals. The measured values of the contrast C [9] in Eq. (5.1) for the NOT and AND logic gates were both 66 (=18 dB), which are larger than the value in the first row of Table 5.2, confirming the high performance of these devices.

There are technical problems with conventional fabrication methods, e.g., molecular beam epitaxy, including the need for large equipment with high energy consumption. In order to solve these problems, a novel method has been developed to control the sizes and separations of the QDs by confining them in polymer microspheres.

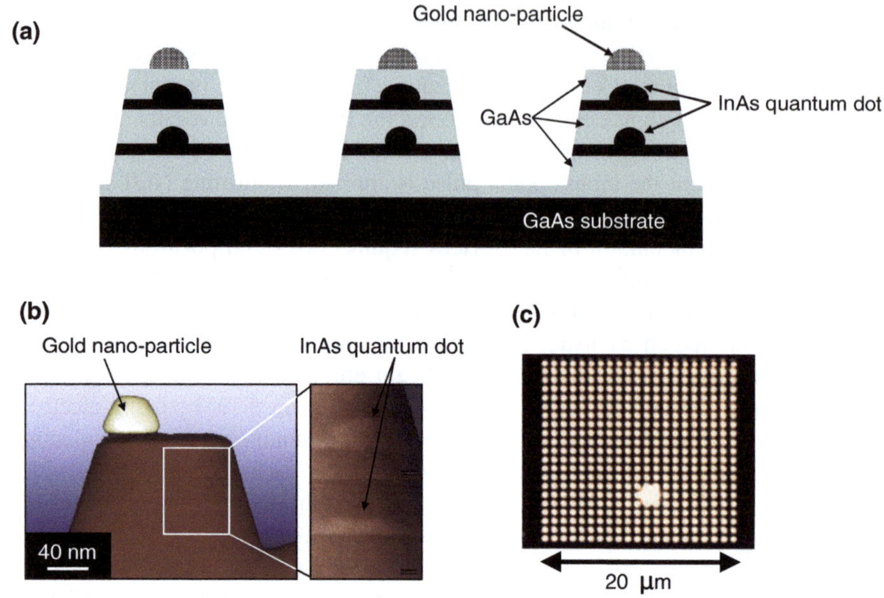

Fig. 5.7 Room temperature-operation NOT logic gate composed of InAs quantum dots. **a** Cross-sectional structure of the mesa-shaped devices. **b** Scanning transmission electron microscope image of the cross-sectional profile. **c** Optical microscope image of a two-dimensional array of fabricated devices

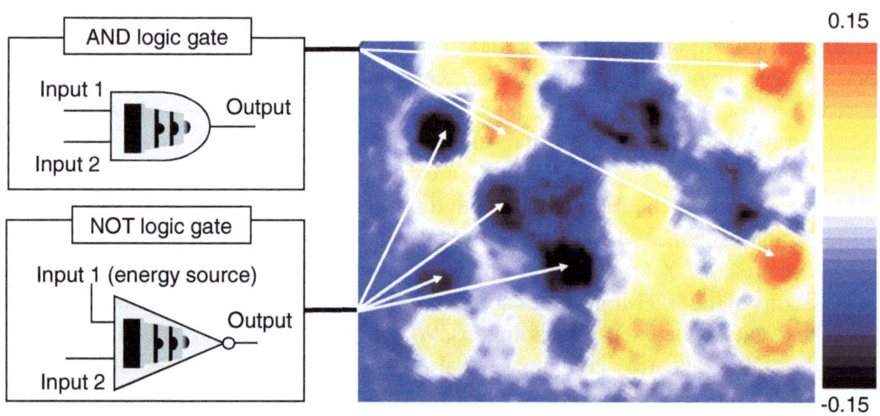

Fig. 5.8 Measured spatial distribution of the output signal intensity from a two-dimensional array of fabricated devices composed of InAs quantum dots

This method autonomously hardens a UV-cured resin by using the energy transferred between QDs [10]. The method is expected to be applied not only to the production of DP devices but also to a variety of technologies such as energy conversion (Chap. 7) and information security (Sects. 8.1–8.3).

By applying the above operating principles of the NOT and AND logic gates (as well as the XOR logic gate reviewed in Chap. 3), other devices can be realized, such as OR, NOR, and NAND logic gates, as shown in Figs. 5.9a–c, respectively. With these devices, a complete set of logic gates can be realized.[2] In summary, although it has been difficult to fabricate and operate logic gates using conventional propagating light, as well as to realize a complete set of logic operations, this has been made possible by using DPs.

Fan-out, F, is a key quantity for judging whether these devices can be used for optical integrated circuits. This quantity represents the number of devices that can be connected in parallel to the output terminal of the device and is expressed as $F = 1/2\gamma_{rad}T$ in the cases of the optical switch and the NOT gate described in Parts (a) and (b) above. Here, γ_{rad} is the radiative relaxation rate, and T is the time constant for energy transfer, which is inversely proportional to the interaction energy given by Eq. (2.76). In the case of AND and NOT logic gates composed of CuCl QDs, for example, the value of F is estimated to be 10 because $\gamma_{rad} = 5.0 \times 10^8\,\mathrm{s}^{-1}$ and $T = 100\,\mathrm{ps}$. Also, if these logic gates are composed of InAs QDs, F is 5 because $\gamma_{rad} = 1.0 \times 10^9\,\mathrm{s}^{-1}$ and $T = 100\,\mathrm{ps}$. These values are sufficiently large to be used for practical optical integrated circuits.

(c) Nano-optical condenser

A nano-optical condenser converts propagating light to DPs with high efficiency [11] and can be used as an input interface to inject propagating light into a DP device. To construct the nano-optical condenser, as shown in Fig. 5.10a, a large number of small QDs are used, and one large QD is placed at the center. Also, medium-sized QDs are placed in the spaces between them. Since the sizes of these QDs are tuned so that they are resonant with each other, the energies are transferred from small QDs to the medium-sized QDs if excitons are created in the small QDs by applying propagating light. After this transfer, relaxation promptly occurs in the medium-sized QDs. Subsequently, the energies are transferred from the medium-sized QDs to the large QD. After relaxation in the large QD, the output signal is generated. Since the energies of the large number of small QDs are tuned to the photon energy of the incident propagating light, almost all the incident propagating light energy can be absorbed by these small QDs. Furthermore, since the energy loss in this system is due to dissipation by relaxation in the medium-sized QDs and the large QD, its magnitude is very small, as with the cases of the DP devices described in Parts (a) and (b) above. As a result, it is expected that the propagating light will be converted to DPs with high efficiency.

Figure 5.10b shows a near-field optical microscopic image of the spatial distribution of the DP energy when applying propagating light with a wavelength of 384–386 nm to cubic CuCl QDs in a NaCl host crystal. The bright spot at the center corresponds to the place to which the incident light energy was condensed at the

[2] If all of the logical functions expressing the relations between M inputs and N outputs ($M, N = 1, 2, 3, \ldots$) can be given by the combination of a few basic logical functions, the set of these basic logical functions is called a complete set. All digital operations become possible with a complete set.

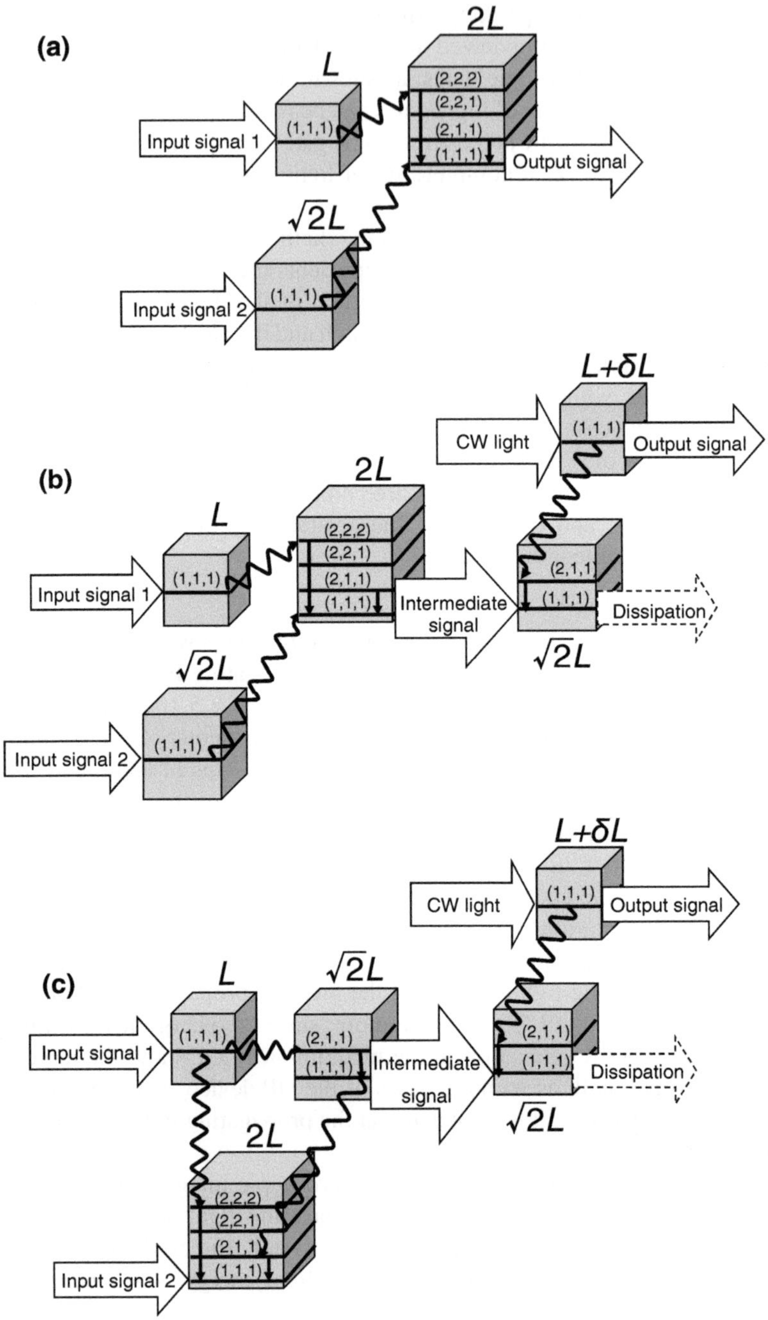

Fig. 5.9 Applications of the principles of logic gate operation. **a** OR logic gate. **b** NOR logic gate. **c** NAND logic gate

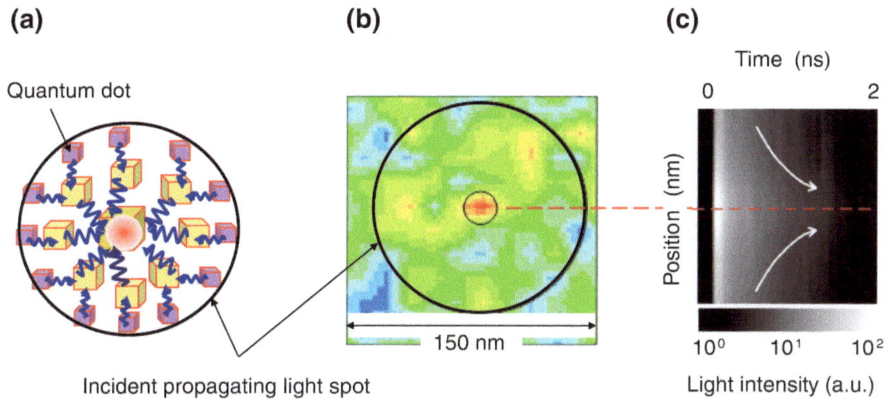

Fig. 5.10 Nano-optical condenser. **a** Structure of the device. **b** and **c** show the spatial distribution and temporal evolution of the light intensity emitted from CuCl quantum dots, respectively

large QD, having a side length of 8 nm. Its diameter is about 20 nm, including the size of the probe apex used for the microscope, which governs the resolution of the measurement. The light power in this spot is more than five times that of the light power emitted from this largest QD when it is isolated from the other QDs. From these results, it is confirmed that this device works as a high-efficiency optical condenser at a scale beyond the diffraction limit. This novel device has also been called an optical nano-fountain [11].

The high performance of this device can be confirmed by comparing it with a convex lens. When propagating light is focused by a convex lens, the theoretical spot diameter in the focal plane, which corresponds to the diffraction limit of focusing, is expressed as λ/NA, where λ is the wavelength of the incident light, and *NA* is a parameter called the numerical aperture, which depends on the shape and material of the lens, and is smaller than unity. By substituting the spot diameter shown in Fig. 5.10b into this formula, one can derive that *NA* is more than 40, which is much larger than that of the conventional convex lens.

Figure 5.10c shows the measured spatial and temporal evolutions of the light intensity. The horizontal axis at the top of the figure represents time, and the vertical axis represents the radial position of the polar coordinate centered at the large QD. The gradation of the brightness is proportional to the number of emitted photons, from which one can find that the energy is condensed to the large QD with a time constant as short as 2 ps.

A practical stacked-layer nano-optical condenser operating at room temperature has been developed by using InAs QDs [12]. Its cross-sectional profile is shown in Fig. 5.11. Large QDs are grown in the central layer, which is sandwiched by 10 layers in which small QDs are grown. These QDs have oblate hemispherical shapes. The average diameter and height of the large QDs are 48 and 4.5 nm, respectively, and the average separation between adjacent QDs in the layer is 15 nm. The average diameter, average height, and separation of the small QDs are 4.3, 2.7, and 20 nm, respectively.

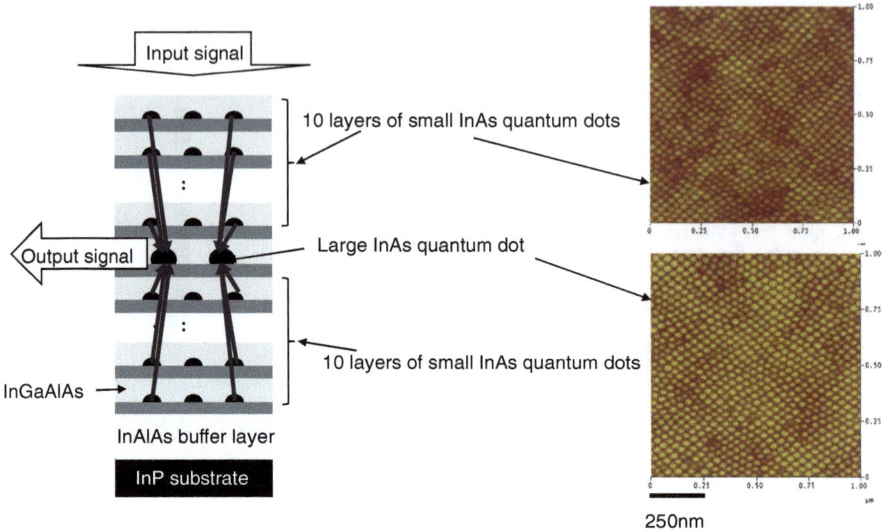

Fig. 5.11 Stacked layered nano-optical condenser composed of InAs quantum dots. Cross-sectional structure of multi-layered quantum dots (*left*), and atomic force microscope images of small and large quantum dots (*right*)

The average vertical separation of the QDs is 15 nm. These values satisfy the resonant conditions for energy transfer. As a result of the energy transfer from the small QDs in the upper and lower layers to the large QD in the central layer, the emitted light intensity from the large QD was confirmed to be 7.7 times that of the isolated large QD at a temperature of 300 K.

(d) Digital-to-analog converter

Figure 5.12a shows the structure of a three-bit digital-to-analog (DA) converter [13], in which three different-sized QDs (QD0, QD1, and QD2) are used as three input terminals. A large QD (QD-O) is used as an output terminal. Its principle of operation is similar to that of the nano-optical condenser in Part (c) above. In this device, however, the sizes of the QDs are tuned so that the energy levels (1, 1, 1) in QD0, QD1, and QD2 are resonant with the energy levels (2, 2, 2), (2, 2, 1), and (2, 1, 1) in QD-O, respectively. Here, it is assumed that the magnitudes of the interaction energy, $\hbar U$ (Eq. (2.76)), are 2.5×10^{-25}, 4.8×10^{-25}, and 2.1×10^{-24} J, respectively, which is made possible by adjusting the separation between QD0 and QD-O to be the largest and that between QD2 and QD-O to be the smallest. With these assumptions, the temporal evolutions of the occupation probabilities of the exciton in QD-O, after the input signals are applied to QD0, QD1, and QD2, are calculated based on the formulation in Chap. 3. The three curves in Fig. 5.12b show the results of the calculations. The total probabilities of energy transfer from QD0, QD1, and QD2 to QD-O are estimated by integrating the areas under these curves for the time range of 0–5 ns. The ratio of their values is found to be 1:2:4. (As a matter of fact, the

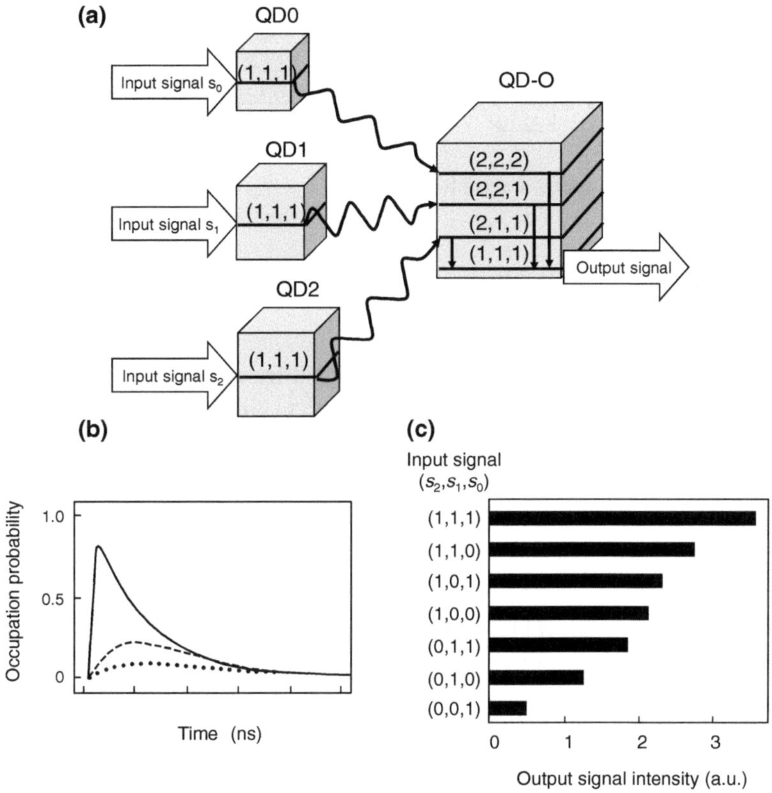

Fig. 5.12 Three-bit digital-to-analog converter. **a** Structure. **b** Calculated occupation probability of exciton in QD-O. *Dotted*, *broken*, and *solid curves* correspond to the input signals s_1, s_2, and s_3 in (**a**), respectively. **c** Measured light intensity emitted from the CuCl QD-O

magnitudes of the interaction energies, $\hbar U$, given above were set in order to achieve this ratio.) By denoting the output signal intensity from QD-O by d and the numbers of excitons in QD0, QD1, and QD2 by s_0, s_1, and s_2 (=0 or 1), respectively, it is found that their relation can be expressed as

$$d = 2^0 s_0 + 2^1 s_1 + 2^2 s_2. \tag{5.2}$$

This relation represents the function of converting a digital signal to an analog one, by which the three-bit DA converter operation is confirmed.

Figure 5.12c shows the experimental results confirming the DA conversion operation by using cubic CuCl QDs. The side lengths of QD0, QD1, and QD2 are tuned to 1.0, 3.1, and 4.1 nm, respectively. The wavelengths of the incident light, to be resonant with these QDs, are 325, 376, and 381 nm, respectively. The values of s_0, s_1, and s_2 in Eq. (5.2) take 1 or 0 depending on whether or not the incident light is applied to these QDs. The side length of QD-O is 5.9 nm. Figure 5.12c shows the

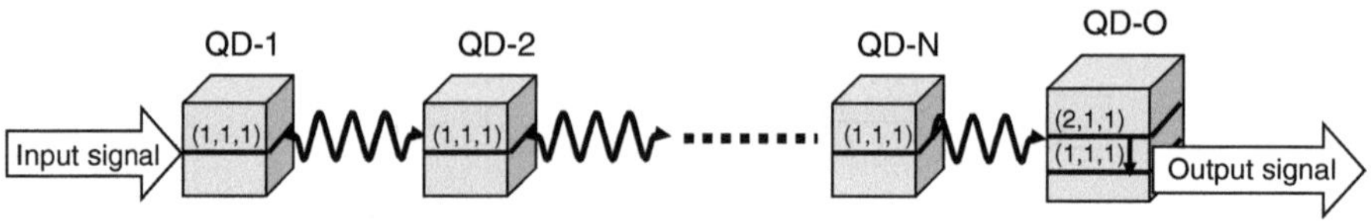

Fig. 5.13 Structure of an energy transmitter

intensity of the light emitted from QD-O, which agrees with Eq. (5.2), confirming the DA conversion operation.

(e) Energy transmitter

An energy transmitter is used to transmit a signal from one DP device to another, which corresponds to a metallic wire in an electrical circuit or an optical waveguide in a conventional optical integrated circuit. The energy transmitter should meet the following two requirements:

(1) Signal reflection from the DP device connected to the end of the transmitter must be avoided to achieve stable uni-directional energy transmission.
(2) Transmission loss must be sufficiently low for a long transmission length.

Figure 5.13 shows a novel DP energy transmitter that has been developed to meet these requirements. It is composed of an array of N QDs of the same size (QD-1 – QD-N). A larger QD (QD-O) is placed at the end of this array. In the case of using cubic QDs, as an example, an exciton is created in the energy level (1, 1, 1) in QD-1 by applying an input signal. This energy is transferred to the energy level (1, 1, 1) in QD-N, and nutation occurs among the N QDs. As a result, these QDs are coupled with each other. If the size of QD-O is tuned so that its electric dipole-forbidden level (2, 1, 1) is resonant with the energy level of this coupled state of the N QDs, the nutating energy is transferred to the upper energy level (2, 1, 1) in QD-O, and the energy is dissipated by subsequent relaxation to the lower energy level (1, 1, 1). The light emitted after this relaxation is used as the output signal.

This device operation has been confirmed experimentally and theoretically by using spherical CdSe QDs [14], whose energy eigenvalues are given by Eqs. (F.15) and (F.16) in Appendix F. By using 2.8 nm-diameter QDs for QD-1 – QD-N and a 4.1 nm-diameter QD for QD-O, the energy levels S in QD-1 – QD-N are resonant with the upper energy level L_u in QD-O, as is schematically explained by Fig. 5.14a. These QDs are dispersed on a SiO_2 substrate with an average separation between adjacent QDs adjusted to 3 nm. By applying pulsed incident light (2 ps pulse width, 0.6 mW optical power), excitons are created in QD-1 – QD-N. Figure 5.14b shows the measured spectral profiles emitted from these QDs. By decreasing the temperature, the intensity of the light emitted from the energy level S in QD-1 – QD-N decreases (the spectral peak at the wavelength of 540 nm is denoted by P_s in this figure), whereas that from the lower energy level L_l in QD-O increases (the spectral peak at the wavelength of 595 nm is denoted by P_{L1} in this figure). This anti-correlated variation in the intensities is because the efficiency of energy transfer from the energy level S to the upper energy level L_u in QD-O increases due to the decrease in the

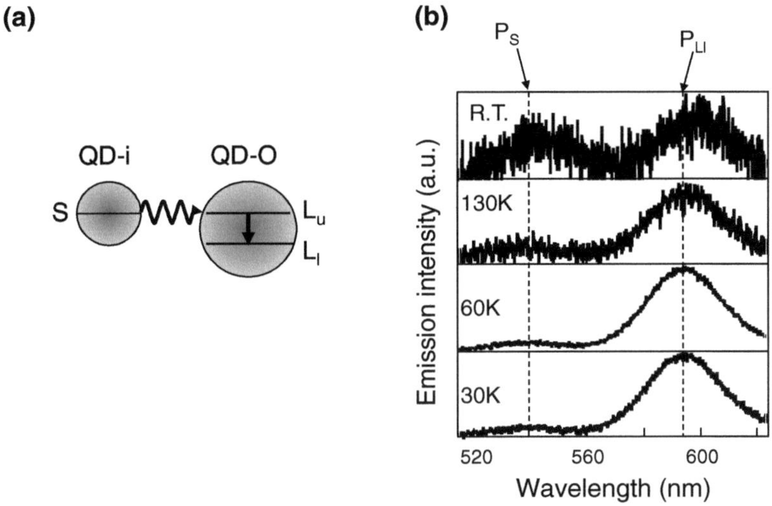

Fig. 5.14 Energy transfer from a small QD, QD-*i* ($i = 1 - N$), to a large QD, QD-O. **a** Energy levels. **b** Measured temperature dependence of the spectra of emitted from the CdSe quantum dots. P_S and P_{L1} represent the spectral peaks emitted from the energy level S in the small quantum dot and from the lower energy level L_l in the large quantum dot, respectively

non-radiative relaxation rate γ with decreasing temperature. From the measured results of the temporal variation of the optical intensity at the peak wavelength, the time constant for the energy transfer from QD-1 – QD-N to QD-O was estimated to be 135 ps.

The energy transmitter of Fig. 5.13 meets requirement (1) above because the exciton cannot be excited to the upper energy level L_u in QD-O even if the exciton is created in the lower energy level L_l in QD-O by back-transfer of the signal from the DP device installed at the next stage after QD-O. Thus, the energy is not back-transferred from QD-O to QD-1 – QD-N. In order to check whether requirement (2) is met, the transmission loss and transmission length are evaluated by using the rate equations, which are approximated equations of the quantum master equations in Chap. 3. For this evaluation, it is assumed that QD-1 – QD-N for the practical device are dispersed on a substrate instead of being arranged linearly and that QD-O is placed among these dispersed QDs.

As schematically illustrated in Fig. 5.15a, small QDs are dispersed along the x-, y-, and z-axes and are used as QD-1 – QD-N. The numbers of rows, columns, and layers are denoted by N_x, N_y, and N_z, respectively [15]. QD-1 and QD-O are respectively denoted by QD_{in} and QD_{out} in this figure. The rate equations representing interactions between two arbitrary QDs are expressed as

$$\frac{dn_i}{dt} = \sum_{j=1}^{N+1} U_{ji}n_j - \sum_{i=1}^{N+1} U_{ij}n_i - \gamma_{rad,i}n_i. \tag{5.3}$$

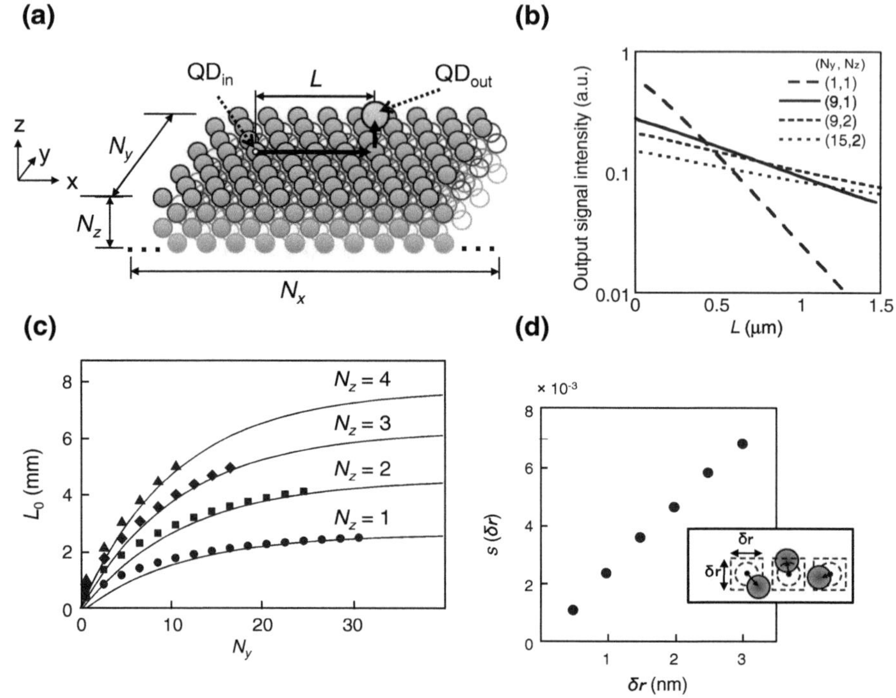

Fig. 5.15 Calculated results for QDs dispersed on a planar substrate. **a** Arrangement of multiple small quantum dots and one larger quantum dot. QDin and QDout represent QD-1 and QD-O in the text, respectively. **b** Dependence of the output signal intensity I_{out} on the distance L between the input and output terminals. **c** Dependence of the energy transmission length L_0 on the numbers, N_y and N_z, of small quantum dots along the y- and z-axes on the substrate. **d** Dependence of the standard deviation $\sigma\ (\delta r)$ of I_{out} on the fluctuation δr of the separation between adjacent quantum dots

Here, n_i is the occupation probability of the exciton in QD-i, where $i = N + 1$ corresponds to QD-O; U_{ij} corresponds to U in the second term of Eq. (3.2), which is proportional to the interaction energy between QD-i and QD-j; and $\gamma_{rad,i}$ is the radiative relaxation rate of the exciton in QD-i. The output optical signal intensity I_{out} emitted from QD-O is given by

$$I_{out} = \int \gamma_{rad,N+1} n_{N+1} dt. \tag{5.4}$$

Corresponding to the experimental conditions using spherical CdSe QDs, the diameters of QD-1 – QD-N are assumed to be 2.8 nm, and that of QD-O is assumed to be 4.1 nm. The energy $\hbar U$ of the interaction between QD-N and QD-O is assumed to be 7.8×10^{-25} J, which corresponds to a QD separation of 7.3 nm. Relaxation rates are assumed to be $\gamma_{rad,i} = 4.5 \times 10^8 \mathrm{s}^{-1}$ $(i = 1 \sim N)$ and $\gamma_{rad,N+1} = 1.0 \times 10^9 \mathrm{s}^{-1}$ [16]. Figure 5.15b shows the calculated dependence of I_{out} on the distance L between the input terminal QD-1 and the output terminal QD-O, where N_x is fixed to 317, and (N_y, N_z) are (1,1), (9,1), (9,2), and (15,2). It is found from this figure that I_{out} decreases in a complicated manner with increasing L, i.e., I_{out} is larger for smaller values of N_y and N_z in the region where the value of L is small. On the other hand, in the region where the value of L is large, the value of I_{out} decreases more slowly for larger values of N_y and N_z. Therefore, in order to maintain I_{out} sufficiently large, a small number of QDs is more advantageous for short-span energy transmission, whereas it is more advantageous to increase the number of QDs for long-span energy transmission. By defining the energy transmission length L_0 using the value of L at which the electric field amplitude $(\propto \sqrt{I_{out}})$ of the output optical signal is -times the input signal, the dependencies of L_0 on the values of N_y and N_z are calculated numerically and are shown in Fig. 5.15c. This figure shows that L_0 increases with increasing N_y and N_z, and moreover, it saturates with increasing N_y. This feature is expressed as

$$L_0\left(N_y, N_z\right) = L_{\max}\left\{1 - \exp\left(-\frac{N_y}{C}\right)\right\}, \tag{5.5}$$

where $L_{\max}$ represents the maximum energy transmission length. The solid curves in this figure are the results of fitting this equation to the experimental results, by which the values of $L_{\max}$ for $N_z = 1$ and 4 are found to be 2.85 and 7.92 μm, respectively.

In the above discussion, fluctuations δr of the separation between QDs were neglected. Figure 5.15d shows the calculated dependence of the standard deviation $\sigma(\delta r)$ of the output signal intensity I_{out} on δr. From this figure, it is found that the value of $\sigma(\delta r)$ is maintained as low as 6.8×10^{-3} even when δr is as large as 3 nm. Since the value of δr can be maintained smaller than 3 nm by employing recent nano-fabrication technologies, it is concluded that this energy transmitter can be produced and operated reliably.

In order to experimentally evaluate the transmission length, spherical CdSe QDs were dispersed on a SiO_2 substrate, with the average separation between adjacent QDs set to 7.3 nm. Moreover, the thicknesses of the QD layers, H, were fixed at 10, 20, and 50 nm, which are proportional to the number of rows N_z of QD-1 – QD-N along the z-axis. These devices are denoted by A, B, and C, respectively. While applying CW light with a wavelength of 473 nm, the relations between L and I_{out} were measured for these devices. The results are shown in Fig. 5.16a, b. The value of the energy transmission length L_0 can be estimated from the fitted exponential curves, represented by broken lines in Fig. 5.16b. That is, the values of L_0 for the devices A, B, and C are estimated to be 1.92, 4.40, and 11.8 μm, respectively. These values agree with the calculated values, as shown in Fig. 5.16c, and increase with increasing H, i.e., with increasing N_z.

Experiments similar to that of Fig. 5.16 have been carried out by using CdCe QDs dispersed in a narrow channel formed in a SiO_2 substrate in order to evaluate the energy transmission length L_0 [17]. Moreover, the selective use of the electric dipole-forbidden transition has been proposed as a means of greatly increasing the value of L_0. Results of calculations have confirmed that the value of L_0 increased to as long as several millimeters [18].

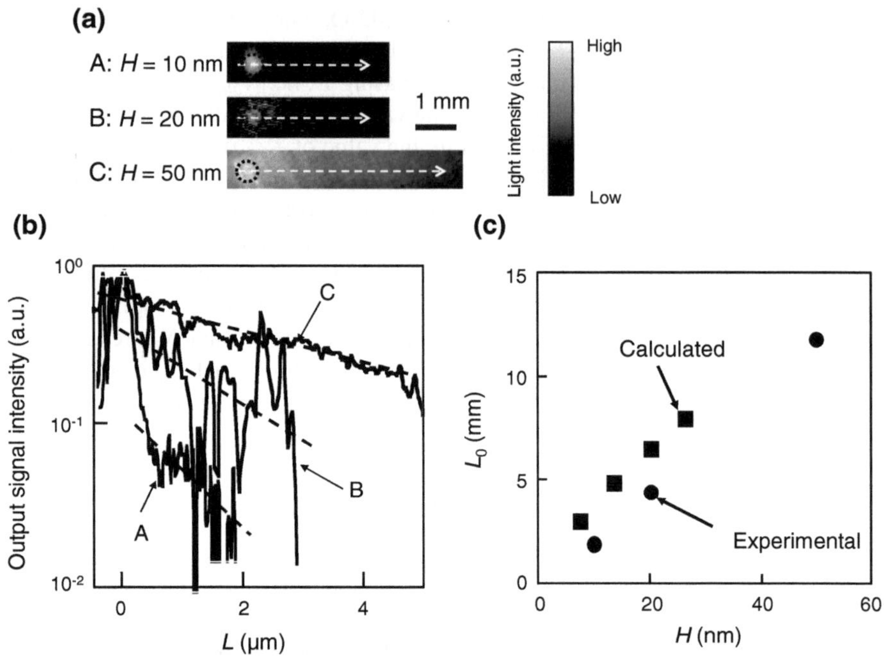

Fig. 5.16 Experimental results for spherical CdSe QDs. **a** Spatial distributions of the light intensities emitted from devices *A*, *B*, and *C*. **b** Dependence of the output signal intensity I_{out} on the distance *L* between the input and output terminals. Curves *A*–*C* correspond to the devices *A*–*C*, respectively. *Broken lines* represent exponential functions fitted to the experimental values. **c** Dependence of the energy transmission length L_0 on the thickness *H* of the small quantum dot layers

(f) Frequency up-converter

In the case of the DP devices described in Parts (a)–(e), uni-directional energy transfer is guaranteed by the relaxation of the exciton from an upper energy level to a lower energy level in a QD. Therefore, the photon energy of the output signal is lower than that of the input signal, by an amount that depends on the energy dissipated by relaxation. In other words, the optical frequency of the output signal is lower than that of the input signal, and therefore, these devices correspond to frequency down-converters.

In contrast, it is possible to realize an optical frequency up-converter, as shown in Fig. 5.17, in which a small QD (QD-A) and a large QD (QD-B) are used. Although the device structure is similar to that of the NOT logic gate described in Part (b), this device uses a high-intensity optical pulse as an input signal in order to create multiple phonons in the heat bath and excite the exciton in QD-B from its lower energy level to its higher energy level.

Creation and excitation of excitons have been experimentally confirmed [19]. Applying a high-intensity optical pulse that is resonant with the lower energy level B_1 of QD-B induces an interaction between the photons and phonons, and multiple

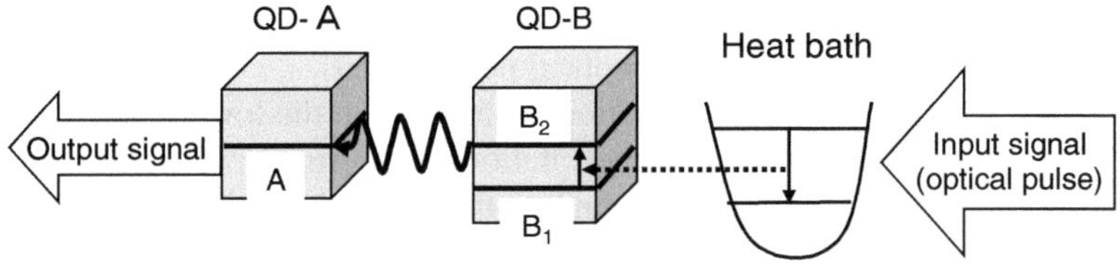

Fig. 5.17 Structure of an optical frequency up-converter using QDs

phonons are created in the heat bath. The number of phonons created follows the Bose–Einstein distribution

$$n\left(\omega_0\right) = \frac{1}{\exp\left[\hbar\omega_0/k_B T\left(S\right)\right] - 1}. \tag{5.6}$$

Here, ω_0, k_B, and T represent the angular frequency of the incident light, the Boltzmann constant, and the heat bath temperature, respectively. It is assumed, for simplicity, that the heat bath temperature is proportional to the time-integrated optical pulse power S. Based on this assumption, the temporal evolution of the occupation probability of the exciton was calculated, and the result is shown in Fig. 5.18. Solid, broken, and dotted curves represent the results for the $\pi/2$-, π-, and $3\pi/2$-pulses of the incident light, respectively. Figure 5.18a shows the occupation probability of the exciton in the lower energy level B_1 in QD-B. On the other hand, Fig. 5.18b is the occupation probability in the energy level of QD-A that is resonant with the upper energy level B_2 in QD-B. This probability takes a nonzero value in this figure, confirming frequency up-conversion; i.e., the photons are emitted from the energy level

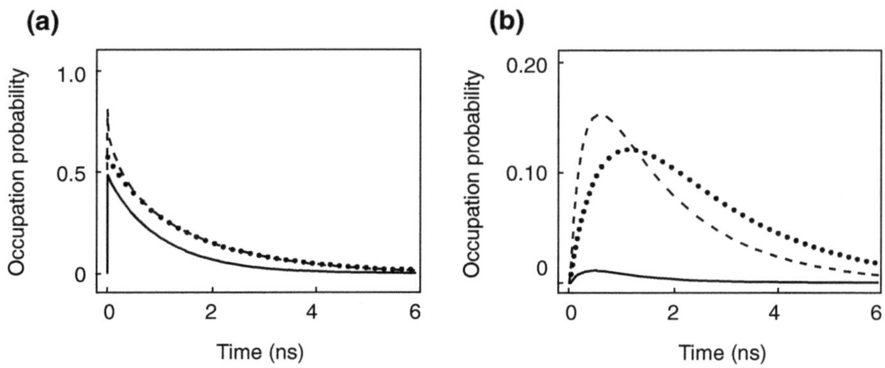

Fig. 5.18 Calculated temporal evolutions of the occupation probabilities of the exciton. **a** and **b** represent occupation probabilities in the lower energy level B_1 in QD-B, and in the energy level in QD-A, respectively. *Solid*, *broken*, and *dotted curves* represent the results for $\pi/2$-, π-, and $3\pi/2$-pulses, respectively

in QD-A after energy transfer from QD-B, and the optical frequency of the output signal is higher than that of the input optical pulse signal.

Figure 5.18a shows that the occupation probability in the lower energy level B_1 takes the maximum value with -pulse excitation. Furthermore, the temporal evolutions of the occupation probability depend on the non-radiative and radiative relaxation rates (γ, γ_{rad}) of the lower energy level B_1. On the other hand, the features of the temporal evolution of the occupation probability in QD-A are more complicated; namely, its maximum value depends on the occupation probability of the lower energy level B_1 in QD-B, and the number of created phonons increases with increasing optical pulse intensity. Therefore, the time constant of the decreasing probability shown by the dotted curve in Fig. 5.18b is longer than that shown by the broken curve. This is because the duration during which the optical pulse is applied is longer for the dotted curve than for the broken curve.

Based on the principle described above, an optical frequency up-converter has been developed by using two closely spaced quantum wells (QW) grown in a ZnO nanorod of 80 nm diameter. As shown in Fig. 5.19a, b, these QWs can be regarded as thin-disk QDs, and the separation between the thinner QW (QW-A, 3.2 nm thickness) and the thicker QW (QW-B, 3.8 nm thickness) is several nm. Figure 5.19c schematically explains the relevant energy levels of the QWs and the energy transfer. The light (3.435 eV photon energy) emitted from the energy level E_{A1} in QW-A, being resonant with the upper energy level E_{B2} in QW-B, can be observed by increasing the optical power of the input signal pulse (3.425 eV photon energy, 10 ps pulse width), which is resonant with the lower energy level E_{B1} in QW-B. Curves A–D in Fig. 5.20a represent the spectral profiles of the light emitted from QW-A for incident light powers of 1, 2, 10, and 60 mW, respectively. The vertical broken line at 3.435 eV represents the emission from the energy level E_{A1}. The curves B–D show spectral peaks at this broken line, from which frequency up-conversion is confirmed. Figure 5.20b shows magnified spectral profiles at around 3.435 eV, to which Lorentzian curves are fitted (broken lines).

Figure 5.20c shows the areas under the Lorentzian curves, which are shown as a function of the incident light power. Closed circles in this figure represent the experimental results. Curves W–Z are the calculated results of the occupation probability of the exciton in the energy level E_{A1} in QW-A, derived by using the quantum master equations of Chap. 3: It is assumed that the heat bath temperature T is proportional to the time-integrated incident optical pulse power S. Then, by inserting $T(S) = \eta S$ (η: proportional constant) into Eq. (5.6) and setting the values of $\hbar\omega_0/k_B\eta$ to 0.1, 0.2, 0.5, and 1.0, the calculated results are fitted to the experimental results of the closed circles, yielding the curves W–Z, respectively. Open squares in this figure represent the calculated results for zero phonon number, which represents zero occupation probability of the exciton. Since the calculated results agree well with the experimental results in this figure, it is confirmed that the optical frequency is up-converted by the coupling of the excitons in the QDs and phonons in the heat bath. This frequency up-converter is expected to find applications in novel light sources, DP-propagating light converters, high-efficiency photodetectors, energy converters, and so on.

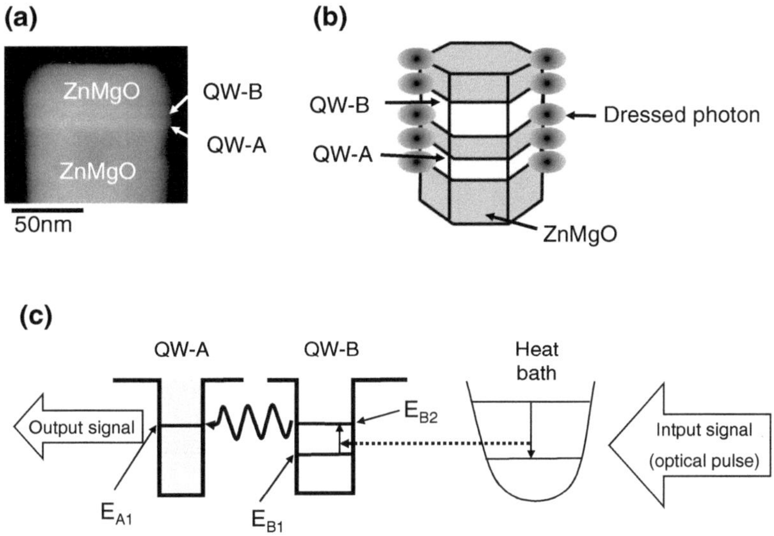

Fig. 5.19 Optical frequency up-converter using quantum wells in a ZnO nanorod. **a** and **b** show a transmission electron microscopic image and the structure of the specimen, respectively. **c** Relevant energy levels of the quantum wells and the energy transfer

Although the structure is more complicated than that of Figs. 5.17, and 5.21 shows another type of optical frequency up-converter that has been proposed, which utilizes the optical switch (AND logic gate) described in Part (a) [20]. Here, the largest QD is used as the input terminal. By applying CW light and the output signal (photon energy: $h\nu_2$) from a front-stage DP device 1 to this device, the photon energy of the output signal is up-converted to $h\nu_1$, which is then applied to the back-stage DP device 2.

(g) Delayed-feedback optical pulse generator

The intensity of the output signal from an active device such as a laser or an optical amplifier can fluctuate or pulsate when the output signal is injected back into the device with some time delay [21]. By using this delayed feedback, it is possible to realize a pulse generator [22]. In the case of using cubic QDs, as illustrated in Fig. 5.22, the size ratio between the small QD and the large QD is arranged to be $1 : \sqrt{2}$, resulting in resonance between the energy level $(1, 1, 1)$ of the exciton in the small QD and the upper energy level $(2, 1, 1)$ in the large QD. The present DP device utilizes several QDs under this resonant condition.

System 1 in Fig. 5.22 corresponds to the active device described above. It is composed of a small QD (QD-C) and a large QD (QD-G), which are resonant with each other. The incident CW light, being resonant with the energy level C_1 in QD-C, is used as an energy source for the pulse generation. The magnitude of the near-field optical interaction between QD-C and QD-G is denoted by $\hbar U_{CG}$.

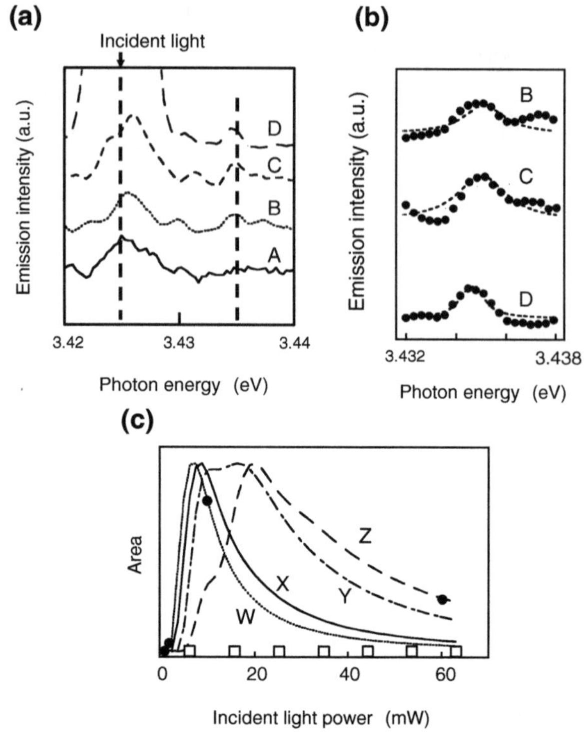

Fig. 5.20 Experimental results of optical frequency up-conversion. **a** Spectral profiles of the output signals. Curves A, B, C, and D represent the results for incident light powers of 1, 2, 10, and 60 mW, respectively. **b** Magnified spectral profiles of curves B, C, and D in **a**. *Broken curves* are Lorentzian curves fitted to the experimental values. **c** Relation between the incident light power and the area under the fitted Lorentzian curves. *Closed circles* represent the experimental values. Curves W, X, Y, and Z are the calculated results fitted to the closed circles by setting $\hbar\omega_0/k_B\eta$ to 0.1, 0.2, 0.5, and 1.0, respectively. *Open squares* represent the calculated results for zero phonon number

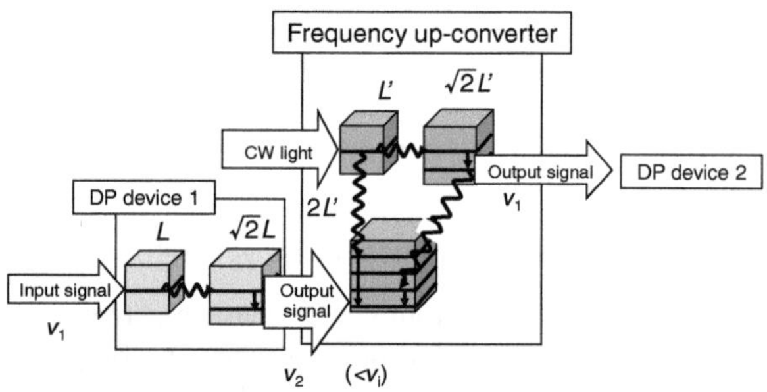

Fig. 5.21 Structure of an optical frequency up-converter using QDs

System 2, on the other hand, corresponds to a delayed-feedback device. It is composed of a small QD (QD-A) and a large QD (QD-B), which are also resonant with each other. By applying the light generated from the energy level C_1 in QD-C to QD-A after a time delay Δ, an exciton is created in the energy level A_1 in QD-A. Here, in order to provide a time delay, several methods can be utilized, for example, using optical nano-fountains [11], multiple QDs of different sizes [15], and excitation recycling [23]. Afterward, the energy is transferred from the energy level A_1 in QD-A to the upper energy level B_2 in QD-B, where the energy of their interaction is denoted by $\hbar U_{AB}$. Subsequent relaxation to the lower energy level B_1 in QD-B and

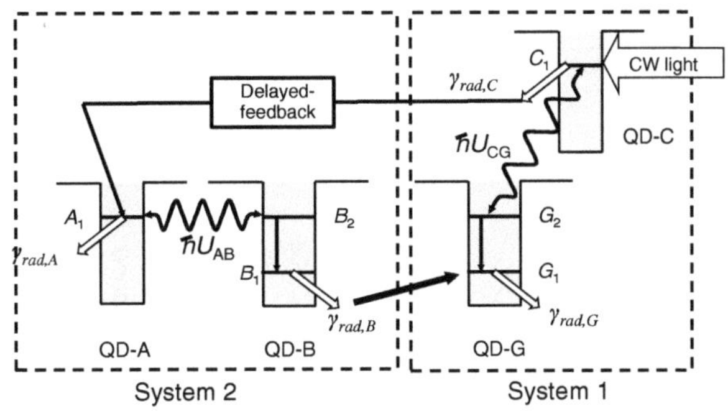

Fig. 5.22 Structure of a delayed-feedback optical pulse generator. The *thick solid arrow* pointing from QD-B to QD-G represents the action of changing the occupation probability of the exciton in the energy level G_1 in QD-G using the output signal from QD-B

annihilation of the exciton generate the output signal, which changes the occupation probability of the exciton in the lower energy level G_1 in QD-G of System 1, as is represented by the right-pointing thick arrow in this figure.

The energy transfer described above serves as the delayed feedback in System 2. The radiative relaxation rate from each energy level is denoted by $\gamma_{rad,i}$ (i = A, B, C, G) in this figure. The total delay time is governed by Δ because the value of Δ provided above can be adjusted to be much longer than the time constants given by U_{AB}, γ, and $\gamma_{rad,i}$.

The temporal evolution of the occupation probability of the exciton in each energy level in the four QDs can be derived by using the quantum master equation of Chap. 3 [24]. System 1 contains three energy levels (C_1 in QD-C, and G_1 and G_2 in QD-G), and the quantum master equations of the density matrix $\rho^{Sys1}(t)$ for the occupation probabilities in these energy levels contain the perturbation Hamiltonian $H_{ext}^{CW}(t)$ due to the CW light incident on QD-C. The value of this Hamiltonian is proportional to that of the electric field $E_{CW}(t)$ of the CW light and is expressed as

$$H_{ext}^{CW}(t) \propto E_{CW}(t) . \tag{5.7}$$

Since the occupation probability of the lower energy level G_1 in QD-G is changed by irradiating the light emitted from the lower energy level B_1 in QD-B, the perturbation Hamiltonian $H_{ext}^{G}(t)$ for QD-G is proportional to the density matrix element $\rho_{B_1}^{Sys2}(t)$ representing the occupation probability in the lower energy level B_1, which is expressed as

$$H_{ext}^{G}(t) = \alpha_G \rho_{B_1}^{Sys2}(t) . \tag{5.8}$$

Here, α_G is the proportionality constant representing the efficiency of the energy transfer from System 2 to System 1.

On the other hand, System 2 contains three energy levels (A_1 in QD-A, and B_1 and B_2 in QD-B). The occupation probability of the energy level A_1 is changed if QD-A is irradiated with the light emitted from the energy level C_1 in QD-C. Therefore, the quantum master equations for the density matrix $\rho^{Sys2}(t)$ representing the occupation probabilities in each energy level of System 2 contain the perturbation Hamiltonian

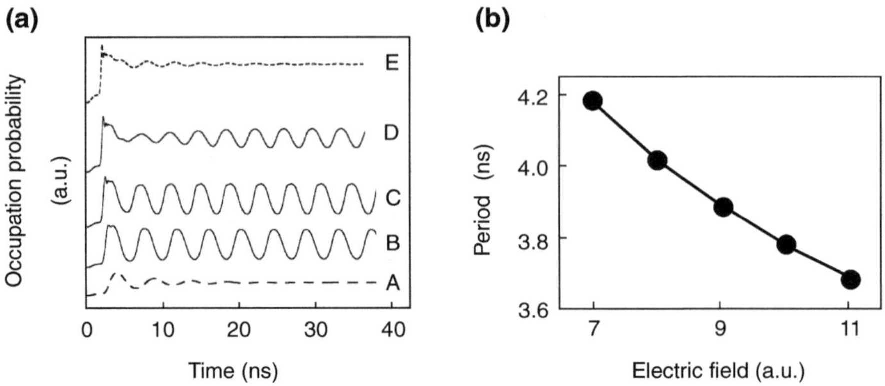

Fig. 5.23 Calculated occupation probabilities in the lower energy level G_1 of QD-G. **a** Temporal evolution of the occupation probability. Curves *A*, *B*, *C*, *D*, and *E* represent the results when the electric field of the CW light has values 5, 7, 9, 11, and 13 (arbitrary units), respectively. **b** Relation between the electric field of the incident light and the period of pulsation

$H_{ext}^{A}(t)$ for QD-A, whose value is proportional to the density matrix element $\rho_{C_1}^{Sys1}(t)$ for the occupation probability of the energy level C_1 in QD-C. Here, by noting that there exists a time delay Δ in the feedback, this Hamiltonian is expressed as

$$H_{ext}^{A}(t) = \alpha_A \rho_{C_1}^{Sys1}(t - \Delta), \tag{5.9}$$

where α_A is the proportionality constant representing the efficiency of the energy transfer from System 1 to System 2. Using Eqs. (5.8) and (5.9), the simultaneous quantum master equations for Systems 1 and 2 are solved to derive the occupation probability of excitons in each energy level.

By taking ZnO QDs as an example [25], the physical quantities are fixed to $\hbar U_{CG} = \hbar U_{AB} = 7.3 \times 10^{-25}$J, $\gamma = 1 \times 10^{11}$s^{-1}, $\gamma_{rad,C} = \gamma_{rad,A} = 2.3 \times 10^9$s^{-1}, $\alpha_G = 0.1$, and $\alpha_A = 0.01$. Figure 5.23a shows the results of numerical calculations of the temporal evolutions of the occupation probabilities of the exciton in the lower energy level G_1 in QD-G, where the value of the electric field $E_{CW}(t)$ of the CW light applied to QD-C is used as a parameter. No pulsation is seen when the value of $E_{CW}(t)$ is too small or too large, as shown by curves A and E, respectively. However, in the case of moderate values of $E_{CW}(t)$, curves B–D clearly show pulsation. Since this pulsation can be regarded as a pulse train, the DP device of Fig. 5.22 is called a delayed-feedback optical pulse generator. As shown in Fig. 5.23b, the period of the pulsation decreases with increasing $E_{CW}(t)$, which is due to the increase in the occupation probability of the exciton in QD-G.

Other features of this optical pulse generator are summarized as follows:

(1) Pulsation does not occur when the value of the delay time Δ is too short. This is because, in System 1, the energy is promptly transferred from the energy level C_1 in QD-C to the upper energy level G_2, and System 1 reaches a steady state within

a sufficiently short duration. The period T of pulsation is proportional to Δ and is expressed as $T = a\Delta + b$, where a does not depend on $E_{CW}(t)$. However, b decreases with increasing $E_{CW}(t)$, which agrees with the results shown in Fig. 5.23b. It should be pointed out that the pulse width depends mainly on the delay time Δ.

(2) Pulsation does not occur when the value of the interaction energy $\hbar U_{CG}$ between QD-C and QD-G is too small because hardly any of the energy is transferred from the energy level C_1 in QD-C to the upper energy level G_2 in QD-G. On the other hand, when the interaction energy is too large, pulsation does not occur either. Pulsation occurs when $\hbar U_{CG}$ takes a moderate value between these two extreme cases.

(3) Pulsation does not occur when the value of $\gamma_{rad,C}$ in QD-C is too large because the duration $\gamma_{rad,C}^{-1}$ in which the energy is transferred from QD-C to QD-G is too short. However, it should be noted that the occupation probability in the lower energy level G_1 increases with decreasing $\gamma_{rad,C}$ because the duration $\gamma_{rad,C}^{-1}$ of the energy transfer increases.

5.1.2 Devices in Which Coupling with Propagating Light is Controlled

In addition to the DP devices reviewed in Sect. 5.1.1, another type of DP device is one in which the coupling between DPs and propagating light is controlled. This short subsection reviews two examples of such devices.

(a) Buffer memory

As described by Eq. (3.7), the total electric dipole moment in the anti-symmetric state is zero because the two electric dipole moments are anti-parallel to each other. Therefore, this state does not absorb or emit propagating light; that is to say, it does not couple with propagating light. By utilizing this feature, a buffer memory device has been proposed [26]. It is composed of three QDs (QD-A, QD-B, and QD-C), as illustrated in Fig. 5.24a. Since the two smaller QDs, QD-A and QD-B, are the same size, their energy levels, with energy eigenvalues denoted by $\hbar\Omega$, are resonant with each other. The large QD, QD-C, is placed at the asymmetric position with respect to QD-A and QD-B. If excitons are created in both QD-A and QD-B by the incident light signal, they occupy the symmetric and anti-symmetric states formed by these two QDs. If the energy $\hbar\Omega_{C2}$ of the upper energy level C_2 in QD-C is tuned to $\hbar(\Omega + U)$ (where $\hbar U$ is the energy of the interaction between QD-A and QD-B), this energy level is resonant with the symmetric state. This means that the exciton energy in the symmetric state is transferred to the upper energy level C_2 in QD-C, subsequently relaxes to the lower energy level C_1, and is dissipated to the heat bath, as is shown by the left part of Fig. 5.24a.

After this energy transfer, the anti-symmetric state is still occupied by the exciton, which is not resonant with the upper energy level C_2 in QD-C. Therefore, the energy

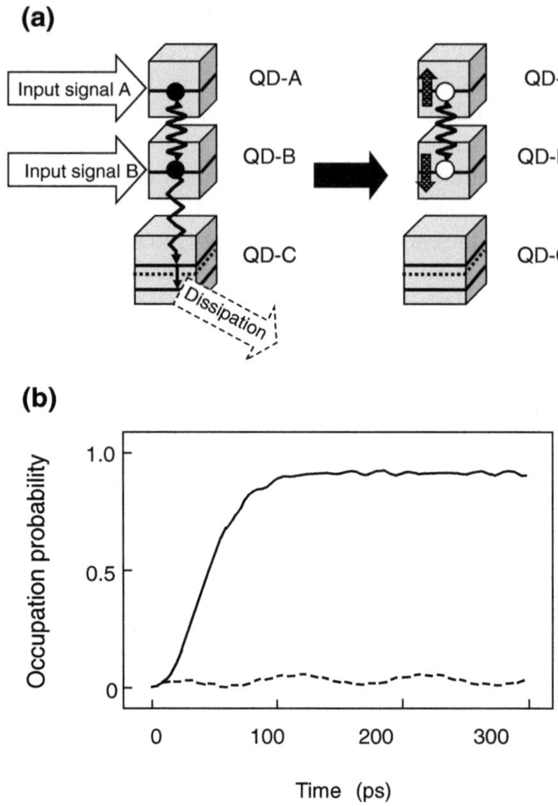

Fig. 5.24 Buffer memory using three QDs. **a** The *left* figure shows that QD-A and QD-B are both occupied by an exciton, as indicated by *closed circles*. On the other hand, the *right* figure shows that an exciton remains either in QD-A or in QD-B, with an occupation probability of 0.5, as indicated by *open circles*. *Upward* and *downward thick arrows* in this figure represent two electric dipole moments that are anti-parallel to each other. **b** Calculated temporal evolutions of the occupation probabilities of the exciton. *Solid* and *broken curves* show values when the excitons occupy the anti-symmetric and symmetric states, respectively

does not transfer to QD-C, as is shown by the right part of Fig. 5.24a. Moreover, since this state is electric dipole-forbidden, no propagating light is emitted. As a result, the input signal energy is stored in the anti-symmetric state, which represents the operation of a buffer memory.

Figure 5.24b shows the calculated results of the dynamic behavior of the device operation [27], where the value of the interaction energy between QD-A and QD-B, $\hbar U$, is 1.1×10^{-23} J, and that between QD-B and QD-C, $\hbar U'$, is 2.1×10^{-24} J. On the other hand, it is assumed that QD-A and QD-C do not interact with each other because QD-C is placed at the asymmetric position, as shown in Fig. 5.24a. Solid and broken curves represent the time evolutions of the occupation probabilities of the exciton in the anti-symmetric and symmetric states formed by QD-A and QD-B, respectively. Comparing these curves, it is found that the value of the solid curve asymptotically approaches unity, whereas that of the broken curve remains zero, from which the buffer memory operation is confirmed.

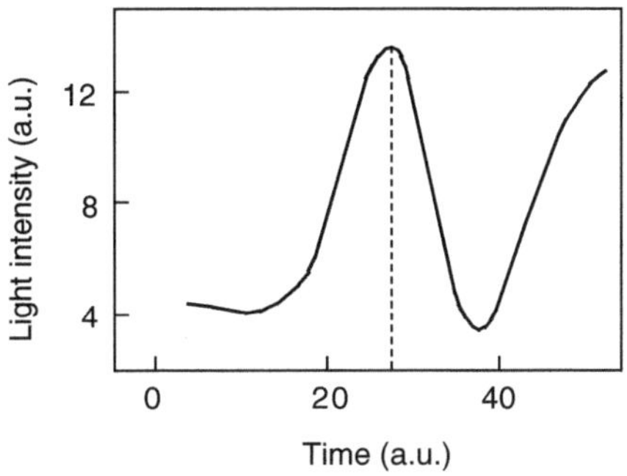

Fig. 5.25 Calculated time evolution of the light intensity emitted from a super-radiant optical pulse generator. The intensity takes the maximum at the time represented by the *vertical broken line* in this figure

(b) Super-radiant optical pulse generator

By utilizing the symmetric state in which two electric dipole moments are parallel to each other, a phenomenon similar to Dicke's super-radiance can occur [28, 29]. An ultrashort optical pulse generator has been proposed by using this phenomenon [30]. This phenomenon is one type of cooperative phenomenon, in which the electric dipole moments generated in multiple QDs oscillate synchronously in an in-phase manner, and the total intensity of the pulsed propagating light emitted from these electric dipole moments is proportional to the square of the number of the QDs, N, and the pulse width is inversely proportional to N.

Figure 5.25 shows the calculated time evolution of the emitted light intensity. Here, the initial condition at time $t = 0$ is that the exciton occupies the upper and lower energy levels in the $2i + 1$th and $2i$th QDs, respectively ($i = 0, 1, 2, \ldots$). The total number of QDs, N, is assumed to be 8. This figure shows that the intensity of the emitted propagating light takes the maximum at the time represented by the vertical broken line because the phases of the oscillations of the electric dipole moments in the QDs coincide with each other at this moment. This maximum value is as high as that ($N(N/2 + 1)/2$) of Dickefs super-radiance. This phenomenon has been experimentally confirmed by using multiple QWs in a ZnO nanorod [31]. The number and thickness of the QWs are 9 and 3.25 nm, respectively. The separation between adjacent QWs is 9 nm. It was confirmed that four QWs coupled coherently to produce super-radiance.

5.2 Characteristics of Dressed Photon Devices

This section reviews the unique characteristics of DP devices, which are important for application to novel optical information transmission and processing systems.

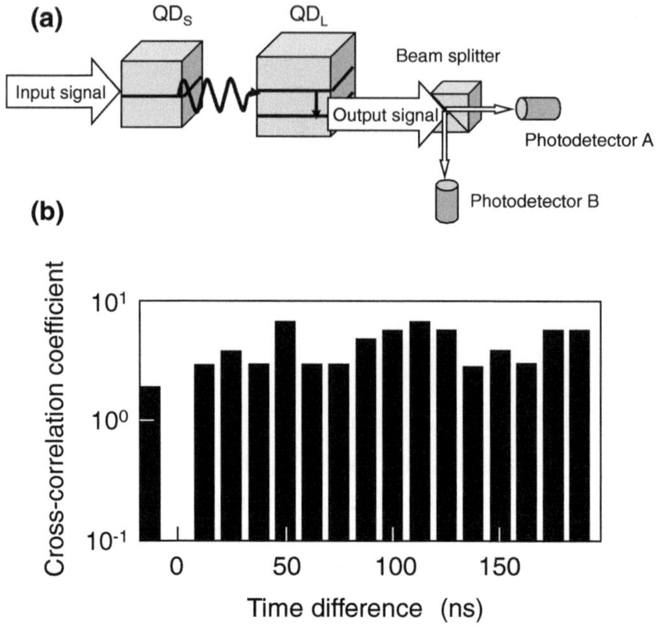

Fig. 5.26 Experimental results for single-photon operation. **a** Setup of photon correlation experiment. **b** Measured dependence of the cross-correlation coefficient on the time difference between the detections by two photodetectors

5.2.1 Low Energy Consumption

(a) Single-photon operation

An exciton is created by injecting a photon to the input terminal of the DP device, and subsequent energy transfer and relaxation generate a photon from the output terminal. Thus, the DP device is driven by a single photon and emits a single photon. This single-photon emission has been confirmed by photon correlation experiments using two cubic QDs [32], as shown in Fig. 5.26a. As was the case of the DP devices described in Sect. 5.1, the ratio of the side lengths of the small and large QDs is adjusted to be $1 : \sqrt{2}$ so that the energy level (1, 1, 1) in the small QD (QD_S) and the upper energy level (2, 1, 1) in the large QD (QD_L) are resonant with each other.

Figure 5.26b shows the experimental results obtained by using two cubic CuCl QDs at a temperature of 15 K [33]. The horizontal and vertical axes represent the time difference between the detections by two photodetectors and the cross-correlation coefficient between the two detected light intensities. When the time difference is zero, the value of the cross-correlation coefficient is zero,[3] from which single-photon emission was estimated to have a probability as high as 99.3 %.

[3] When the value of the cross-correlation coefficient is smaller than unity at zero time difference between the detections, the quantum state of the photon is called an anti-bunching state.

This high probability of single photon emission is due to the following blockade mechanisms: If two excitons are created in QDs, the energy of the level (1, 1, 1) decreases by about 30 meV. Since this decrease is as high as the energy required for coupling the two excitons, the energy level (1, 1, 1) in the input terminal is detuned from the input signal. Therefore, the input signal, the energy level (1, 1, 1) in QDs, and the upper energy level (2, 1, 1) in QD_L become mutually off-resonant, from which it is concluded that only one exciton is created in QDs and, as a result, only one photon is emitted from QD_L.

(b) Magnitude of energy dissipation

This part starts with a discussion of the energy dissipation of conventional electronic devices. Electrical wires are required to connect to a conventional electronic device, as shown in Fig. 5.27a. This means that the magnitude of the energy dissipation is mainly determined not by the electronic device itself but by other elements, including wires, load resistances, and so on, which consume a large amount of energy. As an example of an electronic device, a single-electron tunneling device is considered. The device is connected to an electrical power supply via a load impedance, as shown in Fig. 5.27b. In order to realize single-electron tunneling, the electrostatic energy $E_c = e^2/2C$ (e: electron charge, C: capacitance of the tunneling junction) must be larger than the thermal energy $k_B T$ (k_B: Boltzmann constant, T: temperature) of the heat bath. Furthermore, a high load impedance must be used in order to suppress electron number fluctuations; that is, if the load impedance is an inductance L, the requirement

$$L \gg \frac{\hbar^2 C}{e^4} \tag{5.10}$$

must be met [34]. Due to these requirements, the energy consumed in the circuit of Fig. 5.27b is large.

In contrast, the DP device does not require electrical or optical wires, and the energy is dissipated only in the DP device due to relaxation from a higher energy level to a lower energy level in a nanomaterial. The rate of this relaxation is given by Eq. (3.16), and is about $1.0 \times 10^{11}\,\mathrm{s}^{-1}$ in the case of a CuCl QD.

In order to estimate the magnitude of the energy dissipation in the DP device, two closely spaced QDs are considered, as shown in Fig. 5.27c [35]. By adjusting the size of the QDs, the energy level S in the small QD (QDs) is resonant with the upper energy level L_2 in the large QD (QD_L). By applying propagating light to them, the energy is transferred from S to L_2, and subsequent relaxation from L_2 to the lower energy level L_1 fixes the value of the output signal intensity. If the energy difference between the energy levels L_2 and L_1 in QD_L is small, the magnitude of the energy dissipation is also small. However, the possibility of energy transfer from S to L_1 without passing through L_2 also increases, which would cause malfunction of the DP device. Therefore, the error rate of the DP device depends on the energy dissipation, which is proportional to the energy difference between the energy levels L_2 and L_1. In order to estimate these values, Fig. 5.28a shows the calculated occupation probabilities of the exciton in the lower energy level L_1 in QD_L as a function of the

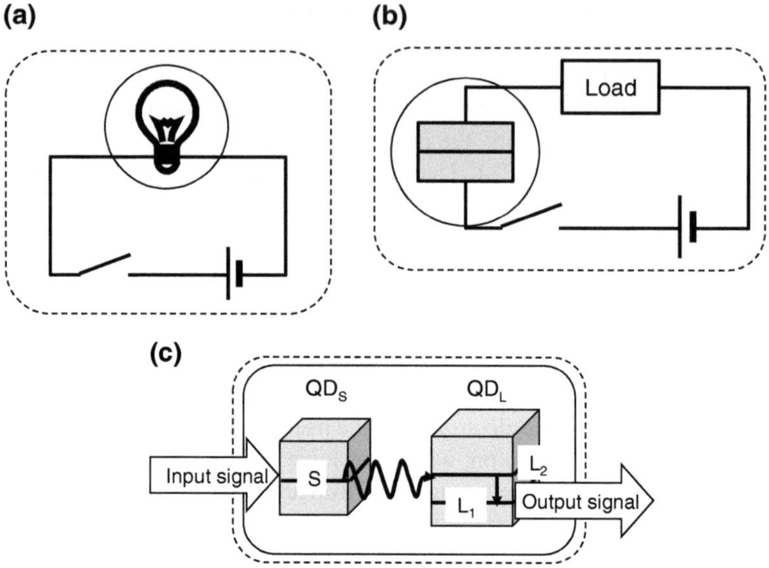

Fig. 5.27 Comparison between systems using electronic devices and dressed photon devices. *Solid* and *broken squares* represent the size of the device and system, respectively. **a** and **b** show a system using electronic devices, namely, a light bulb and a single-electron tunneling device, respectively. **c** The system using dressed photon devices

magnitude of the energy dissipation. In the case of curve A in this figure, the energy is appropriately transferred from QDs to QD_L because their separation is sufficiently short, and thus, this curve corresponds to the value of the output signal intensity. On the other hand, for the curves B–D, the magnitudes of the energy transfer are small because the separation between the two QDs is too long, and thus, these curves represent the magnitudes of the energy dissipation due to the relaxation from level L_2 to level L_1, which correspond to the values of the error intensities. Therefore, the ratios of the value of curve A to those of curves B–D give the error rates. Curves A–G in Fig. 5.28b represent the relations between the values of the error rate and the energy dissipation, which were derived by calculating the ratios described above. For comparison, curve H represents the value of the energy dissipation required for a bit-flip in a CMOS logic gate, as an example of a conventional electronic device [36]. By comparing the values for the DP device and the CMOS logic gate, the energy dissipation of the DP device is found to be extremely low, namely, as low as 10^4 times smaller than that of the CMOS logic gate.

The energy transfer process described above is similar to that observed in a photosynthetic bacteria [37] and, because of its high energy transfer efficiency, is receiving attention as a novel system function that is inherent to complex systems in nano-scale space [38, 39].

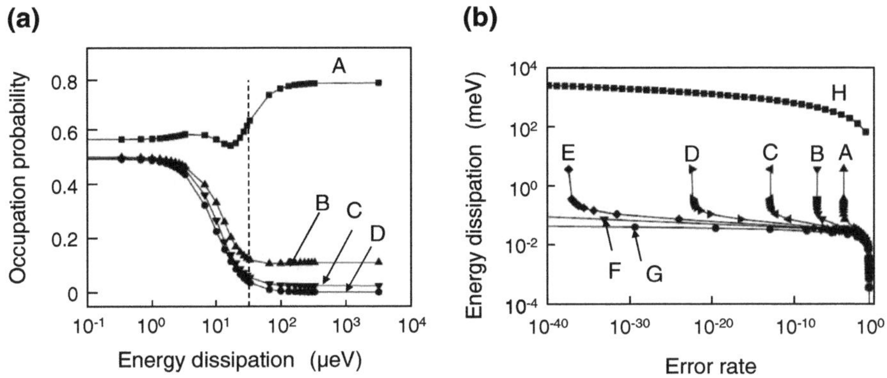

Fig. 5.28 Calculated results for energy dissipation by a dressed photon device. **a** Relation between the magnitude of energy dissipation and the occupation probability of the exciton. Curve *A* is proportional to the value of the output signal intensity. Curves *A*–*D* represent the results when the values of the interaction energy, $\hbar U$, between the two quantum dots are 1.1×10^{-24}, 2.1×10^{-25}, 1.1×10^{-25}, and 1.1×10^{-26} J, respectively. **b** Relation between the error rate and the energy dissipation. Curves *A*–*G* represent the results when the values of the interaction energy, $\hbar U$, between the two quantum dots are 2.1×10^{-25}, 1.8×10^{-25}, 1.5×10^{-25}, 1.3×10^{-25}, 1.2×10^{-25}, 1.1×10^{-25}, and 1.1×10^{-26} J, respectively. For comparison, curve *H* represents the value of the energy dissipation required for a bit-flip in a CMOS logic gate

(c) Magnitude of energy consumption

The magnitude of the energy required to drive the DP device is very low because it can be operated by a single photon, as was confirmed in Part (a). Moreover, the energy dissipation is also very low, due to only the relaxation between energy levels, as was confirmed in Part (b). However, in order to estimate the magnitude of the energy consumption, the magnitude of the driving energy and dissipated energy must be estimated in a more quantitative manner from the viewpoint of transmitting significant information to the receiver [40].

For this estimation, a basic optical information transmission system is considered, as illustrated in Fig. 5.29a. It is composed of an input interface (the nano-optical condenser of Part (c) in Sect. 5.1.1) to convert the propagating light of the input signal to the DP, a NOT logic gate (Part (b)), and an output interface. The former two devices are composed of InAs QDs, and the latter is composed of a gold nanoparticle, as was shown in Fig. 5.7. The generated DP output signal is converted to propagating light by the gold nanoparticle and reaches a photodetector, where it is converted to an electrical signal. The intensity of the propagating light (the output signal) must be sufficiently high in order to achieve detection with a sufficiently large signal-to-noise ratio for definitely recognizing the transmitted information. By noting this requirement, the magnitude of energy consumption is estimated in the following:

(1) Efficiency, η_{in}, of conversion from the propagating light signal input to the DP by the input interface: The nano-optical condenser has an extremely low energy dissipation due to the relaxation from the upper energy level to the lower energy

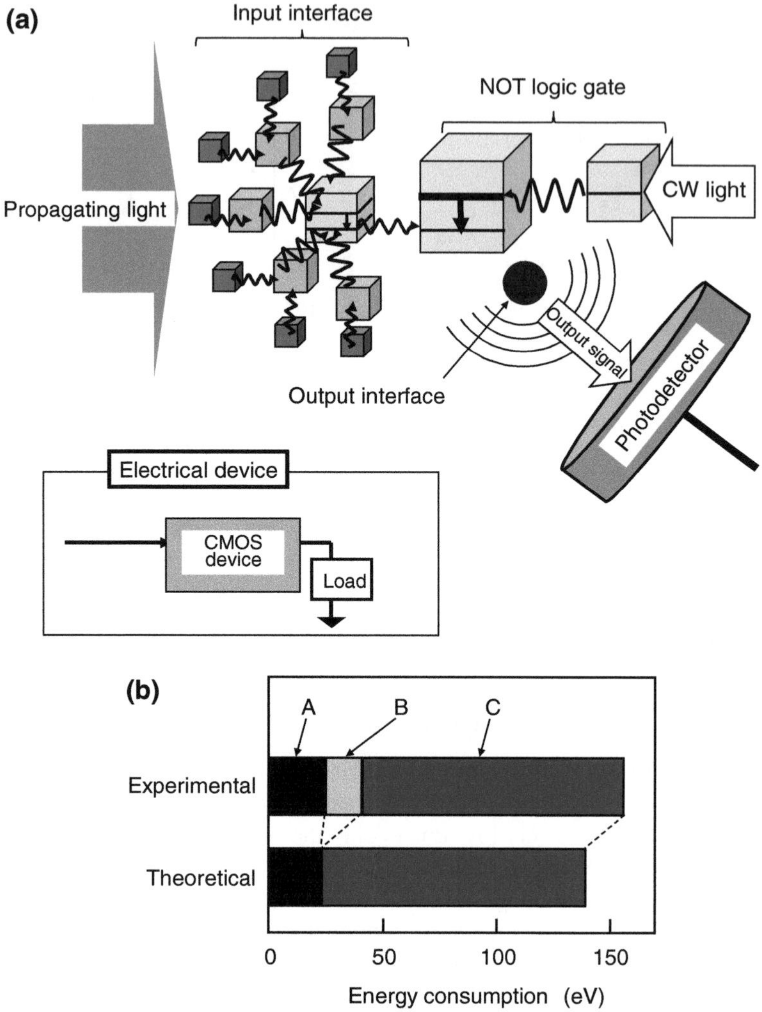

Fig. 5.29 A system composed of an input interface, a NOT logic gate, an output interface, and a photodetector. **a** Structure of the system. For comparison, that of the system composed of a conventional electronic CMOS device is shown in the inset. **b** The magnitudes of the energy consumption. *A*, *B*, and *C* represent the values of the energy consumed in the input interface, the NOT logic gate, and the output interface, respectively

level in the QD. Here, an experimentally evaluated efficiency, η_{in}, of 0.9 is employed for estimation.

(2) Energy dissipation, E_d, in the NOT logic gate: A theoretical energy dissipation, $E_{d(th)}$, of 25 μeV, due to the relaxation from the upper energy level to the lower energy level in the QD of the output terminal [35], is employed for the estimation. Also, an experimental energy dissipation, $E_{d(\mathrm{exp})}$, of 65 meV is employed [7].

(3) Efficiency, η_{out}, of the output interface: In the NOT logic gate of Fig. 5.7, InAs QDs are buried in GaAs layers. The output optical power, P_{emit}, from the QD of the output terminal and the output optical power, $P_{extract}$, extracted outside of the DP device are related by the Fresnel reflection at the surface of the mesa-shaped DP device, expressed as

$$P_{extract} = \left(\frac{n_G - 1}{n_G + 1}\right)^2 P_{emit}, \tag{5.11}$$

where (=3.5) is the refractive index of GaAs. Moreover, in the system of Fig. 5.29a, of the optical power $P_{extract}$, only the forward-scattered light propagating upward in the mesa reaches the photodetector. That is, the output power P_{mesa} from the mesa-shaped DP device is given by

$$P_{mesa} = 0.5 P_{extract}. \tag{5.12}$$

The gold nanoparticle fixed on the upper surface of the mesa-shaped DP device enhances the forward-scattering, and as a result, the scattered optical power, P_m, was measured to be [7]

$$P_m = 3.0 P_{mesa}. \tag{5.13}$$

By combining the numerical values given above, the efficiency of the output interface, η_{out}, is estimated to be

$$\eta_{out} = \frac{P_m}{P_{emit}} = \frac{P_m}{P_{mesa}} \cdot \frac{P_{mesa}}{P_{extract}} \cdot \frac{P_{extract}}{P_{emit}}. \tag{5.14}$$

Inserting the values of Eqs. (5.11)–(5.13) into the three terms on the right-hand side of this equation gives $\eta_{out} = 0.45$.

(4) Total energy consumption, $E_{c,total}$: The number of photons, n_p, required to recognize a one-bit signal with a conventional receiver (composed of an optical amplifier and an avalanche photodiode) is 100 [41], and the required photon energy is expressed as $E_{\mathrm{det}} = n_p h\nu$ (ν: optical frequency). Therefore, the photon energy of the input signal that must be applied to the input interface is expressed by using the efficiencies described in (1)–(3) above:

$$E_{in} = \frac{n_p (h\nu + E_d)}{\eta_{in}\eta_{out}}. \tag{5.15}$$

By using the theoretical and experimental values of the energy dissipation for light with a wavelength of 1.3 μm (photon energy $h\nu = 0.95$ eV) given in (2), the theoretically and experimentally estimated values of E_{in} are found to be 235 and 251 eV, respectively. Therefore, the theoretical and experimental magnitudes of the total energy consumption,

$$E_{c,total} = E_{in} - E_{\mathrm{det}} \tag{5.16}$$

in the system of Fig. 5.29a are 140 and 156 eV, respectively.

In order to summarize the estimated results given above, the magnitudes of the energy consumption in the input interface, the NOT logic gate, and the output interface are compared and shown in Fig. 5.29b. From this figure, it is found that the magnitude of the energy consumption at the output interface is the largest, whereas that in the input interface is smaller. Furthermore, that in the NOT logic gate is extremely small and can be neglected. From these results, it is concluded that increasing the efficiency of the output interface would be effective in further reducing the total energy consumption.

For comparison, here the magnitude of the energy consumption of a CMOS logic gate to which a load impedance is connected is estimated. The experimental value of the energy consumption is 6.3 MeV [42]. Since the experimental value of $E_{c,total}$ described above (156 eV: 25 aJ) is about 10^4 times smaller, it is confirmed that the energy consumption of the system of Fig. 5.29a is extremely low.

Finally, the signal processing rate of the NOT logic gate in the system of Fig. 5.29a is estimated. Although the time, τ, required to transfer the energy from the input QD to the output QD is as short as 50 ps [7], the signal processing rate B of the NOT logic gate depends not only on τ but also on n_p and η_{out} and is expressed as $1/\left(n_p\tau/\eta_{out}\right)$ per bit. By inserting the numerical values given in (3) and (4) into this expression, B is estimated to be 90 Mb/s. By using identical devices in parallel, in other words, device redundancy, the minimum duration for a single bit of information could be shortened, allowing a higher operating speed.

From the estimations described above, it is confirmed that the energy for driving DP devices and the energy consumed by DP devices are both extremely low, which means that a higher degree of integration of these devices can be expected as compared with the integration of conventional electronic devices. Also, the much higher degree of integration compared with conventional optical devices will enable the construction of novel integrated systems that are not possible as long as conventional electrical or optical devices are used. That is, one can be released from the common view in conventional technology that "light should be used for communication because of its high propagation speed, while electrons should be used for computing because of their small size." As an example of the kind of novel system that can be constructed when released from this commonly held view, a DP computer using DP devices has been proposed [43]. Here, it should be noted that DP computing is completely different from conventional optical computing [44], which carries out digital information processing using several technologies based on spatially parallel processing utilizing the wave optical properties of propagating light, for example, holography. In contrast, DP computing carries out digital processing of time-sequential signals, which has never been possible by using conventional optical devices and propagating light. Furthermore, the low energy consumption of DP devices is extremely useful for improving energy efficiency in order to solve the serious problem of energy management now being faced in rapidly developing communication networks [45].

5.2.2 Tamper-Resistance

Since the electrical device in Fig. 5.27b dissipates energy through the electrical power supply or the load impedance, the information transmitted through the circuit can be intercepted by detecting the magnitude of this energy dissipation. This signal interception is called tampering and is a serious problem in information security. In contrast, the DP devices transmit signals without requiring any wires, and moreover, the magnitude of the dissipated energy is extremely low. Therefore, tampering, and thus, side-channel attacks, are difficult in practice [46], suggesting that DP devices have high tamper-resistance.

The possibility of signal tampering is studied here. As illustrated in Fig. 5.30a, a QD (QD_{mon}) is brought close to a DP device composed of two QDs (QD_S and QD_L, for the input and output terminals, respectively). Figure 5.30b shows the result of numerical calculations, where the energy of the interaction between QD_L and QD_{mon} ($\hbar U = 1.0 \times 10^{-24}$ J) is assumed to be equal to that between QD_S and QD_L. The non-radiative relaxation rate, γ, in QD_L is set at $2.0 \times 10^{11} s^{-1}$. The dotted curve in this figure represents the occupation probability of the exciton in the lower energy

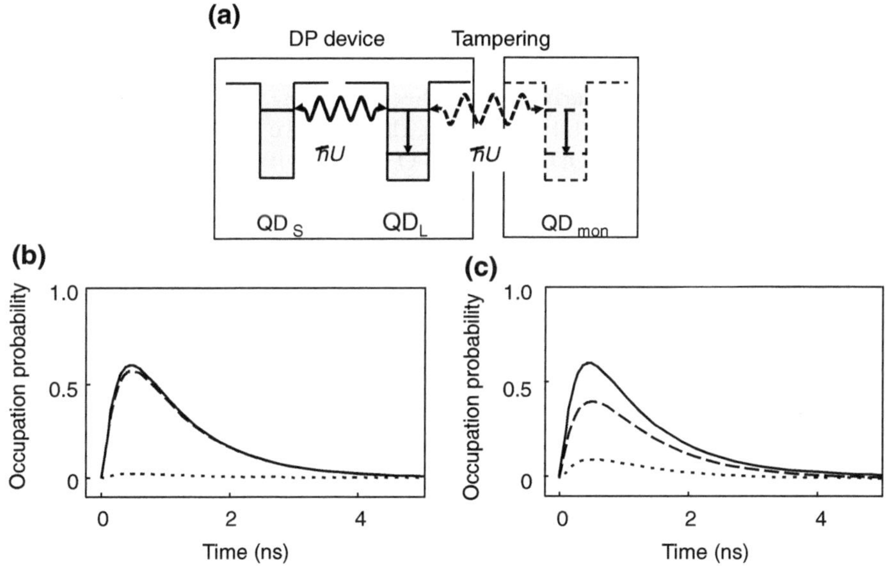

Fig. 5.30 Calculated results of tamper resistance. **a** Dressed photon devices composed of two quantum dots (QD_S and QD_L). The quantum dot (QD_{mon}) is used for tampering. **b** and **c** show the temporal evolutions of occupation probabilities of the exciton in the quantum dots, where the values of the interaction energy, $\hbar U$, are 1.0×10^{-24} and 2.0×10^{-24} J, respectively. The value of the non-radiative relaxation rate, γ, is 2.0×10^{11} s^{-1}. *Solid curves* represent the occupation probabilities in the upper energy level of QD_L in the absence of QD_{mon}. *Broken* and *dotted curves* represent the occupation probabilities in the lower energy level of QD_L and the lower energy level of QD_{mon}, respectively, with QD_{mon} placed in close proximity

level in QD_{mon}, which is very small, and thus, tampering is not possible. On the other hand, Fig. 5.30c shows the result for a larger interaction energy ($\hbar U = 2.0 \times 10^{-24}$ J). Although the dotted curve shows that the occupation probability in QD_{mon} is larger than that of Fig. 5.30b, the value of the broken curve is smaller than that of the solid curve. This means that the occupation probability in QD_L decreases by bringing QD_{mon} close to QD_L, and as a result, the tampering is discovered. From these results, the tamper resistance of the DP device is confirmed.

5.2.3 Skew Resistance

It has been found experimentally that the AND logic gate of Fig. 5.2 generated an output signal even though there was a difference between the arrival times of the two input signals [47], which is related to the finite rise and fall times of the output signal described in Part (a) of Sect. 5.1.1. Moreover, the value of the output signal intensity depended on which input signal arrived earlier than the other. This subsection describes such skew dependence, and also, the skew resistance properties of DP devices.

The intensities of the pulsed input signals applied to QD-I and QD-C in the AND logic gate are denoted by $IN_I(t)$ and $IN_C(t)$, respectively. Then, the value of the output signal intensity from QD-O is calculated by solving the quantum master equations for the density matrix given in Sect. 3.2. Here, the energy, $\hbar U$, of the interaction between QDs is 1.1×10^{-24} J. The radiative relaxation rates, $\gamma_{rad,I}$, $\gamma_{rad,C}$, and $\gamma_{rad,O}$, of QD-I, QD-C, and QD-O are $1.2 \times 10^8\,s^{-1}$, $1.0 \times 10^9\,s^{-1}$, and $3.5 \times 10^8\,s^{-1}$, respectively. The non-radiative relaxation rate, γ, is $1.0 \times 10^{11}\,s^{-1}$. The pulse widths of $IN_I(t)$ and $IN_C(t)$ are 2.0 ps.

Figure 5.31 shows the calculated time evolutions of the occupation probability of the exciton in the energy level (1, 1, 1) in QD-O, where the difference in the arrival times of the two input signals, t_s, is used as a parameter. Here, $t_s > 0$ means that the pulse $IN_I(t)$ arrives earlier than $IN_C(t)$. This figure represents the result for $t_s = 0$ (curve A), -300 ps (curve B), and $+300$ ps (curve C). Curve D represents the case when only the signal $IN_I(t)$ arrives. It is found that the occupation probability of curve B is as large as that of curve A. In contrast, the value for curve C is smaller than those for curves A and B, and furthermore, the value pulsates due to the nutation of the energy transfer between QDs (refer to Sect. 3.1).

Curve D takes nonzero values, because the radiative relaxation rate in QD-C takes a nonzero value. However, this feature is disadvantageous for the AND logic gate operation because an output signal is generated even when only a single input signal is applied. In order to investigate these output signal properties, the dependence of the contrast C (refer to the Part (a) in Sect. 5.1.1) on the time difference t_s is estimated. Here, the value of C corresponds to the ratios of two areas: One is the area under curve A, B, or C in Fig. 5.31. The other is the area under curve D. Closed circles and triangles in Fig. 5.32 represent the calculated values of C for $-1\,\text{ns} \le t_s \le 1\,\text{ns}$ and for $-4\,\text{ns} \le t_s \le 4\,\text{ns}$, respectively. The closed circles show that C take values larger than 2 when $-1\,\text{ns} \le t_s \le 0$. Since the input pulse width is 2.0 ps, it is found that this DP device exhibits high skew resistance over a time longer than the input pulse width. On the other hand, in the case of $t_s > 0$, the value of C decreases rapidly with increasing t_s, representing low skew resistance. These characteristics are due to the fact that the magnitude of the energy transferred from QD-I to QD-O depends strongly on the occupation probability of the exciton in QD-C. From the results of calculations, it is found that earlier arrival of $IN_C(t)$ is more advantageous for reliable tamper-resistant operation of the AND logic gate.

To examine skew resistance further, photon correlation experiments (Part (a) in Sect. 5.2.1) were carried out to measure the temporal evolution of the output signal intensity from an InGaAs QD, and the results are shown in Fig. 5.33. The optical wavelength and pulse width of the input signal were 855 nm and 2 ps, respectively. Curves A, B, and C represent the results for $t_s = -1.1, 0,$ and 1.1 ns, respectively. The output signal intensity for curve B is the largest because $t_s = 0$. On the other hand, the intensity for curve A ($t_s = -1.1$ ns) is larger than that for curve C ($t_s = 1.1$ ns), which is consistent with the calculated results of Fig. 5.32. For reference, the closed squares in Fig. 5.32 represent the measured values of C for $-3\,\text{ns} \le t_s \le 3\,\text{ns}$, which agree with the calculated values represented by closed triangles.

As was described above, the DP device exhibits tamper resistance even when the difference in the arrival times of the two input signals is longer than their pulse widths. This nature suggests the possibility of realizing an asynchronous architecture by using unique energy transfer phenomena of DPs [48]. Since this structure is robust with respect to temporal fluctuations of the input signals, it has been recently used to design an asynchronous cellular automaton [49].

5.2.4 Autonomy in Energy Transfer

In the nano-optical condenser (Part (c) in Sect. 5.1.1) and the energy transmitter (Part (e)), energy transfer from multiple small QDs (QD_S) to a larger QD (QD_L) and subsequent relaxation were used. In particular, Fig. 5.15c experimentally demonstrated that

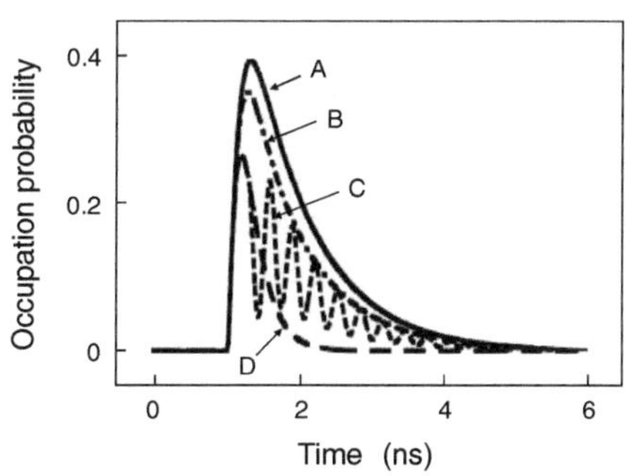

Fig. 5.31 Temporal evolutions of the occupation probability of the exciton in the lower energy level of QD-O used as an output terminal of the AND logic gate. For curves *A*, *B*, and *C*, the differences, t_s, in the arrival times of the two input signals are 0, −300, and +300 ps, respectively. Curve *D* represents the result for one input signal

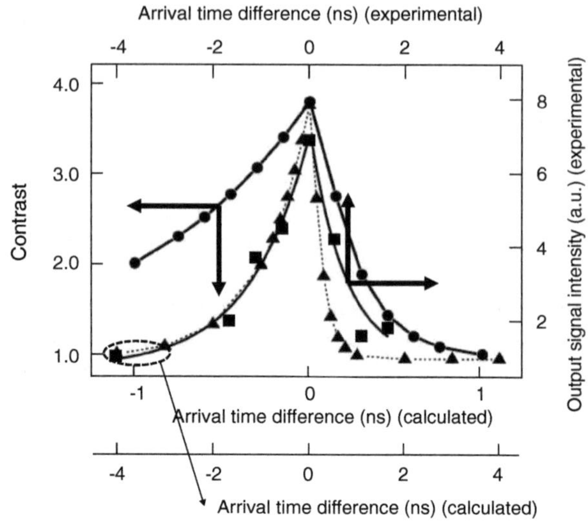

Fig. 5.32 Relation between the difference in the arrival time, and the contrast. *Closed circles* and *triangles* represent the calculated values for $-1\,\text{ns} \leq t_s \leq 1\,\text{ns}$ and $-4\,\text{ns} \leq t_s \leq 4\,\text{ns}$, respectively. *Closed squares* represent the experimental values of the output signal intensities for $-3\,\text{ns} \leq t_s \leq 3\,\text{ns}$, which is proportional to the contrast, C

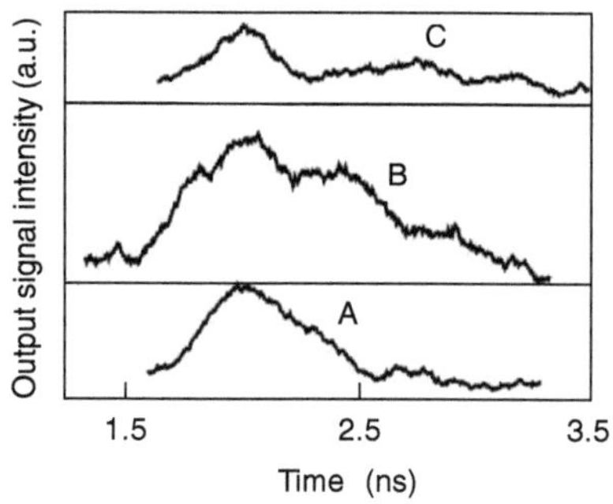

Fig. 5.33 Measured temporal evolutions of the output signal intensities from InGaAs QDs. Curves *A*, *B*, and *C* represent the results for = -1.1, 0, and 1.1 ns, respectively

the energy transmission length, L_0, increases with an increasing number of QDs, N_z. These characteristics suggest that the behavior of the multiple QDs exhibits unique autonomy in the energy transfer to QD_L [50]. This autonomy is examined here.

The above-mentioned energy transfer from the multiple small QDs (QD_S) to one larger QD (QD_L), illustrated in Fig. 5.34, is analyzed. The energy is transferred also between adjacent QD_Ss using their energy levels S, as illustrated in Fig. 5.34b, c. Here, the system S2-L1 of Fig. 5.34d contains two QD_Ss. Similarly, Fig. 5.34e–g show systems that contain 3–5 QD_Ss, called S3-L1 to S5-L1, respectively. The characteristics of the energy transfer from QD_Ss to QD_L in these systems are analyzed in the rest of this subsection [50].

First, the system S_2-L1 is studied. The energy of the interaction between QD_S and QD_L is denoted by $\hbar U_{SL}$, and that between adjacent QD_Ss is denoted by $\hbar U_{S1S2}$. The radiative relaxation rates from the energy level S in QD_{S1} and QD_{S2} are denoted by $\gamma_{rad,S1}$ and $\gamma_{rad,S2}$, respectively, and that from the lower energy level L_l in

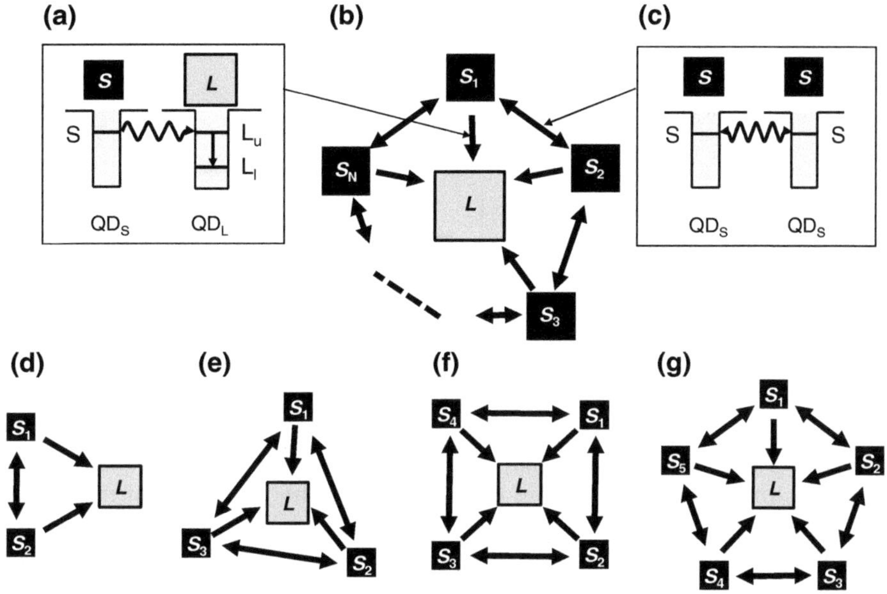

Fig. 5.34 Layout of multiple small quantum dots (QDs) and one large quantum dot (QD_L). **a** Energy transfer from QDs to QD_L and subsequent relaxation. **b** Layout of quantum dots. **c** Energy transfers between adjacent QD_Ss. **d–g** Show systems S2-L1, S3-L1, S4-L1, and S5-L1, respectively

QD_L is denoted by $\gamma_{rad,L1}$. The non-radiative relaxation rate from the upper energy level L_u to the lower energy level L_l in QD_L is denoted by γ. By assuming that each QDs is initially occupied by an exciton, quantum master equations for the density matrix given in Sect. 3.2 are solved by setting the numerical values of the quantities mentioned above to $\hbar U_{SL} = 5.3 \times 10^{-25}$ J, $\hbar U_{S1S2} = 1.1 \times 10^{-24}$ J, $\gamma_{rad,S1} = \gamma_{rad,S2} = 3.4 \times 10^8\,\mathrm{s}^{-1}$, $\gamma_{rad,L1} = 1.0 \times 10^9\,\mathrm{s}^{-1}$, and $\gamma = 1.0 \times 10^9\,\mathrm{s}^{-1}$. The quantum master equations can be similarly solved for systems S_3-L1 to S_5-L1 by assuming that each QDs is initially occupied by an exciton, and one can derive the occupation probability of the exciton in the lower energy level L_l in QD_L. The time-integrated value of this probability corresponds to the output signal intensity.

This intensity is calculated as a function of the number, N, of QD_Ss normalized to the number of QD_Ls, where it is assumed that the areal number density is independent of N. The calculated results are represented by the closed circles in Fig. 5.35 and show that the efficiency in energy transfer is highest when $N \cong 4$. Since $\gamma_{rad,L1}$ of QD_L takes a finite value, the energy is not transferred to QD_L until the exciton in the lower energy level L_l is annihilated, and as a result, the energy is dissipated from QDs if N is too large. Therefore, the output signal intensity does not increase if too many QD_Ss are placed around a QD_L, and as a result, the efficiency of the energy transfer to QD_L decreases when N is larger than 4.

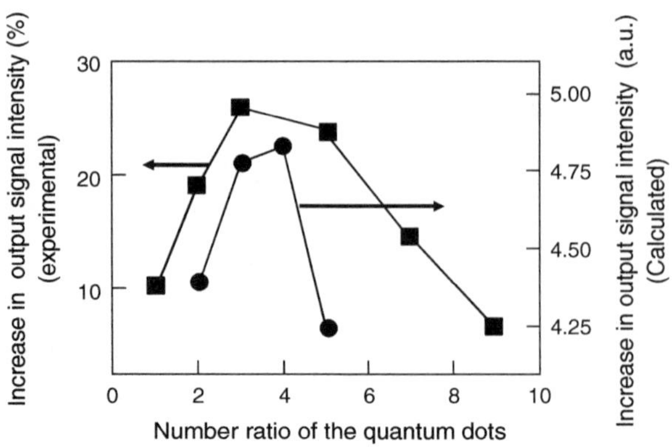

Fig. 5.35 Dependences of the output signal intensities emitted from spherical CdSe QDs on the ratio of the number, N, of QD_Ss to that of QD_L

Second, in an experiment, small and large spherical CdSe QDs (diameters of 2.0 and 2.8 nm, respectively) were used to measure the magnitude of the energy transferred from QDs to QD_L, similar to the case of Part (e) in Sect. 5.1.1. The results are represented by the closed squares in Fig. 5.35,[4] which show that the output signal intensity takes the maximum at $N = 4$.

The dependence of the energy transfer on the number ratio N between the two QDs suggests that the output signal intensity can be controlled by designing the layout of the QDs. Here, this controllability is examined by taking system S_5-L1 of Fig. 5.34g as an example. The values of the interaction energies between QDs and QD_L and that between adjacent QD_Ss are assumed to be $\hbar U_{SL} = \hbar U_{SiSj} = 1.1 \times 10^{-24}$ J. Other numerical values are assumed to be the same as those above.

Let us assume that interactions between some QD_Ss and QD_L can be degraded or lost because their resonant conditions are not satisfied due to, for example, size-detuning of the QDs, fluctuations in the separation between QDs, and deterioration of the QD material. In the case of a pentagonal layout, as shown in Fig. 5.36, there can be eight degraded configurations by referring to the system without any degradations, E0 (equivalent to system S_5-L1 in Fig. 5.34g): System E1 represents the layout in which the interaction between one QDs and QD_L is degraded or lost (represented by the mark × between S_1 and L in this figure); and systems E2 and E2′ in this figure represent that they have two degraded interactions.

Figure 5.37a shows the calculated temporal evolutions of the occupation probabilities of the exciton in the lower energy level L_l in QD_L, which correspond to the

[4] The magnitude of the transferred energy was evaluated by measuring the photocurrent from a photodiode on which these QDs were dispersed [38]. As can be understood from Fig. 5.34a, this optical-to-electrical energy conversion corresponds to the optical frequency down-conversion because the photon energy emitted from the lower energy level L_l in QD_L is lower than that of the incident light, which is resonant with the energy level S in QDs. In the case of the presently used CdSe QDs, it corresponds to conversion from ultraviolet light to visible light. Thus, as an example, dispersing these QDs on the surface of a solar cell is expected to increase its optical-to-electrical energy conversion efficiency, and additionally, the surface of the solar cell can be protected from ultraviolet radiation exposure [38].

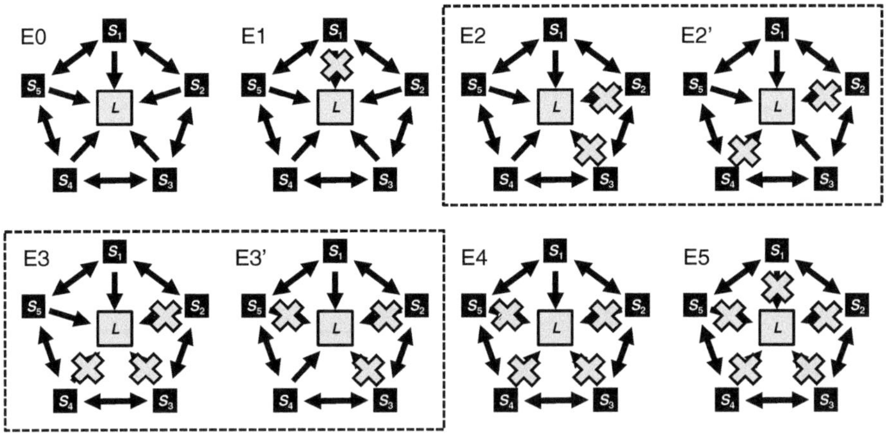

Fig. 5.36 Degraded or lost interactions between some QD_Ss and QD_L, which are represented by the mark $\times$. The symbol En represents the layout of quantum dots in which the number of degraded or lost interactions is n (=1–5). E0 is equivalent to system S5-L1 in Fig. 5.34g

output signal intensity. Figure 5.37b shows the relation between the systems (E0 to E5) and the time-integrated values of the occupation probabilities. This figure shows that the system E5 does not generate any output signals because the interaction between QDs and QD_L is completely lost. In contrast, the output signal intensities from systems E1 to E4 with degraded interactions are larger than that from system E0. In particular, the value of the output signal intensity from system E2 is 1.64 times greater than that from system E0, which is consistent with the fact that the signal intensity takes the maximum at $N = 4$ in Fig. 5.35. Figure 5.37c shows the calculated temporal evolutions of the occupation probabilities of an exciton in the lower energy level L_l in QD_L by assuming that the number of QD_Ss occupied by the exciton is 1–5, as an initial condition. The solid and broken curves represent the cases for systems E0 and E2 in Fig. 5.36, respectively. In the case when only one QDs is initially occupied by the exciton, the output signal intensity from system E2 (curve A_2 in this figure) is much larger than that from system E0 (curve A_0), because the energy is stored in the QD_Ss without any dissipation until it is transferred to QD_L, and therefore, the efficiency of the energy transfer depends on the layout of the QDs.

Moreover, the autonomy in energy transfer can be understood from Fig. 5.38a. This figure shows the temporal evolutions of the occupation probabilities of excitons in energy levels S in five QD_Ss in system E2, in which two interactions are degraded (the interaction between S_2 and QD_L and the interaction between S_3 and QD_L), as was shown in Fig. 5.36. The energy levels S in all of the QD_Ss are initially occupied by the excitons, and for several nanoseconds afterwards, the occupation probabilities in S_2 and S_3 remain high, which means that the energy is efficiently stored in S_2 and S_3 until it is transferred to QD_L. On the other hand, Fig. 5.38b shows the time evolutions of the occupation probabilities in the case of system E0, in which the energy levels in three QD_Ss (S_1, S_3, and S_4) are initially occupied by the excitons. It is found from

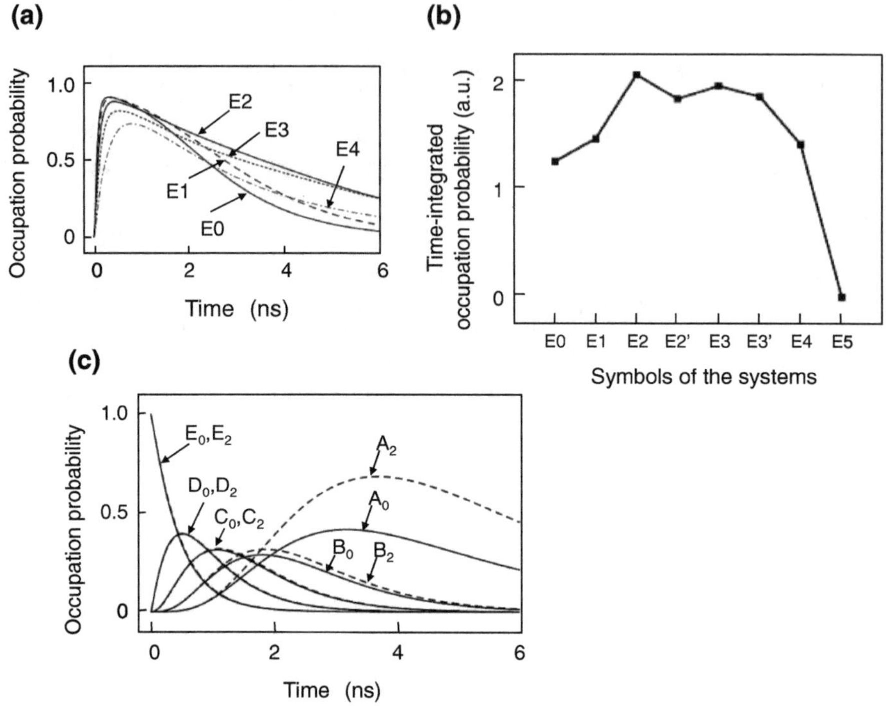

Fig. 5.37 Calculated results of occupation probabilities. **a** Temporal evolutions of the occupation probabilities of the exciton in the lower energy level L_l in QD_L for systems E0 to E4. **b** Time-integrated values of the occupation probabilities for systems E0 to E5. **c** Temporal evolutions of the occupation probabilities of the exciton in the lower energy level L_l in QD_L. *Solid* and *broken curves* represent the results for the systems E0 and E2 in Fig. 5.36, respectively. For curves *A*–*E*, it is assumed that the number of QD_Ss occupied by the exciton is 1–5, respectively

this figure that the occupation probabilities for S_2 and S_5 increase within 2 ns even though they were initially zero. This means that the transferred energy autonomously searches for the unoccupied QD_Ss in the system.

The ability to control the efficiency and autonomy of the energy transfer by arranging the layout of the QDs can be applied to novel information and communications technologies (ICT): First, consider the autonomous behavior observed in the energy transfer. As was demonstrated above, efficient energy transfer is realized even though there is no "central controller" in the system. Such an intrinsic (seemingly intelligent) behavior of the nano-sized physical system may also provide valuable lessons for designing self-organizing, distributed, and complex ICT systems on the Internet scale. The benefits of such distributed systems are that (i) a single point of failure at the server is avoided, (ii) the overall complexity of the system is reduced, and (iii) distributed layouts can more suitably handle suddenly appearing overload conditions or can balance traffic load and energy consumption. A recent trend in communication networks also shows that distributed and cooperative methods inspired by

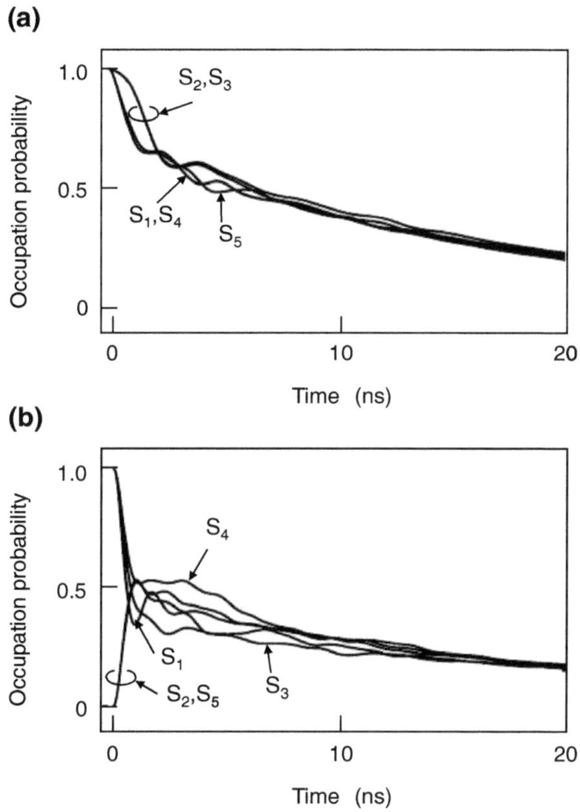

Fig. 5.38 Temporal evolutions of the occupation probabilities of the exciton in the energy levels *S* in five QD_Ss in the system S5-L1. The curves *S1* to *S5* correspond to the QD_Ss (*S1* to *S5*) in this system. **a** Calculated results for system E2 in which two interactions are degraded. **b** Calculated results for system E0 without any degraded interactions

biological [51, 52] and physical [53] phenomena have gained much attention as flexible and robust mechanisms for autonomous network management and control. These nature-inspired autonomous mechanisms have great benefits in terms of sustainability and reliability under unknown or changing conditions, similar to the autonomous and efficient energy transfer in the systems involving missing or failing links, as was shown in Fig. 5.36. Since the phenomenon of autonomous energy transfer between QDs is similar to the inherent behavior of amoeba exploited in bio-computing, several simulations have been carried out for application to novel, non-Von Neumann type computing systems for solving constraint satisfaction problems [54], Boolean satisfiability problems [55], and decision making problems [56].

Second, increases in the output signal intensity are observed, which is due to the degraded interactions, as was shown in Fig. 5.37b. This indicates the robustness against errors occurring in the system. Such behavior is also of great importance for future communication networks. Since new-generation networks are expected to accommodate a large number of heterogeneous end-devices, access technologies, network protocols/services, and traffic characteristics, some measures against failures or sudden fluctuations in performance will be essential. Since DP devices have the potential to provide superior behavior in the presence of errors, helpful guidelines and principles for constructing efficient ICT systems can be derived by studying the inherent robustness of DP devices.

Finally, increasing the energy efficiency in optical information transmission or processing systems is a key issue for future ICT technologies. It was demonstrated in Part (b) in Sect. 5.2.1 that a single process of energy transfer is about 10^4 times more efficient compared with the single bit-flip energy required in conventional electronic device. On the other hand, energy transfer in light harvesting antennas exhibits superior efficiency [57, 58], and these structures have similarities with nanostructures networked via near-field optical interactions. In summary, the studies described here will be extremely helpful for developing advanced DP devices with higher performance.

References

1. M. Ohtsu, K. Kobayashi, T. Kawazoe, T. Yatsui, M. Naruse, *Prinicples of Nanophotonics* (CRC Press, Bica Raton, 2007), pp. 29–108
2. T. Kawazoe, K. Kobayashi, S. Sangu, M. Ohtsu, Appl. Phys. Lett. **82**, 2957 (2003)
3. S. Sangu, K. Kobayashi, T. Kawazoe, A. Shojiguchi, M. Ohtsu, Trans. Materials Res. Soc. Jpn. **28**, 1035 (2003)
4. T. Yatsui, S. Sangu, T. Kawazoe, M. Ohtsu, S.J. An, J. Yoo, G.-C. Yi, Appl. Phys. Lett. **90**, 223110 (2007)
5. T. Kawazoe, K. Kobayashi, K. Akahane, M. Naruse, N. Yamamoto, M. Ohtsu, Appl. Phys. B **84**, 243 (2006)
6. R. Heitz, F. Guffarth, I. Mukhametzhanov, M. Grundmann, A. Madhukar, D. Bimberg, Phys. Rev. B **62**, 16881 (2000)
7. T. Kawazoe, M. Ohtsu, S. Aso, Y. Sawado, Y. Hosoda, K. Yoshizawa, K. Akahane, N. Yamamoto, M. Naruse, Appl. Phys. B **103**, 537 (2011)
8. V. Zwiller, H. Blom, P. Jonsson, N. Panev, S. Jeppesen, T. Tsegaye, E. Goobar, M.E. Pistol, L. Samuelson, G. Bjork, Appl. Phys. **78**, 2476 (2001)
9. Y. Francis, J. Suganda, Y. Shizuhuo (ed.), *Introduction to Information Optics* (Academic Press, San Diego, 2001), p. 202
10. N. Tate, Y. Liu, T. Kawazoe, M. Naruse, T. Yatsui, M. Ohtsu, Appl. Phys. B **110**, 39 (2013)
11. T. Kawazoe, K. Kobayashi, M. Ohtsu, Appl. Phys. Lett. **86**, 103102 (2005)
12. K. Akahane, N. Yamamoto, M. Naruse, T. Kawazoe, T. Yatsui, M. Ohtsu, Jpn. J. Appl. Phys. **50**, 04DH05 (2011)
13. M. Naruse, T. Miyazaki, F. Kubota, T. Kawazoe, K. Kobayashi, S. Sangu, M. Ohtsu, Opt. Lett. **30**(2), 201–203 (2005)
14. W. Nomura, T. Yatsui, T. Kawazoe, M. Ohtsu, J. Nanophotonics **1**, 011591 (2007)
15. W. Nomura, T. Yatsui, T. Kawazoe, M. Naruse, M. Ohtsu, Appl. Phys. B **100**, 181 (2010)
16. P. Guynot-Sionnest, M. Shim, C. Matranga, M. Hines, Phys. Rev. B **60**, R2181 (1999)
17. W. Nomura, T. Yatsui, T. Kawazoe, M. Naruse, Appl. Phys. B **107**, 257 (2012)
18. Y. Liu, T. Morishima, T. Yatsui, T. Yonezawa, M. Washizu, H. Fujita, M. Ohtsu, *Extended Abstracts* (*The Annual Meeting, 2009*) (Soc. Instrument and Control Eng., Tokyo, 2009) paper number 3B5-7
19. T. Yatsui, S. Sangu, K. Kobayashi, T. Kawazoe, M. Ohtsu, J. Yoo, G.-C. Yi, Appl. Phys. Lett. **94**, 083113 (2009)
20. M. Ohtsu, in Progress in Nanophotonics I, ed. by M. Ohtsu,*Nanophotonics: Dressed Photon Technology for Qualitatively Innovative Optical Devices, Fabrication, and Systems*, (Springer, Berlin, 2011), pp. 15–18

21. M. Ohtsu, *Highly Coherent Semiconductor Lasers* (Artech House, Boston, 1992), pp. 40–43
22. M. Naruse, H. Hori, K. Kobayashi, T. Kawazoe, M. Ohtsu, Appl. Phys. B **102**, 717 (2011)
23. T. Franzl, T.A. Klar, S. Schietinger, A.L. Rogach, J. Feldman, Nano Lett. **4**, 1599 (2004)
24. H.J. Carmichael, *Statistical Methods in Quantum Optics I* (Springer, Berlin, 1999)
25. T. Yatsui, H. Jeong, M. Ohtsu, Appl. Phys. B **93**, 199 (2008)
26. S. Sangu, K. Kobayashi, M. Ohtsu, IEICE Trans. Electron. **E88-C**, 1824 (2005)
27. S. Sangu, K. Kobayashi, in Handbook of Nanophysics, ed. by K.D. Sattler, *Operation in Nanophotonics* (CRC Press, Boca Raton, 2010), pp. 33/11–33/12
28. R.H. Dicke, Phys. Rev. **93**, 99 (1954)
29. M. Gross, S. Haroche, Phys. Rep. **93**, 301 (1982)
30. A. Shojiguchi, K. Kobayashi, S. Sangu, K. Kitahara, M. Ohtsu, J. Phys. Soc. Jpn. **72**, 2984 (2003)
31. T. Yatsui, A. Ishikawa, K. Kobayashi, A. Shojiguchi, S. Sangu, T. Kawazoe, M. Ohtsu, J. Yoo, G.-C. Yi, App. Phy. Lett. **100**, 233118 (2012)
32. R. Hambury Brown, R.Q. Twiss, Nature **178**, 1447 (1956)
33. T. Kawazoe, S. Tanaka, M. Ohtsu, J. Nanophotonics **2**, 029502 (2008)
34. G.-L. Ingold, Y.V. Nazarov, in Single Charge Tunneling, ed. by H. Grabert, M.H. Devoret. *Charge Tunneling Rates in Ultrasmall Junctions* (Plenum Press, New York, 1992), pp. 21–107
35. M. Naruse, H. Hori, K. Kobayashi, P. Holmstrom, L. Thylen, M. Ohtsu, Opt. Express **18**, A544 (2010)
36. L.B. Kish, IEEE Proc. Circ. Dev. Syst. **151**, 190 (2004)
37. H. Imahori, J. Phys. Chem. B **108**, 6130 (2004)
38. M. Naruse, T. Kawazoe, R. Ohta, W. Nomura, M. Ohtsu, Phys. Rev. B **80**, 125325 (2009)
39. N. Johnson, *Simply Complexity* (Oneworld Publications, Oxford, 2007)
40. M. Naruse, P. Holmstrom, T. Kawazoe, K. Akahane, N. Yamamoto, L. Thylen, M. Ohtsu, Appl. Phys. Lett. **100**, 241102 (2012)
41. Y. Yamamoto, *Lecture Notes on Fundamentals of Noise Processes*, online text, http://www.qis.ex.nii.ac.jp/qis/lecturenotes.html
42. F. Moll, M. Roca, E. Isern, Microelectron. J. **34**, 833 (2003)
43. M. Ohtsu, *Extended Abstracts (RLNR/Tokyo-Tech International Symposium on Nanoscience and Nanotechnology on Quantum Particles, 2003)* (Tokyo Inst. Tech, Tokyo, 2003) paper number I-3
44. N. Streibl, K.-H. Brenner, A. Huang, J. Jahns, J.L. Jewell, A.W. Lohmann, D.A.B. Miller, M. Murdocca, M.E. Prise, T. Sizer, Proc. IEEE **77**, 1954 (1989)
45. A. Qureshi, R. Weber, H. Balakrishnan, J. Guttag, B. Maggs, ACM SIGCOMM Comput. Commun. Rev. **39**, 123 (2009)
46. M. Naruse, H. Hori, K. Kobayashi, M. Ohtsu, Opt. Lett. **32**, 1761 (2007)
47. M. Naruse, F. Pepper, K. Akahane, N. Yamamoto, T. Kawazoe, N. Tate, M. Ohtsu, ACM J. Emerg. Technol. Comput. Syst. **8**, 4-1 (2012)
48. S. Hauck, Proc. IEEE **83**, 69 (1995)
49. J. Lee, S. Adachi, F. Peper, S. Mashiko, J. Comput. Syst. Sci. **70**, 201 (2005)
50. M. Naruse, K. Leibnitz, F. Peper, N. Tate, W. Nomura, T. Kawazoe, M. Murata, M. Ohtsu, Nano Commun. Netw. **2**, 189 (2011)
51. S. Balasubramaniam, D. Botvich, J. Mineraud, W. Donnelly, N. Agoulmine, IEEE Netw. **24**, 20 (2010)
52. S. Balasubraianiam, K. Leipnitz, P. Lio, D. Botvich, M. Murata, IEEE Commun. Mag. **49**, 44 (2011)
53. S. Tumps, Comput. Netw. **52**, 360 (2008)
54. M. Naruse, M. Aono, S.-J. Kim, T. Kawazoe, W. Nomura, H. Hori, M. Hara, M. Ohtsu, Phys. Rev. B **86**, 125407 (2012)
55. M. Aono, S.-J. Kim, L. Zhu, M. Naruse, M. Ohtsu, H. Hori, M. Hara, in *Proceedings of The 2012 International Symposium on Nonlinear Theory and Its Applications* (The Inst. Electron, Info. and Commun. Eng., Tokyo, 2012) pp. 586–589

56. S.-J. Kim, M. Naruse, M. Aono, M. Ohtsu, M. Hara, *Technical Digest of The 1st International Workshop on Information Physics and Computing in Nano-scale Photonics and Materials* (The Opt. Soc. Jpn., Tokyo, 2012) paper number IPCN1-14
57. A. Olaya-Castro, C.F. Lee, F.F. Olsen, N.F. Johnson, Phys. Rev. B **78**, 085115 (2008)
58. H. Tamura, J.-M. Mallet, M. Oheim, I. Burghardt, J. Phys. Chem. C **113**, 7458 (2009)

Chapter 6
Fabrication Using Dressed Photons

Longum iter est per praecepta, breve et efficax per exempla.
Lucius Annaeus Seneca, *Epistulae,* VI, 5

This chapter reviews novel nano-fabrication methods that have been made possible based on the principles described in Chap. 4. These methods use a novel excitation process that originates from the phonons in the dressed-photon–phonons (DPPs), called a phonon-assisted process.

6.1 Molecular Dissociation by Dressed-Photon–Phonons

As a first example of fabrication technology, this section reviews the dissociation of molecules and subsequent deposition of nanomaterial onto the substrate by using DPPs. Theories and experiments on the dissociation process are described, followed by a review of deposition experiments.

6.1.1 Comparison Between Experiments and Theories

When a gas molecule comes flying towards the apex of a fiber probe, energy is transferred from the probe to the molecule, mediated by DPPs. This excites molecular vibrations that are forbidden in the case where propagating light is used. This subsection compares the experimental results of dissociating DEZn molecules (Fig. 4.2) with theoretical predictions in order to confirm that the molecular dissociation is explained by the principles described in Chap. 4.

Figure 6.1 shows electronic states and vibrational states involved in dissociation, together with their energy levels. This figure is a simplification of Figs. 4.1 and 4.3. The three solid lines represent the ground state and two excited states of an

M. Ohtsu, *Dressed Photons*, Nano-Optics and Nanophotonics,
DOI: 10.1007/978-3-642-39569-7_6, © Springer-Verlag Berlin Heidelberg 2014

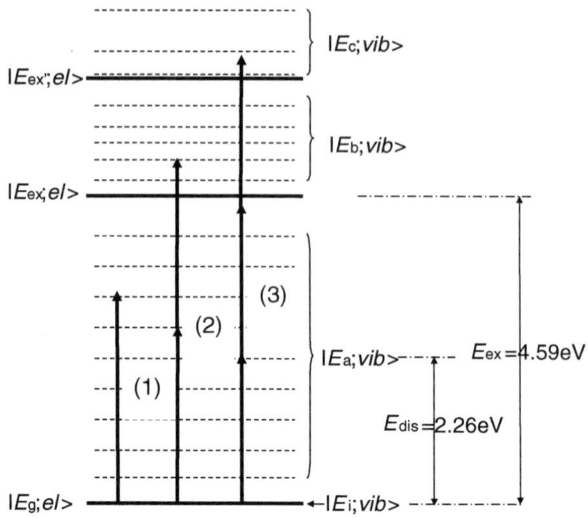

Fig. 6.1 Electronic states and vibrational states involved in dissociation of a molecule, together with their energy levels

electron, and the numerous broken lines represent the molecular vibrationmolecular vibrational states associated with each electronic state. Via the DPPs, the molecule can be excited to a higher vibrational state in the electronic ground state, as is shown by arrow (1). For this excitation, the molecular vibrationmolecular vibrational state contributes as a sideband of the modulated electronic states, as was described in Sect. 4.4. The molecule can be dissociated if the energy of the higher vibrational state is larger than the dissociation energy E_{dis}. Note that this excitation is not possible by using propagating light because the transition in the electronic ground state is electric-dipole forbidden. After this excitation, the molecule can be successively excited to the vibrational states in the electronic excited states by multi-step excitation, as shown by arrows (2) and (3). Note that this excitation is possible not only by using the DPPs but also by propagating light because the transition from the electronic ground state to the excited state is electric-dipole allowed. After this excitation, the molecule transitions to the anti-binding excited state in Fig. 4.3 and dissociates.

In order to study the dissociation process described above, consider the following state vectors, which are composed of the state of the probe and the electronic and vibrational states of the molecule. They are respectively expressed as

$$|i\rangle = |N; probe\rangle \otimes \left|E_g; el\right\rangle \otimes |E_i; vib\rangle \,, \tag{6.1a}$$

$$|f_1\rangle = |N-1; probe\rangle \otimes \left|E_g; el\right\rangle \otimes |E_a; vib\rangle \,, \tag{6.1b}$$

$$|f_2\rangle = |N-2; probe\rangle \otimes |E_{ex1}; el\rangle \otimes |E_b; vib\rangle \,, \tag{6.1c}$$

$$|f_3\rangle = |N-3; probe\rangle \otimes |E_{ex2}; el\rangle \otimes |E_c; vib\rangle \,, \tag{6.1d}$$

where the symbol $\otimes$ represents the direct product. The state vector $|i\rangle$ represents the initial state. The vectors $|f_1\rangle$ to $|f_3\rangle$ are the final states after single-step, two-step, and three-step excitations, respectively, because one to three quanta of the DPPs are annihilated. These excitations are shown by arrows (1) to (3) in Fig. 6.1. The first term, $|N;\, probe\rangle$, in each equation represents the state of the probe, which is an eigenstate of the Hamiltonian of Eq. (4.35) and is identified by the number N of DPPs. The second and third terms represent the states of the molecule flying towards the apex of the probe. The second term, $|E_\alpha;\, el\rangle$, represents the electronic state in the molecule, where E_α is its eigenenergy, and $\alpha = g$, $\alpha = ex1$, and $\alpha = ex2$ represent the ground state, the first-excited state, and the second-excited state, respectively. Since the adiabatic approximation is not valid, the molecular vibrationmolecular vibrational states have to be considered in addition to the electronic states. The third term, $\left|E_\beta;\, vib\right\rangle$, represents such vibrational states, where E_β is its eigenenergy, $\beta = i$ and $\beta = a$ are respectively the vibrational ground state and excited state in the electronic ground state, and $\beta = b$ and $\beta = c$ are respectively the excited vibrational states in the first-excited state and second-excited state of the electron.

The probabilities of excitation from the initial state of Eq. (6.1a) to the final state of Eqs. (6.1b)–(6.1d) are expressed as

$$P_1\left(\omega_p\right) = \frac{2\pi}{\hbar}\left|\langle f_1|\, \hat{H}_{\rm int}\, |i\rangle\right|^2, \tag{6.2a}$$

$$P_2\left(\omega_p\right) = \frac{2\pi}{\hbar}\left|\langle f_2|\, \hat{H}_{\rm int}\, |i\rangle\right|^2, \tag{6.2b}$$

$$P_3\left(\omega_p\right) = \frac{2\pi}{\hbar}\left|\langle f_3|\, \hat{H}_{\rm int}\, |i\rangle\right|^2, \tag{6.2c}$$

where $\hat{H}_{\rm int}$ is the interaction Hamiltonian between the probe and molecule and is given by

$$\hat{H}_{\rm int} = -\int \hat{\boldsymbol{p}}\,(\boldsymbol{r}) \cdot \hat{\boldsymbol{D}}^{\perp}\,(\boldsymbol{r})\, d^3 r. \tag{6.3}$$

Although the details of the interaction Hamiltonian are given by Eq. (C.25) in Appendix C, Eq. (6.3) is given here in order to describe the transition probability analytically. The processes $\langle f_2|\, \hat{H}_{\rm int}\, |i\rangle$ and $\langle f_3|\, \hat{H}_{\rm int}\, |i\rangle$ represent the two- and three-step excitations, respectively. They can be resolved into each step as follows:

$$\langle f_2|\, \hat{H}_{\rm int}\, |i\rangle = \langle f_2|\, \hat{H}_{\rm int}\, |f_1\rangle\, \langle f_1|\, \hat{H}_{\rm int}\, |i\rangle\,, \tag{6.4a}$$

$$\langle f_3|\, \hat{H}_{\rm int}\, |i\rangle = \langle f_3|\, \hat{H}_{\rm int}\, |f_2\rangle\, \langle f_2|\, \hat{H}_{\rm int}\, |f_1\rangle\, \langle f_1|\, \hat{H}_{\rm int}\, |i\rangle\,. \tag{6.4b}$$

In the interaction Hamiltonian of Eq. (6.3), $\hat{\boldsymbol{p}}\,(\boldsymbol{r})$ is the operator of the electric dipole moment in the molecule, which is the sum of the operators originating from the electronic excitation $\hat{\boldsymbol{p}}^{el}$ and the molecular vibrational excitation $\hat{\boldsymbol{p}}^{vib}$. They are expressed by using the annihilation and creation operators of the electron and the molecular vibration ($\hat{e}_l$, $\hat{e}_l^\dagger$, $\hat{v}_{ib}$, $\hat{v}_{ib}^\dagger$):

$$\hat{\boldsymbol{p}}^{el} = \boldsymbol{p}^{el}\left(\hat{e}_l^{\dagger} + \hat{e}_l\right), \tag{6.5a}$$

$$\hat{\boldsymbol{p}}^{vib} = \boldsymbol{p}^{vib}\left(\hat{v}_{iv}^{\dagger} + \hat{v}_{iv}\right), \tag{6.5b}$$

$$\hat{\boldsymbol{p}} = \hat{\boldsymbol{p}}^{el} + \hat{\boldsymbol{p}}^{vib}. \tag{6.5c}$$

Here, $\hat{\boldsymbol{D}}^{\perp}(\boldsymbol{r})$ is the operator of the electric displacement vector of the DPP, whose absolute value is expressed as

$$\left|\hat{\boldsymbol{D}}^{\perp}(\boldsymbol{r})\right| = \sqrt{\frac{\hbar\omega_p \Phi\left(\omega_p\right)}{V_{DPP}}}. \tag{6.6}$$

Here, ω_p is the angular frequency of the propagating light injected into the base of the probe, $\Phi\left(\omega_p\right)$ is the number of DPPs generated per second, and V_{DPP} is the volume of the DPP field of the DPP at the apex of the probe. Using these quantities, the excitation probabilities can be derived as follows:

(1) Excitation from $|i\rangle$ to $|f_1\rangle$: Since this is the transition between the molecular vibrationmolecular vibrational states in the ground state of the electron, $\hat{\boldsymbol{p}}^{vib}$ of Eq. (6.5b) is involved, and one obtains:

$$\langle f_1|\,\hat{H}_{\text{int}}\,|i\rangle = p^{vib}\sqrt{\frac{\hbar\omega_p\Phi}{V_{DPP}}}, \tag{6.7}$$

where $p^{vib} \equiv \left|\hat{\boldsymbol{p}}^{vib}\right|$.

(2) Excitation from $|f_1\rangle$ to $|f_2\rangle$: Since this is the transition from the electronic ground state to the first-excited state of the electron, $\hat{\boldsymbol{p}}^{el}$ of Eq. (6.5a) is involved, and one obtains:

$$\langle f_2|\,\hat{H}_{\text{int}}\,|f_1\rangle = p^{el}\sqrt{\frac{\hbar\omega_p\Phi}{V_{DPP}}}\delta\left(\hbar\omega_p - \left(E_{f_2} - E_{f_1}\right)\right), \tag{6.8a}$$

where $p^{el} \equiv \left|\hat{\boldsymbol{p}}^{el}\right|$. The delta function is given by

$$\delta\left(\hbar\omega_p - \left(E_{f_2} - E_{f_1}\right)\right) = \frac{1}{\left|\hbar\omega_p - \left(E_{f_2} - E_{f_1} + i\gamma_{vib}\right)\right|}, \tag{6.8b}$$

where E_{f1} and E_{f2} are the eigenenergies of the states $|f_1\rangle$ and $|f_2\rangle$, respectively. Since the resonant condition is satisfied here, $\hbar\omega_p = E_{f_2} - E_{f_1}$ holds. γ_{vib} is the relaxation rate of the vibrational state.

(3) Excitation from $|f_2\rangle$ to $|f_3\rangle$: Since this is the transition from the first-excited state to the second-excited state of the electron, $\hat{\boldsymbol{p}}^{el}$ of Eq. (6.5a) is involved, and one obtains

$$\langle f_3|\,\hat{H}_{\text{int}}\,|f_2\rangle = p^{el}\sqrt{\frac{\hbar\omega_p\Phi}{V_{DPP}}}\delta\left(\hbar\omega_p - \left(E_{f_3} - E_{f_2}\right)\right), \tag{6.9a}$$

$$\delta\left(\hbar\omega_p - \left(E_{f_3} - E_{f_2}\right)\right) = \frac{1}{\left|\hbar\omega_p - \left(E_{f_3} - E_{f_2} + i\gamma_{vib}\right)\right|}, \tag{6.9b}$$

as was the case of (2) above, where E_{f3} is the eigenenergy of the state $|f_3\rangle$.

By inserting Eqs. (6.8a) and (6.9b) into Eqs. (6.2a) to (6.2c) with the resonant condition

$$\hbar\omega_p = E_l - E_k \quad (k, l = f_1, f_2, f_3), \tag{6.10}$$

the probabilities of the single-step, two-step, and three-step excitations are expressed as

$$P_1\left(\omega_p\right) = \frac{2\pi}{\hbar}\left|\langle f_1|\,\hat{H}_{\text{int}}\,|i\rangle\right|^2 = \frac{2\pi}{\hbar}\left(p^{vib}\right)^2\left(\frac{\hbar\omega_p\Phi}{V_{DPP}}\right), \tag{6.11a}$$

$$\begin{aligned} P_2\left(\omega_p\right) &= \frac{2\pi}{\hbar}\left|\langle f_2|\,\hat{H}_{\text{int}}\,|f_1\rangle\,\langle f_1|\,\hat{H}_{\text{int}}\,|i\rangle\right|^2 \\ &= \frac{2\pi}{\hbar}\left(p^{el}\right)^2\left(p^{vib}\right)^2\left(\frac{\hbar\omega_p\Phi}{V_{DPP}}\right)^2\frac{1}{\gamma_{vib}^2}, \end{aligned} \tag{6.11b}$$

$$\begin{aligned} P_3\left(\omega_p\right) &= \frac{2\pi}{\hbar}\left|\langle f_3|\,\hat{H}_{\text{int}}\,|f_2\rangle\,\langle f_2|\,\hat{H}_{\text{int}}\,|f_1\rangle\,\langle f_1|\,\hat{H}_{\text{int}}\,|i\rangle\right|^2 \\ &= \frac{2\pi}{\hbar}\left(p^{el}\right)^4\left(p^{vib}\right)^2\left(\frac{\hbar\omega_p\Phi}{V_{DPP}}\right)^3\frac{1}{\gamma_{vib}^4}. \end{aligned} \tag{6.11c}$$

It should be noted that these probabilities, $P_i\left(\omega_p\right)$ $(i = 1, 2, 3)$, are for exciting one molecule. On the other hand, the experimentally measurable quantity is the molecular dissociation rate, $R_i\left(\omega_p\right)$, which corresponds to the probability of exciting all of the molecules in the volume of the DPP field, V_{DPP}. This rate is represented by the product of $\left|\langle f|\,\hat{H}_{\text{int}}\,|i\rangle\right|^2$ and ρV_{DPP}, where ρ is the volume density of the molecules. As a result, one derives

$$R_1\left(\omega_p\right) = \rho V_{DPP} P_1\left(\omega_p\right) = 2\pi\rho\left(p^{vib}\right)^2\left(\omega_p\Phi\right), \tag{6.12a}$$

$$R_2\left(\omega_p\right) = \left(\rho V_{DPP}\right)^2 P_2\left(\omega_p\right) = 2\pi\hbar\rho^2\left(p^{el}\right)^2\left(p^{vib}\right)^2\left(\omega_p\Phi\right)^2\frac{1}{\gamma_{vib}^2}, \tag{6.12b}$$

$$R_3\left(\omega_p\right) = \left(\rho V_{DPP}\right)^3 P_3\left(\omega_p\right) = 2\pi\hbar^2\rho^3\left(p^{el}\right)^4\left(p^{vib}\right)^2\left(\omega_p\Phi\right)^3\frac{1}{\gamma_{vib}^4}. \tag{6.12c}$$

Since the dissociation rate $R\left(\omega_p\right)$ is a function of $\Phi\left(\omega_p\right)$, its polynomial expansion with a third-order approximation is expressed, by using Eqs. (6.12a) to (6.12c), as

$$\begin{aligned} R\left(\omega_p\right) &\equiv R_1\left(\omega_p\right) + R_2\left(\omega_p\right) + R_3\left(\omega_p\right) \\ &= a_{\omega_p}\Phi\left(\omega_p\right) + b_{\omega_p}\Phi^2\left(\omega_p\right) + c_{\omega_p}\Phi^3\left(\omega_p\right), \end{aligned} \tag{6.13}$$

where the expansion coefficients are given by

$$a_{\omega_p} = 2\pi\rho\left(p^{vib}\right)^2 \omega_p, \tag{6.14a}$$

$$b_{\omega_p} = 2\pi\hbar\rho^2\left(p^{el}\right)^2\left(p^{vib}\right)^2 \omega_p{}^2 \frac{1}{\gamma_{vib}^2}, \tag{6.14b}$$

$$c_{\omega_p} = 2\pi\hbar^2\rho^3\left(p^{el}\right)^4\left(p^{vib}\right)^2 \omega_p^3 \frac{1}{\gamma_{vib}^4}. \tag{6.14c}$$

In order to compare Eq. (6.13) with the experimental results [1, 2], Fig. 6.2 shows the relations between the photon number flux incident into the base of the probe and the deposition rate of Zn atoms on the sapphire substrate after DEZn molecules

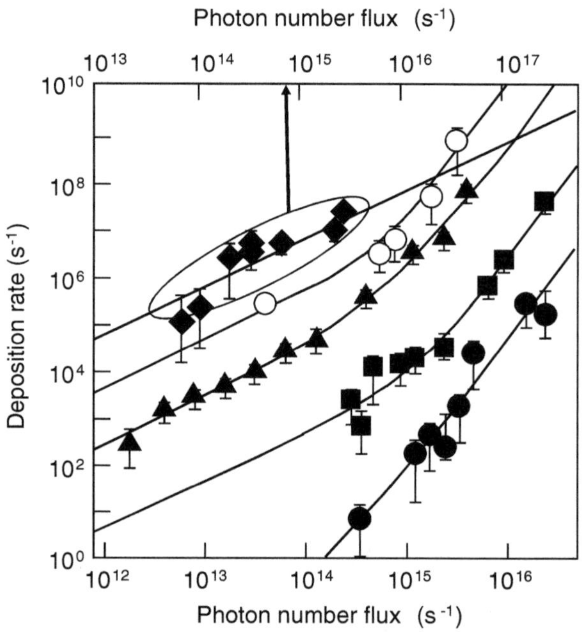

Fig. 6.2 Relation between the photon number flux incident into the base of the probe and the deposition rate of Zn atoms on the sapphire substrate after molecules are dissociated by dressed-photon–phonons. *Closed diamonds, open circles, closed squares*, and *closed circles* represent the experimental results obtained by dissociating DEZn molecules, where the photon energies used are 3.81, 3.04, 2.54, and 1.84 eV, respectively. Closed triangles represents the result of dissociating $Zn(acac)_2$ molecules, where the photon energy is 3.04 eV. *Solid curves* are theoretical results fitted to the experimental values

are dissociated by DPPs. The photon number flux is proportional to $\Phi\left(\omega_p\right)$. On the other hand, since the deposition rate is proportional to the dissociation rate $R\left(\omega_p\right)$, the quantity on the vertical axis can be regarded as $R\left(\omega_p\right)$. Four experimental results are presented in this figure, which depend on the incident photon energy. (The closed triangles in this figure will be explained in Sect. 6.1.2.) They are:

(a) Closed diamonds: Photon energy, 3.81 eV (wavelength, $\lambda = 325$ nm)
(b) Open circles: Photon energy, 3.04 eV (wavelength, $\lambda = 408$ nm)
(c) Closed squares: Photon energy, 2.54 eV (wavelength, $\lambda = 488$ nm)
(d) Closed circles: Photon energy, 1.84 eV (wavelength, $\lambda = 684$ nm)

The solid curves represent Eq. (6.13) fitted to the experimental values. In the cases of (c) and (d), for example, the values of the coefficients in Eq. (6.13) used for the fitting are

(c) $a_{2.54} = 4.1 \times 10^{-12}$, $b_{2.54} = 2.1 \times 10^{-27}$, and $c_{2.54} = 1.5 \times 10^{-42}$,
(d) $a_{1.81} = 0$, $b_{1.81} = 4.2 \times 10^{-29}$, and $c_{1.81} = 3.0 \times 10^{-44}$.

In the case of (c), the first-order coefficient $a_{2.54}$ takes a nonzero value because the photon energy (2.54 eV) is higher than the dissociation energy E_{dis} (2.26 eV) of DEZn, even though it is lower than its excitation energy E_{ex} (4.59 eV). On the other hand, in the case of (d), the coefficient $a_{1.81}$ is zero because the photon energy (1.81 eV) is lower than both E_{dis} and E_{ex}. It is confirmed by these values that the experimental results in this figure represent novel phenomena that originate from the DPPs, because the molecules are not dissociated by propagating light if its photon energy is lower than E_{ex}.

Furthermore, from the values of coefficients given above, it is found that the following unique relation holds:

$$\frac{b_{2.54}}{a_{2.54}} = \frac{c_{2.54}}{b_{2.54}} = \frac{c_{1.81}}{b_{1.81}} \simeq 10^{-15}. \tag{6.15}$$

The values of the ratios in Eq. (6.15) can be theoretically derived: Since it is difficult to estimate the volume density ρ of the molecules experimentally, it is expressed, by using a_{ω_p} of Eq. (6.14a), as

$$\rho = \frac{1}{2\pi\omega_p\left(p^{vib}\right)^2} a_{\omega_p}. \tag{6.16}$$

By inserting this into Eqs. (6.14b) and (6.14c), the ratio between the coefficients can be expressed as

$$\frac{b_{\omega_p}}{a_{\omega_p}} = \frac{c_{\omega_p}}{b_{\omega_p}} = \frac{\hbar}{2\pi}\left(\frac{p^{el}}{p^{vib}}\right)^2 \frac{1}{\gamma_{vib}^2} a_{\omega_p}. \tag{6.17}$$

Here, a_{ω_p} on the right-hand side is replaced by the value $a_{2.54} = 4.1 \times 10^{-12}$, which was used for drawing the solid curve in Fig. 6.2. Furthermore, by using the values $p^{el}/p^{vib} = 1 \times 10^{-4}$, $p^{vib} = 1$ Debye, $p^{el} = 1 \times 10^{-4}$ Debye, and $\gamma_{vib} = 0.1$ eV, one derives [1,2]

$$\frac{b_{2.54}}{a_{2.54}} = \frac{c_{2.54}}{b_{2.54}} \simeq 10^{-15}, \tag{6.18}$$

which agrees with the experimental value of Eq. (6.15). This agreement confirms that the observed novel phenomena of DEZn dissociation are explained by the principles described in Chap. 4.

6.1.2 Deposition by Molecular Dissociation

After dissociating DEZn molecules using DPPs, Zn atoms in the molecules are deposited on a substrate, and the deposited Zn atoms eventually form Zn nanoparticles. This method is called chemical vapor deposition (CVD) using DPPs. Figure 4.2 is an image of the profile of such a Zn nanoparticle acquired by an atomic force microscope (AFM). Figure 6.2 shows the measured rate of deposition of Zn atoms on a sapphire substrate. There are a number of advantages of using DPP-induced molecular dissociation for CVD:

(1) Since the molecules are not dissociated by propagating light even if it leaks from the apex of the probe, the spatial profile of the deposited nanoparticle is not affected by diffraction of the propagating light and depends only on the spatial distribution of the DPP field.
(2) It is possible to dissociate not only conventional optically active molecules but also optically inactive ones because an electric dipole-forbidden transition can be used for the dissociation. This increases the range of materials that can be deposited.
(3) The cost of the CVD equipment decreases because expensive ultraviolet light sources are not required.

As an example of advantage (2), optically inactive zinc-bis(acetylacetonate) molecules ($Zn(acac)_2$) have been dissociated to deposit Zn nanoparticles [3]. Although

---

[1] The value of the electronic polarization component p^{el} is larger than that of the vibrational component p^{vib} in the case where the electron is in a macroscopic material or in vacuum and can move freely. However, in the present case, one has to consider the response of the electron to the DPP, whose spatial extent is smaller than the coherence length of the electron. That is, if the volumes of the spaces in which the electron and nucleus are confined are equal to each other, the value of the polarization induced by the DPP depends on the state density of the electron or nucleus. Furthermore, the state densities of the confined electron and nucleus depend on their effective masses. Here, the electron mass in vacuum is 1×10^{-3} times the nuclear mass, and the effective mass of the electron in semiconductors and dielectrics is 1/10 times the mass of the electron at rest. Therefore, the mass ratio of the electron and nucleus can be estimated to be 1×10^{-4}, and as a result, the ratio p^{el}/p^{vib} of the polarizations is thus 1×10^{-4}.

[2] The value 0.1 eV was used in the text, based on the estimated value of 0.1–0.2 eV from the absorption spectroscopy of gaseous DEZn molecules [2].

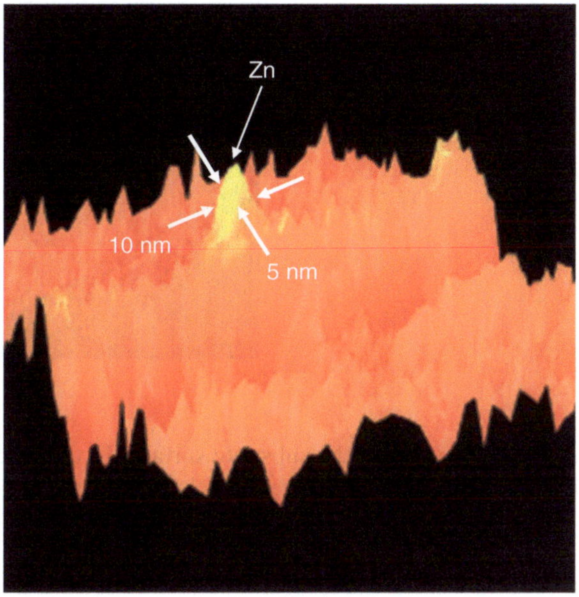

Fig. 6.3 Atomic force microscope image of a Zn nanoparticle deposited on a sapphire substrate by dissociating $Zn(acac)_2$ molecules

this molecule has been popularly used for metal-organic CVD, it has never been used for the conventional CVD using propagating light because of its optical inactivity. However, since this molecular gas is very stable and is not explosive, it can be safely used for CVD using DPPs. Figure 6.3 shows an AFM image of a Zn nanoparticle deposited on a sapphire substrate by dissociating $Zn(acac)_2$ molecules. The height of the nanoparticle is 0.3 nm, which corresponds to the thickness of only two atomic layers of Zn. Its diameter is 5–10 nm, which is much smaller than those shown in Fig. 4.2, making them the smallest reported nanoparticles that have ever been deposited. For reference, the closed triangles in Fig. 6.2 represent the experimental values for $Zn(acac)_2$. Since the rates of dissociation and deposition of this molecule are lower than those of other molecules, these values were accurately measured, as is confirmed by the length of the error bars attached to the closed triangles, which are shorter than those of other experimental results in this figure. As a result of this high accuracy, the theoretical curve was precisely fitted to these experimental values.

It should be pointed out that fine, high-precision formation of nanoparticles is possible because their sizes and positions depend on those of the nanometer-sized apex of the probe on which the DPPs isare generated [4, 5]. Moreover, by scanning the probe on the substrate, it is possible to form unique, high-resolution geometrical patterns that have not been possible with conventional fabrication methods. Furthermore, by successively dissociating several different species of molecules, hybrid deposition of several species of nanoparticles is possible. As an example, nanoparticles of Zn and Al were deposited in close proximity to each other by successively dissociating DEZn and $Al(CH_3)_3$ molecules. Their AFM images are shown in Fig. 6.4 [5]. This method has been improved further from the viewpoint of the interaction between the

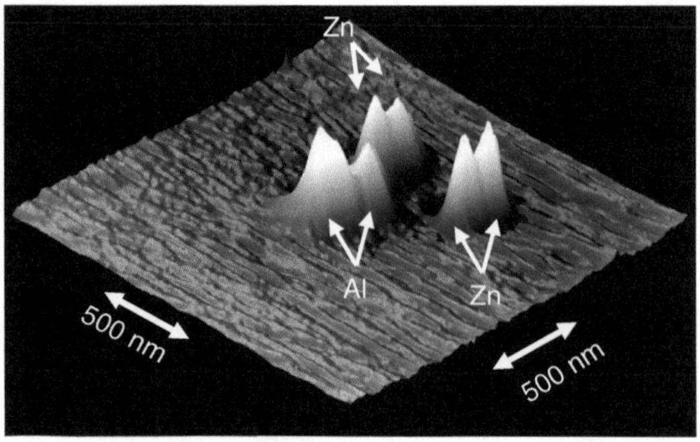

Fig. 6.4 Atomic force microscope image of Zn and Al nanoparticles deposited on a common substrate

deposited atoms and the substrate surface. In the case of Zn deposition, molecules adsorbed on a substrate were dissociated instead of dissociating gaseous molecules, and as a result, the size accuracy of the deposited nanoparticles was improved [6].

By oxidizing Zn, it is possible to create zinc oxide (ZnO) nanoparticles, which is a promising material for DP devices because it emits photons efficiently even at room temperature and is also chemically and thermally stable. Figure 6.5 shows the spatial distribution of the photoluminescence intensity emitted from a ZnO nanoparticle created by oxidizing a Zn nanoparticle after CVD using DPPs [6]. The half-width at the half-maximum of this distribution is 85 nm, which is much smaller than the photoluminescence wavelength (360 nm), confirming that the nanoparticle can act as a small light emitter at a scale beyond the diffraction limit. By improving this method, ZnO nanoparticles with a wurtzite structure have been recently formed [7]. These nanoparticles emit light with a photon energy of 3.29 eV and an emission spectral width as narrow as 140 meV, confirming their superior crystal quality. It should be pointed out that, for this method, a narrow channel was fabricated on the SiO_2 substrate, and the position at which the ZnO nanoparticle was deposited was controlled by the DPPs generated at the edge of the channel without the use of a fiber probe. Such position-controllability is advantageous also in the fabrication of DP devices. Moreover, strong ultraviolet emission has also been observed also from a GaN nanoparticle deposited by this novel CVD method [8].

In order to improve the performance of the CVD method described above, appropriate selection and combination of molecular gases, light sources, and surface processing methods are critical. However, it should be noted that this CVD method can be applied to a variety of molecules. Examples of the applicable molecules are summarized in Table 6.1.

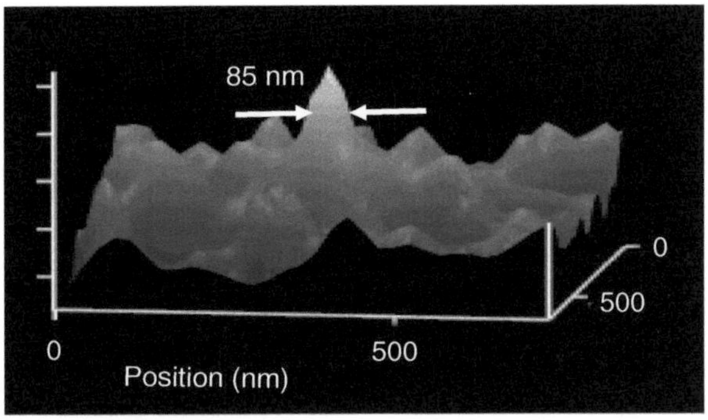

Fig. 6.5 Spatial distribution of the photoluminescence intensity emitted from a ZnO nanoparticle

Table 6.1 Examples of the molecules to which the phonon-assisted CVD can be applied

Material	Molecule	Absorption edge wavelength (nm)[a]
Zn	$Zn(C_2H_5)_2$	270
Al	$Al(CH_3)_3$	250
W	$W(CO)_6$	300
S	H_2S	270
P	PH_3	200
N	NH_3	220
O	O_2	250
Ga	$Ga(CH_3)_3$	260
Si	SiH_4	120
Si	SiH_6	195
P	PH_3	220
Sn	$Sn(CH_3)_4$	225

[a]This wavelength is inversely proportional to the excitation energy E_{ex} of the molecule.

6.2 Lithography Using Dressed-Photon–Phonons

Lithography using propagating light has been employed for practical mass-production to fabricate electronic devices. This method includes the processes of light exposure, etching, doping, and deposition. Light exposure projects a pattern drawn on a photo-mask onto a photo-resist film. Although there have been increasing demands to reduce the pattern size, the achievable fabrication precision, i.e., the resolution, is limited by the diffraction of light (the so-called diffraction limit). This section reviews a novel lithography method that can go beyond the diffraction limit.

Although the energy levels of photo-resist materials are much more complicated than those of the gas molecules considered in Sect. 6.1, a novel lithography method using DPPs has been developed by applying the principles described in Chap. 4. As schematically shown in Fig. 6.6a, a photo-resist film is coated on a substrate, and a photo-mask is placed on it. By illuminating the upper surface of the photo-mask with propagating light, DPPs are generated at the edge of the aperture on the lower surface of the photo-mask. These DPPs expose the photo-resist to form a fine pattern.

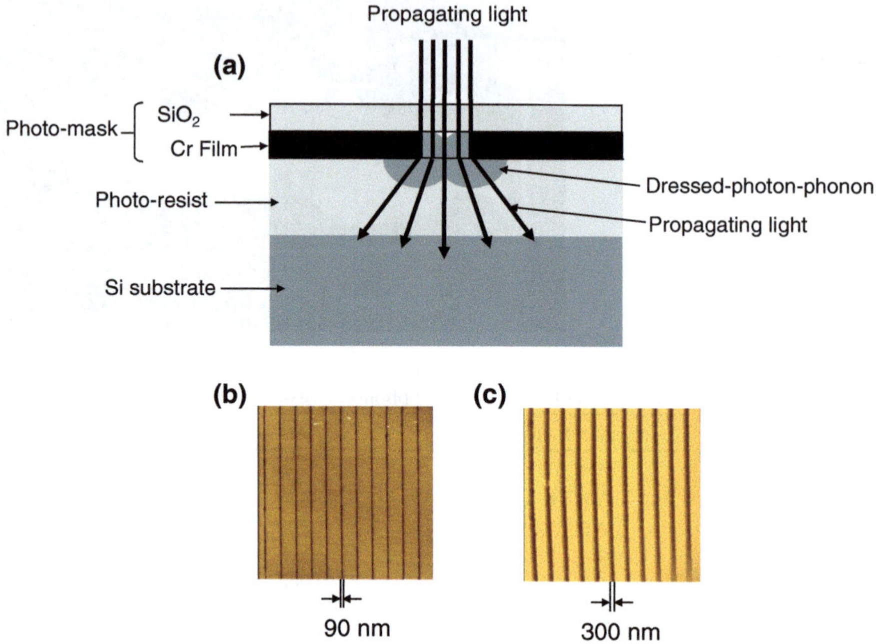

Fig. 6.6 Lithography by using dressed-photon–phonons. **a** Principle. **b**, **c** Atomic force microscope images of photo-resists patterned by using visible and ultraviolet light sources, respectively

The photo-resist material used for conventional lithography is sensitive to short-wavelength ultraviolet propagating light. In the other words, it is not exposed if it is illuminated by propagating light having a lower photon energy, that is to say light having a wavelength longer than the absorption-edge wavelength of the photo-resist. However, even by using such long-wavelength propagating light, the photo-resist can be exposed by the phonon-assisted process if DPPs are generated by the propagating light. As a result, a fine pattern is formed, which is smaller than that formed by conventional lithography using ultraviolet propagating light.

As an example, if red propagating light irradiates a photo-mask having a pattern of parallel narrow lines, the photo-resist under the photo-mask ought not to be exposed. However, as shown by Fig. 6.6b, the photo-resist is exposed, and a pattern of parallel narrow lines is formed [9]. The widths of these lines correspond to the widths of the slits in the photo-mask. This novel pattern formation originates from the DPPs generated on the photo-mask surface. As shown in Fig. 6.7a, two narrow areas of the photo-resist surface are exposed by two DPPs generated at the two edges of the slit in the photo-mask when the exposure time is short. In addition, the inner part of the photo-resist is also exposed, which is a novel phenomenon originating from the constructive interference between the two DPPs generated at the two edges of the slit. By increasing the exposure time, the volumes of the three exposed areas increase, as shown in Fig. 6.7b, and finally, they merge to form a pattern whose width

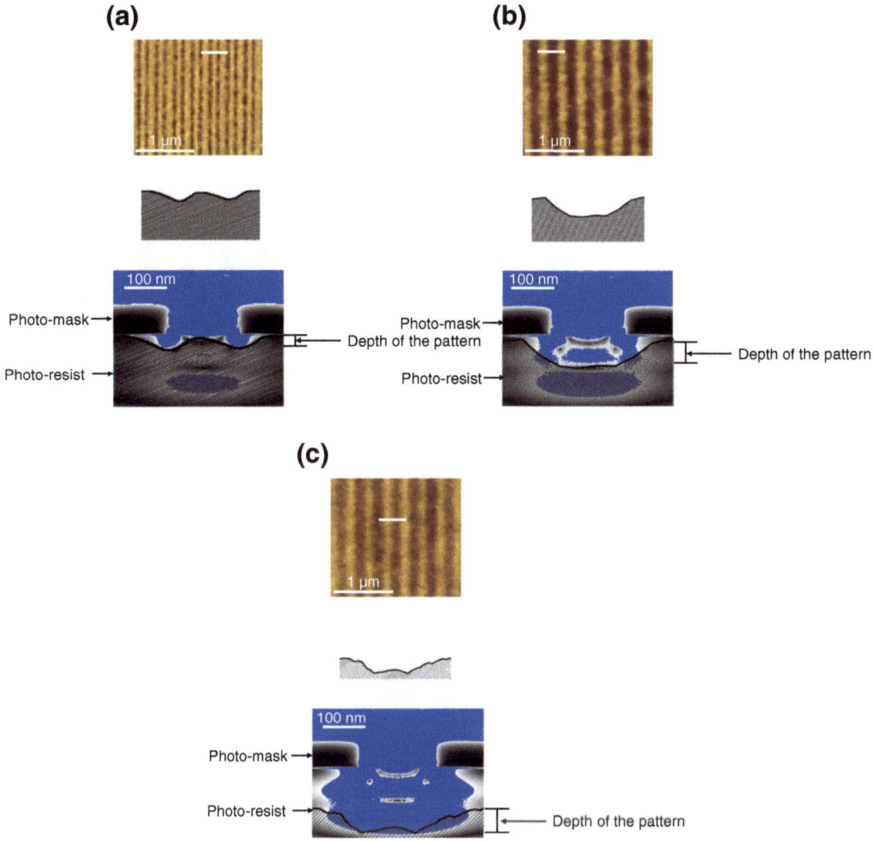

Fig. 6.7 Temporal evolution of the cross-sectional profile of the exposed photo-resist. **a–c** Represent the results for short, medium, and long exposure times, respectively. The *top* images in these figures are atomic force microscope images of the linear-patterned photo-resist. The *middle* images are cross-sectional profiles taken along the white lines in the *top* images. The *bottom* images represent the results of numerical calculations of the cross-sectional profiles, which correspond to the *middle* images

is equivalent to the width of the slit of the photo-mask, as shown in Fig. 6.7c. The experimental relation between the exposure time and the depth of the pattern formed in the photo-resist is shown by closed circles and squares in Fig. 6.8, which are fitted by the solid curve. They clearly show the threshold exposure time required to form a deep pattern as a result of merging the three areas [10]. This threshold feature is also reproduced by a simulation whose results are shown in the bottom images in Fig. 6.7a–c. For comparison, the relation obtained by irradiating the photo-mask with ultraviolet light is represented by open circles, open squares, and broken curves in Fig. 6.8, which does not exhibit any threshold.

Figure 6.6c shows the pattern formed by irradiating the photo-mask with ultraviolet light. In this case, the photo-resist is exposed not only by the DPPs but also by the

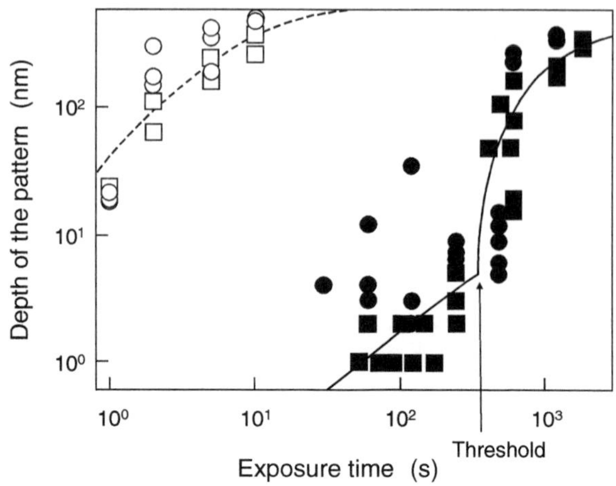

Fig. 6.8 Relation between the exposure time and the depth of the pattern formed on the photo-resist. *Closed circles, closed squares,* and a *solid curve* represent the experimental results obtained by using a visible light source. *Open circles, open squares,* and a *broken curve* are those obtained using an ultraviolet light source

low-intensity propagating light that is transmitted through the slit in the photo-mask. Due to the diffraction of this transmitted propagating light, the width of the formed pattern is broader than that of the slit in the photo-mask and that of the pattern formed by the DPPs. In the case of Fig. 6.6b, on the other hand, since the photo-resist is not exposed by the transmitted low-intensity propagating light but only by the DPPs, the width of the formed pattern is equivalent to that of the slit in the photo-mask, which is narrower than that of Fig. 6.6c.

Lithography using DPPs has several advantages, which are equivalent to advantages (1)–(3) of the CVD method described in Sect. 6.1. Among them, advantage (1) can be easily understood by comparing Fig. 6.6b and c. As an example of advantage (2), an optically inactive resist-film was patterned by DPPs, as shown in Fig. 6.9 [9]. The resist material used for this pattering (ZEP520) is sensitive only to an electron beam and X-rays but not to propagating light, which means that, strictly speaking, it is not a "photo"-resist. However, it was exposed by DPPs, and a two-dimensional array of the disk patterns was formed, as shown in Fig. 6.9. Since this resist material is carefully prepared for nano-fabrication, it has an extremely flat surface when coated on a Si substrate, and thus, Fig. 6.9 shows that the formed disk patterns have sharp edges as well as a flat top surface. Advantage (3) is attained because a conventional exposure apparatus can be used, without requiring an expensive ultraviolet light source or related optical components.

Based on these advantages, some novel patterning methods have been developed:

(1) Pattern duplication: After the photo-resist is coated on a transparent substrate, as shown in Fig. 6.10a, an integrated circuit, as an example, is place on top for use as a photo-mask. The photo-resist is not exposed by the visible propagating light that illuminates the rear surface of the substrate. Therefore, the propagating light is transmitted through the photo-resist and reaches the photo-mask surface, where DPPs are generated. The photo-resist is exposed by these DPPs to duplicate the pattern of the integrated circuit, as shown in Fig. 6.10b.

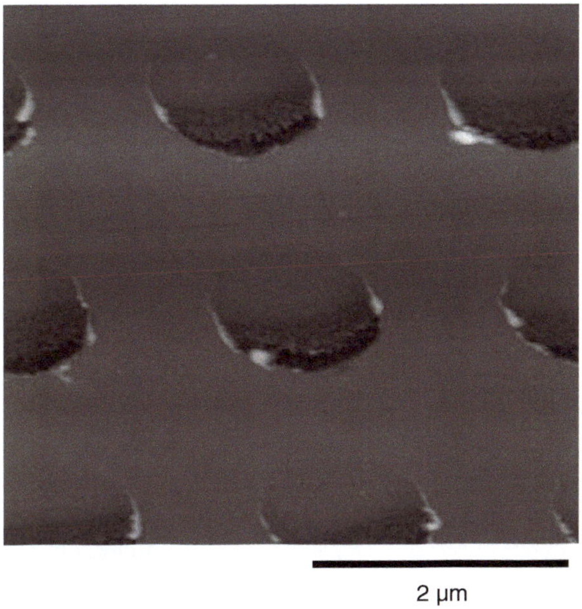

Fig. 6.9 Scanning electron microscope image of a two-dimensional array of circular patterns formed on an optically inactive ZEP520 photo-resist by using dressed-photon–phonons

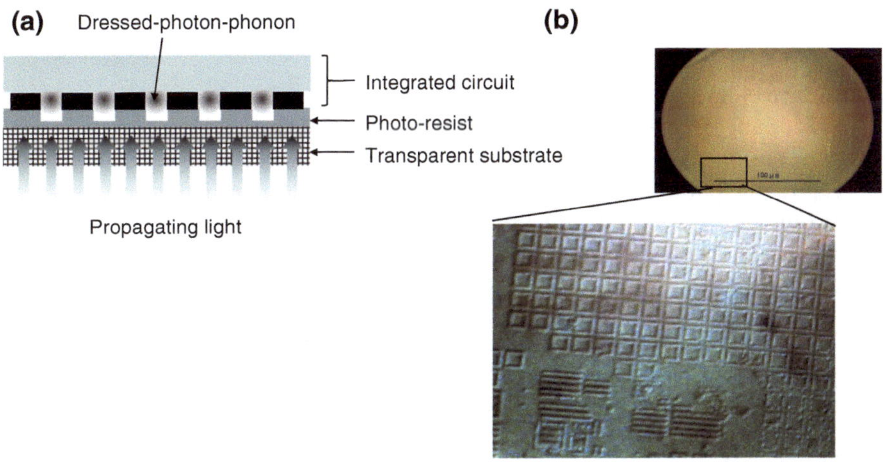

Fig. 6.10 Duplication of an integrated circuit pattern. **a** Principle. **b** Optical microscope images of a pattern duplicated on a photo-resist

(2) Multiple exposure: An array of linear patterns, as an example photo-mask, is placed on a photo-resist, as shown in Fig. 6.11a. A first exposure exposes the photo-resist with the DPPs by irradiating visible propagating light from the rear

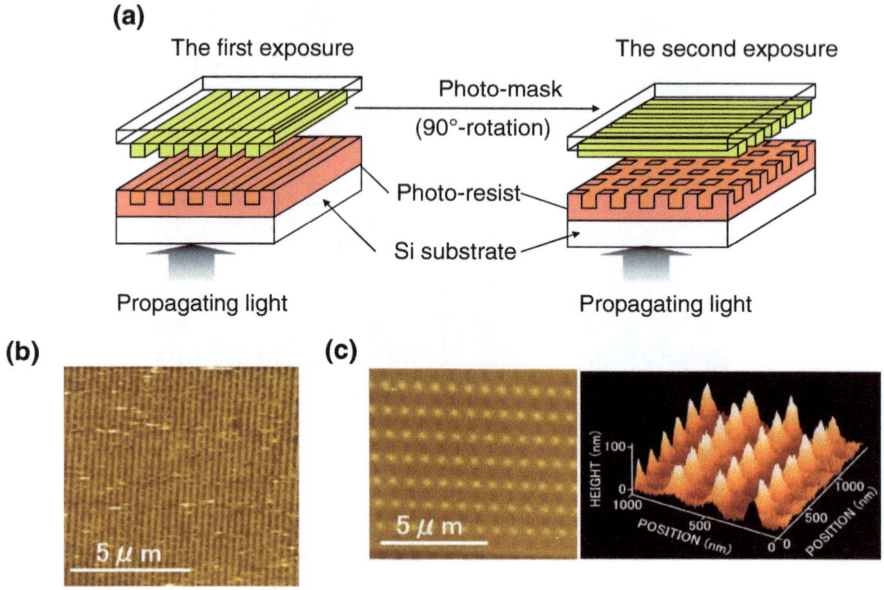

Fig. 6.11 Multiple exposure. **a** Principle. **b** Atomic force microscope image of a linear pattern formed on the photo-resist by the first exposure. **c** Atomic force microscope image of a lattice pattern formed on the photo-resist by the second exposure. The *left* and *right* figures represent a top view and a perspective view, respectively

surface of the transparent substrate, as was the case in (1). As a result, an array of linear patterns is formed, which is shown in Fig. 6.11b. For the second exposure, after the photo-mask is rotated 90 degrees, it is placed on the photo-resist to expose the photo-resist again. As a result, a lattice pattern is formed, as shown in Fig. 6.11c. A high contrast of this pattern can be maintained even when the number of exposures is increased because this photo-resist is optically insensitive to the visible propagating light.

At the early stage of development of these methods, a photo-mask for generating DPPs was fabricated by electron beam lithography. Specifically, an electron beam was raster-scanned on a thin film of Cr on a SiO_2 substrate to form a mask pattern. However, the throughput of this patterning method is very low due to the limited scanning speed of the electron beam. Furthermore, this method of patterning has several problems, including the high cost of the electron beam machine and high energy consumption. In order to solve these problems, electron beam lithography has been recently replaced with lithography using DPPs for high-throughput fabrication of photo-masks.

Lithography using DPPs has been employed to produce a variety of devices. Examples include:

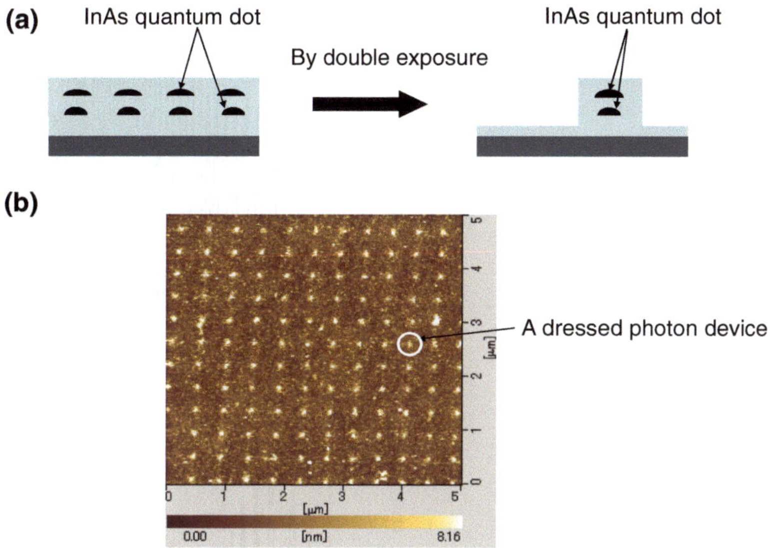

Fig. 6.12 Fabrication of a two-dimensional array of mesa-shaped dressed-photon devices. **a** Principle. **b** Atomic force microscope image of the fabricated two-dimensional array

(1) Two-dimensional array of DP devices [11]: A two-dimensional array of the mesa-shaped DP devices shown in Fig. 5.7 was formed by electron beam lithography and Ar-ion milling. The above-mentioned multiple exposure method can be employed to form the array using less expensive equipment with lower energy consumption. After two different-sized QDs of InAs are grown on the upper and lower layers on the substrate, as shown in Fig. 6.12a, the lattice pattern is formed by double exposure to produce the two-dimensional array of mesas. The size of the mesas is adjusted so that each layer in the mesa contains only one QD. The pair of the QDs in the upper and lower layers constitute a DP device. Figure 6.12b shows an AFM image of the two-dimensional pattern formed, demonstrating that the mesa-shaped DP devices are two-dimensionally aligned.

(2) Fresnel zone plate for soft X-rays [12]: A Fresnel zone plate (FZP) is composed of a large number of concentric circular stripes and has been widely employed to focus soft X-rays. Although an electron beam or focused ion beam has been conventionally employed to fabricate FZPs, lithography using DPPs has been used recently to increase the fabrication throughput. As an example, an FZP composed of 151 concentric circles focuses 0.42 nm-wavelength soft X-rays to a focused spot diameter of 196 nm in a focal length of 50 nm. The diameter and width of the outermost circular stripe are 56 μm and 190 nm, respectively; that is, the diameter of the FZP is 112 μm. In order to produce this FZP, a 180 nm-thick Ta film is deposited on a 200 nm-thick Si_3N_4 film on a Si substrate. This Ta film is coated with photo-resist (FH-SP-3CL), and a photo-mask (Cr film on a Si_3N_4 membrane) is placed on top. A Xe lamp is used as a visible light source. The

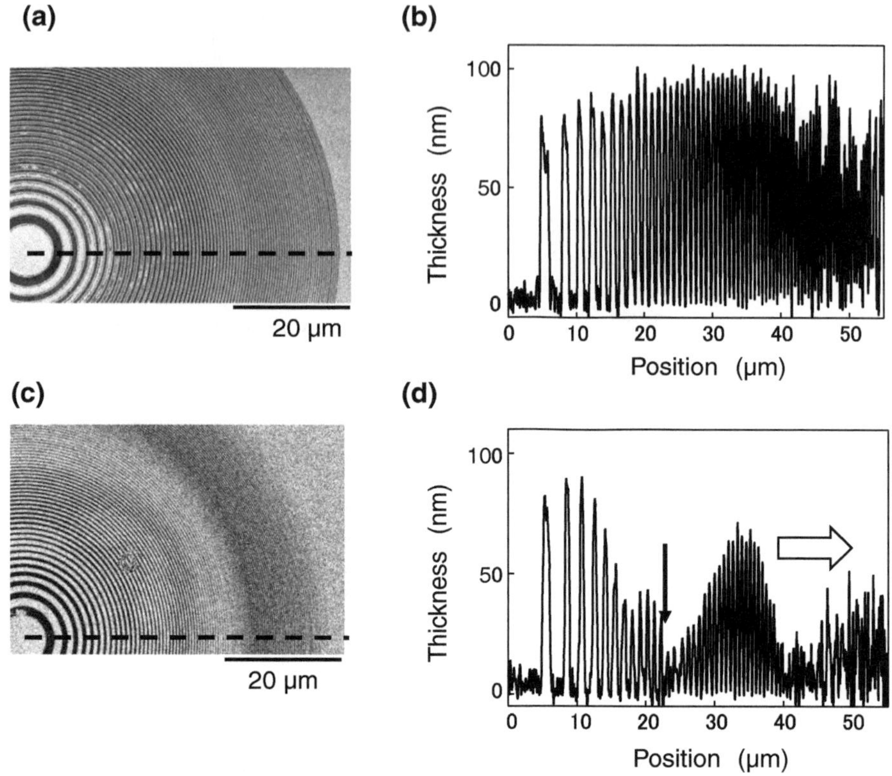

Fig. 6.13 Scanning electron microscope images of a Fresnel zone plate for soft X-rays. **a** Scanning electron microscope (SEM) image of the zone plate pattern fabricated by using a Xe lamp (center wavelength of the emission spectrum: 550 nm) as a light source. **b** Cross-sectional profile taken along the *broken line* in **a**. **c** SEM image of the zone plate pattern fabricated by using a Hg lamp (center wavelength of the emission spectrum: 450 nm) as a light source. **d** Cross-sectional profile taken along the *broken line* in **c**

center wavelength and the half-width of its emission spectrum are 550 and 80 nm, respectively. The optical power density is 100 mW/cm^2. Even when using such long-wavelength propagating light, a pattern is formed by the phonon-assisted process due to the DPPs. After exposure, the Si substrate is removed, and the Si_3N_4 film is thinned. Figure 6.13a shows a scanning electron microscope image of the fabricated FZP. Concentric circular stripes are clearly seen even though the width (190 nm) of the outermost circular stripe is much narrower than the emission wavelength of the Xe lamp. Figure 6.13b is a cross-sectional profile, showing the thickness of the each circular stripe.

For comparison, Fig. 6.13c and d show images of the pattern fabricated by using an Hg lamp as a light source (center wavelength: 450 nm, half-width: 40 nm, optical power density: 100 mW/cm^2). Since its emission wavelength is shorter than the

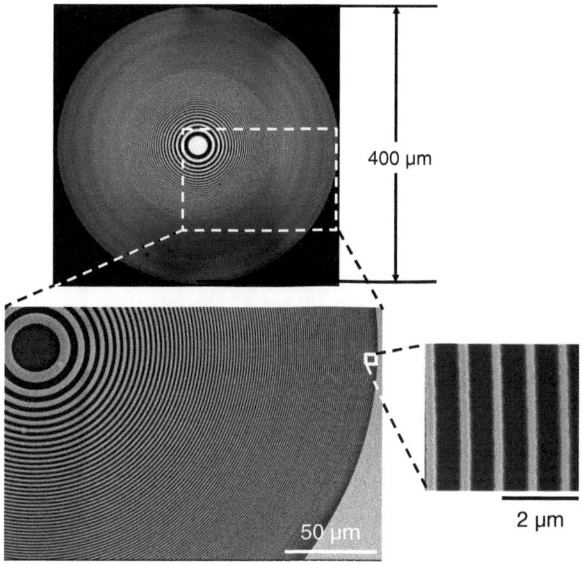

Fig. 6.14 Fresnel zone plate for ultraviolet light. The *upper image* is an optical microscope image. The *lower two images* are scanning electron microscope images

absorption-edge wavelength of FH-SP-3CL, the pattern of the fabricated FZP is exposed by the propagating light. As a result, in the area indicated by the right-pointing thick white arrow in Fig. 6.13d, the circular stripes are thin, which is due to diffraction of the propagating light. Furthermore, at the position indicated by the downward arrow, the thickness of the stripe takes the local minimum. By noting that the width (450 nm) of this stripe is equal to the emission wavelength of the Hg lamp, this local minimum originates from the increase in the light intensity due to the constructive interference between the propagating light transmitted through adjacent concentric circular apertures of the photo-mask. By comparing Fig. 6.13b and d, it is confirmed that use of DPPs is advantageous for producing a high-contrast FZP.

Furthermore, Fig. 6.14 shows an image of an FZP for 325 nm-wavelength ultraviolet light. Even though the diameter of the FZP is as large as 400 µm and the ratio of the widths between the outermost and innermost circular stripes is as large as 60, this figure shows a pattern with very high contrast.

Figure 6.15 shows a photographic image of the photo-mask used for fabrication. 7×7 photo-masks are two-dimensionally aligned in the a square area with the a side length of 7 mm. This means that 49 FZPs can be simultaneously produced, which greatly increases the fabrication throughput for mass-production.

(3) Diffraction grating for soft X-rays [13]: A photo-mask is placed on a photo-resist coated on a Si substrate to form a linear pattern with 7800 lines/mm on the Si substrate. After this pattern formation, the pattern is coated with a thin film of Mo to produce a diffraction grating for soft X-rays. A scanning electron microscope

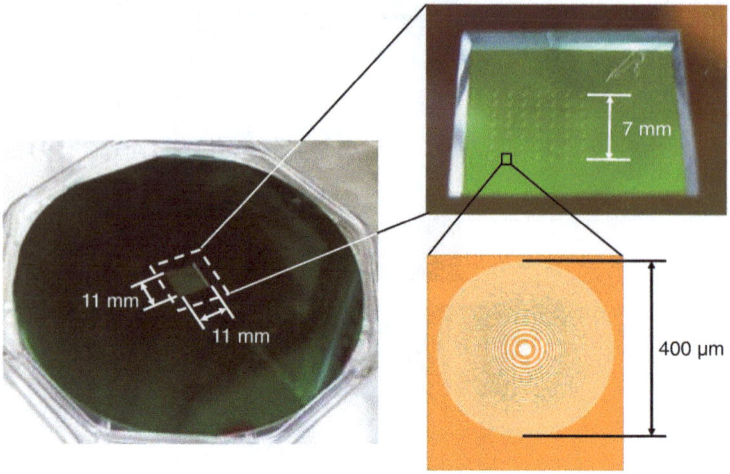

Fig. 6.15 Photo-masks used for fabricating the Fresnel zone plate in Fig. 6.14

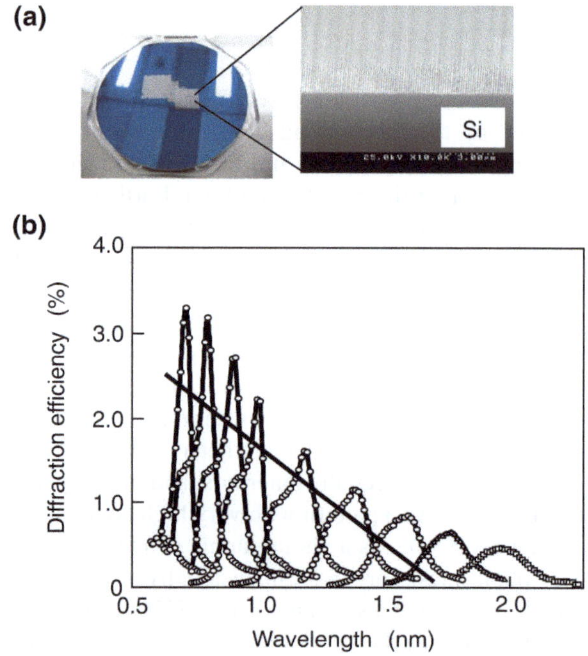

Fig. 6.16 A diffraction grating for soft X-rays. **a** Photograph (*left*) and the scanning electron microscope image (*right*). **b** Measured diffraction efficiency. The *solid line* represents the efficiency of a conventional diffraction grating made by using a KAP crystal lattice

image is shown in Fig. 6.16a. As shown in Fig. 6.16b, the measured diffraction efficiency at a wavelength of 0.6 nm is as high as 3.3 %, which is 1.4-times higher than that of a conventional diffraction grating fabricated by using a KAP crystal lattice structure.

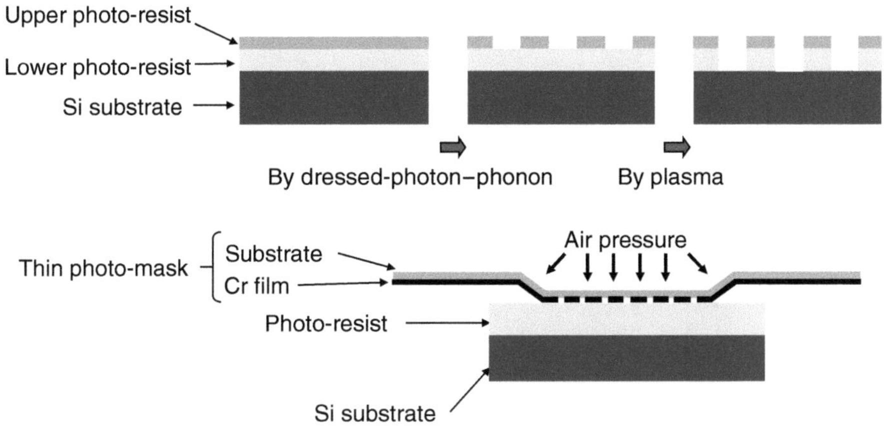

Fig. 6.17 Method of contacting the photo-mask to the photo-resist

In order to produce a variety of devices, including the devices demonstrated above, a practical lithography machine using DPPs has been developed. Several technical ideas were employed in the construction of this machine:

(1) Exploration of photo-resist materials: Photo-resist materials producing a uniform film coated on a substrate are explored. Photo-resists composed of smaller polymer molecules are used to improve the resolution of pattern formation.
(2) Contacting the photo-mask to the photo-resist: By thinning the photo-mask, it can be brought into contact with the photo-resist uniformly over a wide area by using air pressure, as shown in Fig. 6.17. A lubricant film is coated on the photo-mask in order to avoid damaging its surface when it is removed from the photo-resist after light exposure.
(3) Pattern formation over a large area of the photo-resist surface: After exposure with the photo-mask in contact with the photo-resist, the photo-mask is taken off and moved to the adjacent position on the photo-resist surface for successive exposure. By repeating this process, a wide area of the photo-resist surface can be exposed and patterned. This method is called the step-and-repeat method.
(4) Reducing mechanical vibrations: In order to avoid misalignment in the relative positions of the photo-mask and photo-resist caused by mechanical vibrations of the machine, the structure of the machine was carefully designed to incorporate anti-vibration measures.
(5) Removing dust particles: In order to prevent dust particles from adhering to the surfaces of the photo-mask and photo-resist, the degree of cleanness[3] of the chamber in which the system in Fig. 6.17 is installed is maintained as high as

[3] Class X represents the level of the cleanness. In Japan, it is classified according to JIS9920. If the number of dust particles having a diameter smaller than 0.1 μm is less than X in a 1 m^3 volume, the level of cleanness is called class X. Values of X are always powers of 10.

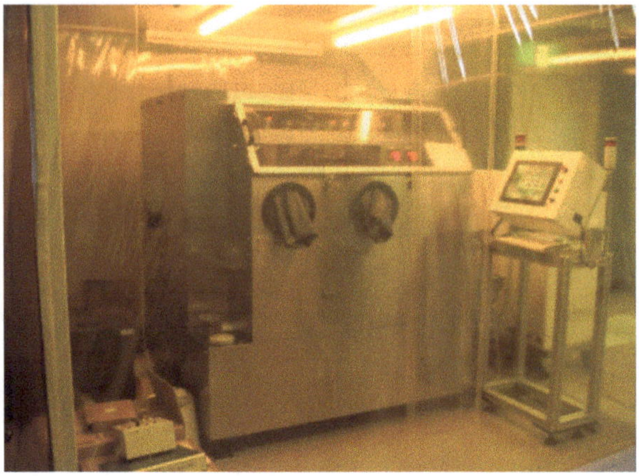

Fig. 6.18 Photograph of a compact lithography machine for practical use

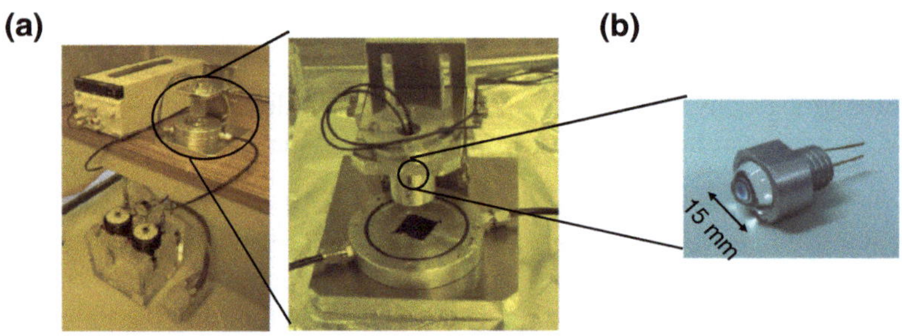

Fig. 6.19 Desktop lithography machine. **a** Photographs of the machine. **b** A light emitting diode used as a light source

class 10. The cleanness of the space around this chamber is maintained to as high as class 100–1000.

Based on these technical ideas, a prototype compact lithography machine has been constructed, as shown in Fig. 6.18 [14]. It occupies a floor area of only 1 m^2. For automatic operation, the specimens, such as Si substrates, are transferred automatically by a robot. The fabrication process, including this transfer, is controlled by a computer. A Xe lamp was used as a light source for generating DPPs; however, it has recently been replaced with a green LED. Since this LED consumes only 4.5 W of electric power, low energy consumption is guaranteed. Moreover, a much more compact, desktop machine was recently developed, as shown in Fig. 6.19, in which the green LED is also used.

Figure 6.20 shows several examples of the linearly aligned stripe patterns formed by using these machines [15]. They are formed over an area of 50 mm × 60 mm by using a photo-mask with an area of 5 mm × 5 mm with the step-and-repeat method. A pattern with a line width of 40 nm and an interval of 90 nm (Fig. 6.20a), a narrow,

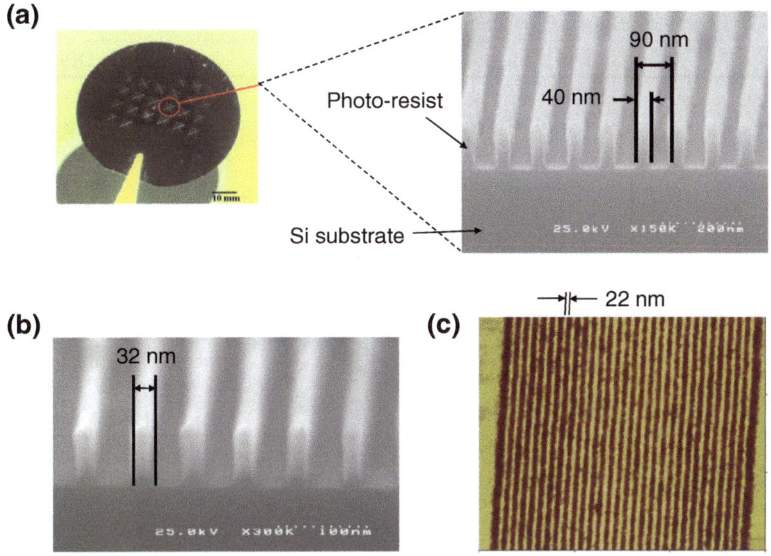

Fig. 6.20 Examples of fabricated linear patterns. **a** A pattern with a line width of 40 nm and a line spacing of 90 nm. **b** A pattern with a large aspect ratio. **c** A pattern with the smallest reported line width of 22 nm

deep pattern with an aspect ratio as large as 3.3 (Fig. 6.20b), and a pattern with the smallest reported line width of 22 nm (Fig. 6.20c) have been successfully formed.

In conventional lithography using propagating light, an ultraviolet light source has been employed to increase the diffraction- limited resolution. To increase the resolution further, lithography systems using extreme ultraviolet light sources and synchrotron radiation sources have recently been under development. However, these light sources are extremely large and consume high energy. Moreover, they require a huge vacuum system and a huge clean room, which also consume a high amount of energy. Enormous costs are required for installing these systems in factories, and this conventional method has now reached stalemate due to the diminishing cost-effectiveness. In contrast, lithography using DPPs does not require any special light sources, such as ultraviolet light sources. Furthermore, high vacuum systems and huge clean rooms are not required either.

To summarize these features, Table 6.2 compares the amount of energy consumed by the DPP lithography system described here and a conventional system. From this table, it is confirmed that the DPP system consumes much less energy. Therefore, it is expected that DPP lithography systems will be capable of producing multiple types of devices in smaller lots (e.g., optical devices, bio-chips, and micro-chemistry chips), which has never been possible with conventional lithography systems. This will allow the development of novel application systems, which will create new markets.

Table 6.2 Energy consumption of lithography systems

	System using propagating light[a,b]	System using DPPs[a]
Light source	ArF laser (9×10^7 kW h/year)	Light emitting diode (1×10^6 kW h/year)
Optical component	Fabrication of optical elements for ultraviolet light (1×10^7 kW h/year)	Fabrication of optical elements for visible light (5×10^6 kW h/year)
Vacuum system	Large-scale high vacuum system ($>3 \times 10^9$ kW h/year)	Simple, low vacuum system (1×10^9 kW h/year)
Environment	Clean room (200 m^2 floor space) ($>>9 \times 10^9$ kW h/year)	Clean booth (2 m^2 floor space) and a small clean room for installing the machine (4.5×10^9 kW h/year)
Total	$>>1.21 \times 10^{10}$ kW h/year	$\cong 5.5 \times 10^9$ kW h/year

[a] 1000 systems are assumed to be in operation
[b] The amount of energy consumed by the conventional lithography system was estimated based on reference [16]

6.3 Fabrication by Autonomous Annihilation of Dressed-Photon–Phonons

Fiber probes or photo-masks have been used to generate DPPs for the novel nano-fabrication methods reviewed in the previous sections. However, the theoretical model of one-dimensional lattice vibration described in Chap. 4 claims that DPPs can be generated in a finite one-dimensional material and localized at its apex. Based on this possibility, this section reviews an autonomous nano-fabrication method in which neither a fiber probe nor a photo-mask is required.

6.3.1 Smoothing a Material Surface by Etching

As the first example of autonomous nano-fabrication, this subsection reviews a novel method of smoothing the surface of a synthetic silica substrate. A variety of materials are smoothed by chemical-mechanical polishing (CMP) by using abrasive materials, such as CeO_2 [17]. In the case of polishing a synthetic silica substrate by CMP, a suspension containing CeO_2 abrasives particles with a diameter of 100 nm and a polishing pad with a surface roughness of 10 μm have been used. However, the achieved minimum roughness, R_a, of the polished surface is as large as 0.2 nm [18]. Furthermore, scratches and digs are formed due to the contact of the CeO_2 particles or impurities in the suspension with the substrate surface in the polishing process. Here, R_a is defined by

$$R_a = \frac{1}{l} \int_0^l f(x)dx, \tag{6.19}$$

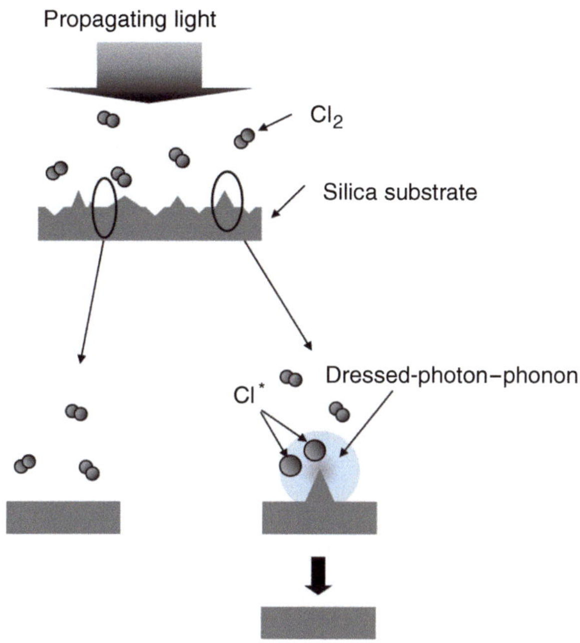

Fig. 6.21 Principle of chemical etching using dressed-photon–phonons, for smoothing a material surface

where $f(x)$ is the thickness of the substrate at position x, l is the total side length of the measurement area on the substrate surface, and dx is the resolution of the measurement.

Rapidly progressing high-power laser technology is requiring high-quality laser mirrors made of synthetic silica that can withstand high optical power. To meet this requirement, the value of R_a of the substrate for the mirror must be decreased to 0.1 nm [19]. To realize this value, a novel chemical etching (CE) method using DPPs has been invented [20]. As illustrated in Fig. 6.21, a planar synthetic silica substrate with a diameter of 30 mm, as an example, is installed in a vacuum chamber filled with Cl_2 molecular gas to a pressure of 100 Pa. This substrate was grown by vapor-phase axial deposition, and the concentration of the OH-group was less than 1 ppm [21]. Although the Cl_2 molecule does not chemically react with silica, the chemically active radical Cl* does. The substrate surface is smoothed as a result of this reaction. In order to create Cl*, the Cl_2 molecule is dissociated by the molecular dissociation described in Sect. 6.1. Here, in order to use DPPs, the substrate is illuminated by propagating light whose photon energy (2.33 eV; 532 nm-wavelength) is lower than the excitation energy E_{ex} (3.10 eV) of Cl_2. The optical power density incident on the entire substrate surface is 0.28 W/cm^2. It should be noted that the Cl_2 molecule is not dissociated by propagating light having such a low photon energy.

If this propagating light is incident on a completely flat substrate surface, the Cl_2 molecule stays in its ground state because of the absence of DPPs. However, if the propagating light is incident on a rough surface, DPPs are generated at the apexes

Fig. 6.22 Photograph of a chemical etching machine

of the bumps, causing dissociation of the Cl_2 due to the phonon-assisted process described above, and Cl* radicals are created. Since the Cl* chemically reacts with the bumps, CE proceeds, gradually varying the shape and size of the bumps. Here, it should be pointed out that the DPPs are not generated in the bottom of digs in the surface because digs are three-dimensional funnel-shaped structures, and thus, the one-dimensional lattice vibration model of Chap. 4 cannot be applied.

The principle of CE described enables autonomous smoothing. That is, selective CE of the bumps spontaneously starts due to light illumination, and the etching progresses autonomously and spontaneously stops when DPPs are no longer generated. In the other words, at the moment the DPPs are annihilated after autonomous variation of the shapes of the bumps, CE no longer occurs. The light power and the Cl_2 molecule pressure must be optimized in order to realize a sufficiently high throughput of autonomous smoothing. Figure 6.22 shows a photograph of a CE machine which has been constructed for practical use.

Figure 6.23a shows an AFM image of the surface of a 30 mm-diameter synthetic silica substrate which was preliminarily polished by CMP. As well as bumps, several linear scratches and digs formed in the CMP process are also seen in this figure. Figure 6.23b shows an AFM image of the surface after CE by DPPs. Comparing it with Fig. 6.23a, it is found that the scratches, digs, and bumps are removed. The surface roughness R_a of Eq. (6.19) was derived by using the results measured by AFM for nine $10\,\mu m \times 10\,\mu m$ square areas on the substrate surface, separated by $100\,\mu m$ (refer to the left part of Fig. 6.23c). Closed squares in the right part of Fig. 6.23c show the average $\bar{R}_a$ of the nine values of R_a measured above, which are plotted as a function of the etching time. The resolution of the measurement dx in Eq. (6.19) is 40 nm, which depends on the apex size of the cantilever of the AFM. This figure shows that the value of $\bar{R}_a$ is 0.21 nm before the CE; however, it decreases to a stationary value of 0.13 nm after the 30 min of CE [22]. Moreover, it was confirmed by auxiliary experiments that the standard deviation of the nine values of R_a was decreased by the CE, from which homogeneous smoothing over the entire substrate surface was

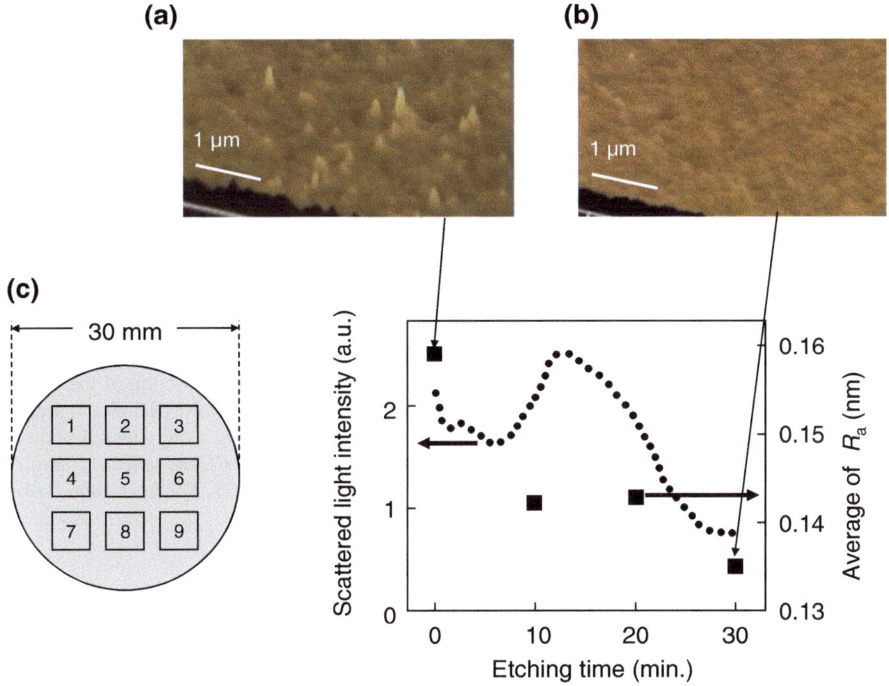

Fig. 6.23 Experimental results of smoothing the surface of a circular synthetic silica substrate. **a** and **b** Atomic force microscope images of the substrate surface before and after the chemical etching, respectively. **c** *Nine squares* in the *left* figure represent the areas in which the surface roughnesses were measured by using an atomic force microscope. The *closed squares* in the *right* figure represent the average of the nine measured roughnesses. The *dotted curve* represents the scattered light intensity

confirmed. The dotted curve in the right part of Fig. 6.23c shows the scattered light intensity of supplementary laser light which illuminated the substrate surface to monitor the surface roughness as CE proceeded. Its photon energy was lower than E_{ex} (wavelength: 632 nm, light power density: 0.13 W/cm^2, light beam diameter: 1.0 mm) [22]. The temporal evolution of this scattered light intensity monitored in real-time is equivalent to those represented by the closed squares. This figure also shows that the values of $\bar{R}_a$ and the scattered light intensity temporarily increased at around 10 min. after the CE started. From spatial power spectral density analysis of the AFM images, it was confirmed that these temporary increases occur because a single tall bump was split into several lower bumps by CE. However, as CE proceeded further, the values of $\bar{R}_a$ and scattered light intensity decreased gradually. Precise power spectral analysis of these temporal evolutions has been carried out [23], which can be explained by the mathematical science model given in Sect. 8.4.

A laser mirror was produced by coating a highly reflective film on a smoothed synthetic silica substrate surface, and its ability to withstand high optical power was evaluated using a standard test method (ISO 11254-022 on 1). The closed diamonds in

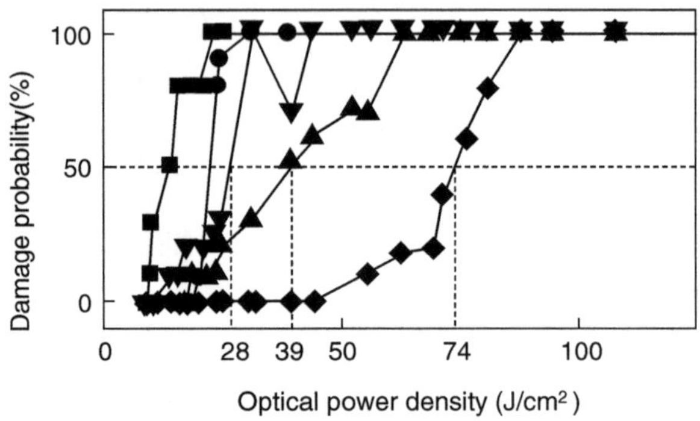

Fig. 6.24 Measured damage probabilities of mirror surfaces, plotted as a function of the optical power density of the irradiated laser light. *Closed diamonds* and downward triangles represent the results for mirrors using substrates smoothed by the chemical etching using dressed-photon–phonons and conventional chemical–mechanical polishing, respectively. *Closed upward triangles, circles*, and *squares* represent the values for commercially available mirrors. The *horizontal broken line* represents the damage threshold

Fig. 6.24 show the evaluated results [24], which represent the probability of damaging the mirror surface by irradiating it with 355 nm-wavelength pulsed laser light. For comparison, the values for mirrors using substrates polished by conventional CMP are also shown. The damage threshold, defined as the 50 % probability of causing damage, is found to be as high as 74 J/cm^2 after CE by the DPPs (closed diamonds). Since the damage thresholds were only 28 J/cm^2 before the CE (closed downward triangles) and 39 J/cm^2 for the strongest commercially available mirror polished by CMP (closed upward triangles), the excellent performance of the present CE method is confirmed.

The mechanism involved in mirror damage has been studied by pump-probe spectroscopy using 750 nm-wavelength pulsed laser light [25]. Immediately after irradiating a substrate with a high-intensity pump pulse in order to damage its surface, a low-intensity probe pulse was radiated to measure the spectra of the Raman signals generated from the damaged surface. As a result, it was found that amorphous synthetic silica was crystallized with a 300 fs-time constant by radiating the pump pulse, followed by breaking of the Si–O bonds with a 200 fs-time constant, inducing damage. After the CE, the threshold (6.9 kJ/cm^2) for inducing this damaginge was 1.3-times higher than that before the CE. This result confirmed that CE was effective in increasing the ability to withstand high optical power.

This CE method can also be applied to a variety of surface shapes [26]. They include convex and concave surfaces, and the inner surfaces of cylinders. Furthermore, the surfaces of multiple substrates can be simultaneously smoothed even when they are stacked. That is, by inserting spacers between adjacent substrates, their surfaces can be chemically etched autonomously because the propagating light and Cl_2 molecules can reach the surfaces through the gaps formed by the spacers. As another example, Fig. 6.25 schematically explains how to etch the side surface of a diffraction

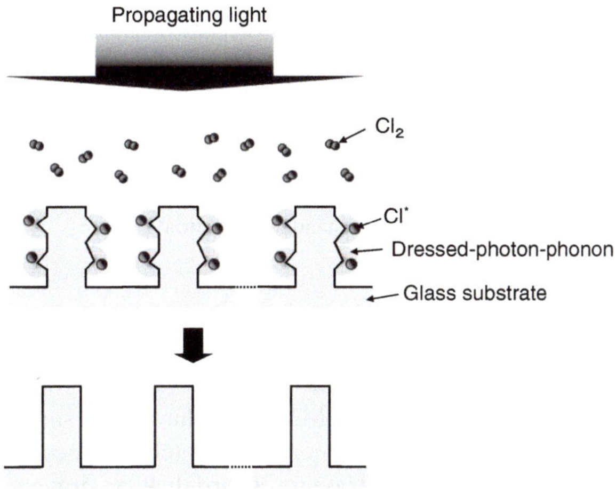

Fig. 6.25 Principle of chemical etching of the side surfaces of the corrugations of a diffraction grating

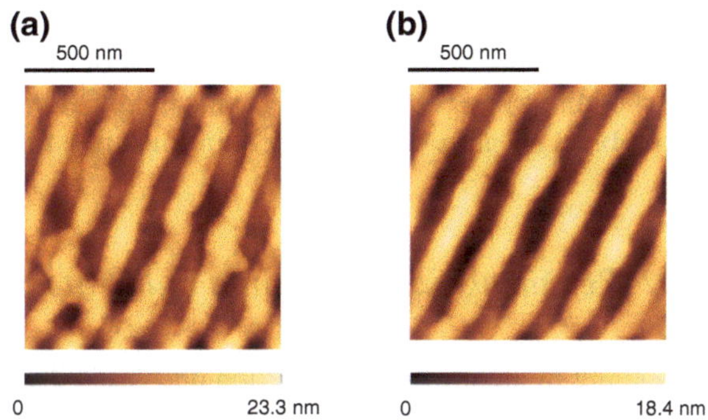

Fig. 6.26 Atomic force microscope images of a diffraction grating formed on a soda lime glass substrate. **a** and **b** Atomic force microscope images acquired before and after the chemical etching

grating composed of parallel linear corrugated patterns. This diffraction grating was made from a soda lime glass substrate by using a thermal nano-imprinting method [27]. The average height and interval of the corrugated pattern are 13.5 and 175 nm, respectively. Since the propagating light can reach the side surfaces of the corrugations, CE is possible by using visible propagating light (wavelength: 532 nm, power density: 0.28 W/cm^2) and Cl_2 gas at a pressure of 100 Pa. Figure 6.26a shows an AFM image before the CE, in which corrugations meander due to their rough side surfaces. In contrast, Fig. 6.26b shows the AFM image after the 30 minutes of CE, in which the meanders decreased. Table 6.3 summarizes the roughness of the side and bottom surfaces of the corrugations before and after the CE, which were evaluated by using AFM images. The table also shows that the average and the standard deviation of the width of the corrugations decreased after CE. From the values shown in this table, it is confirmed that the side surface was successfully smoothed.

This CE method has also been applied to a glass substrate for a disk in a magnetic storage system [28], and also to a photo-mask for conventional lithography [29]. Moreover, it can be applied to a variety of materials, such as crystals, plastics, and

Table 6.3 Surface roughness of each part before and after chemical etching (CE)

	Before etching	After etching
Average width of corrugation	94.4 nm (20.7 nm)[a]	89.8 nm (17.6 nm)[a]
Roughness R_a of side surface[b]	0.68 nm	0.36 nm
Roughness R_a of bottom surface[c]	0.76 nm	0.26 nm

[a]Value in parentheses is the standard deviation
[b]Etching speed: 0.64 nm/h
[c]Etching speed: 1.0 nm/h

metals, as long as they react with radical atoms. For example, the surfaces of GaN [30], diamond [31], and PMMA [32] have been smoothed with this method. Besides Cl_2, F_2 and O_2 have also been used as etching gases. Moreover, in order to smooth the surface of a glass substrate for a hard disk, an organic film is coated on the substrate surface instead of using the molecular gases, and high-throughput smoothing has been demonstrated in air by irradiating the substrate with propagating light [28]. Since conventional CMP uses a large amount of CeO_2 abrasives, it has the problem of depletion of rare resources. In contrast, the CE method described here can solve this problem because it does not require such rare materials.

6.3.2 Repairing Scratches on a Substrate Surface by Deposition

Smoothing in the previous subsection decreased the roughness by etching the substrate surface. In contrast to this, it is possible to decrease the roughness by depositing material. As an example, this subsection reviews a method for repairing scratches in a poly-Al_2O_3 crystal surface by material deposition using DPPs [33]. A variety of transparent materials have been used as laser media, optical windows, and so on [34–38]. Among them, a transparent poly-Al_2O_3 crystal has been used to construct a ceramic laser for application to automobile spark plugs [39]. Although a smooth surface of this crystal is essential for maintaining its optical transparency, it has many scratches created by the preliminary polishing. Even though CMP can be used for single-crystals or amorphous materials, it is not effective for poly-Al_2O_3 crystals because of the anisotropic interaction between the poly-crystalline surface and the abrasive.

In order to repair a scratched surface, Al_2O_3 particles are deposited by a novel sputtering method. In conventional sputtering, as illustrated in Fig. 6.27a, it has been found that the migration length of the Al_2O_3 particles on the substrate surface depends on the height of the Schwöbel barrier of the free energy [40, 41]. Since this barrier is high at the ridgeline of a scratch, it is difficult for the Al_2O_3 particles to climb over the ridgeline in order to migrate to the slope and bottom of the scratch. As a result, the Al_2O_3 particles are deposited selectively at the ridgeline and do not fill the scratch, with the result that the surface cannot be smoothed. In contrast, as illustrated

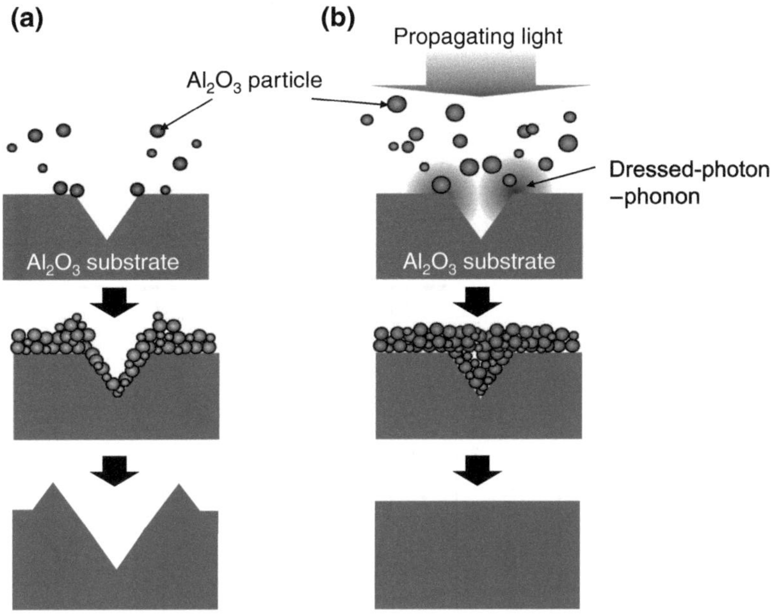

Fig. 6.27 Principle of repairing scratches on a poly-Al_2O_3 crystal surface by sputtering Al_2O_3 particles. **a** Conventional sputtering. **b** Sputtering with light irradiation

in Fig. 6.27b, DPPs are generated at the ridgeline by irradiating the substrate surface with propagating light in the sputtering process. The Al_2O_3 particles sputtered to this ridgeline are activated by the DPPs due to the phonon-assisted process, and as a result, the particles climb over the Schwöbel barrier and migrate to the slope and bottom of the scratch. Consequently, the scratches are filled with the Al_2O_3 particles, and the surface is smoothed.

In order to confirm this, a poly-Al_2O_3 crystal was preliminary polished by CMP using diamond particles of 0.5 μm-diameter as abrasives. After the CMP, Al_2O_3 particles were deposited by rf-magnetron sputtering. The substrate surface was simultaneously irradiated with propagating light (wavelength: 473 nm, light power density: 2.7 W/cm^2) whose photon energy (2.62 eV) was lower than the excitation energy E_{ex} (4.96 eV) of Al_2O_3 [42]. After 30 mins of sputtering, a 100 nm-thick Al_2O_3 layer was formed. Figure 6.28a, b, and c shows AFM images of the substrate surface acquired before the sputtering, after the sputtering without light irradiation, and after the sputtering with light irradiation, respectively. Comparing Fig. 6.28c with a and b, the scratches were successfully repaired by the sputtering with light irradiation. Figure 6.28d, e, and f shows cross-sectional profiles of the scratches, taken along the white lines in Fig. 6.28a, b, and c, respectively. From these profiles, the depths of the scratches were 4.0, 4.4, and 1.8 nm, confirming the effectiveness of light irradiation. For more quantitative evaluation, the surface roughness R_a of Eq. (6.19) was evaluated by using the AFM images in Fig. 6.28a, b, and c. The results are summarized in

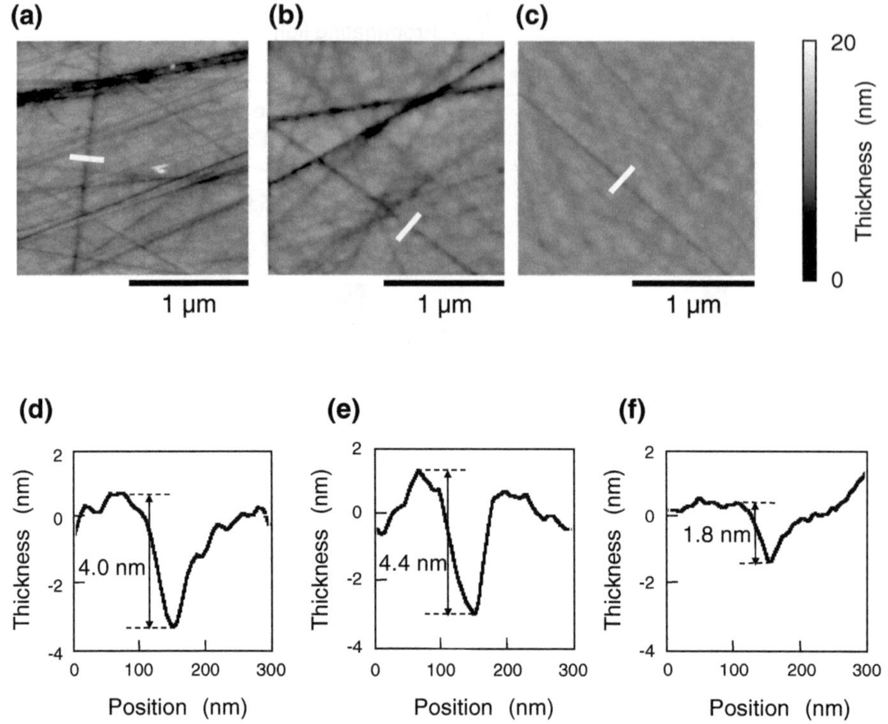

Fig. 6.28 Results of repairing the scratches on a poly-Al_2O_3 crystal surface. **a** Scanning electron microscope (SEM) images of the surface before the sputtering. **b**, **c** SEM images after the conventional sputtering and after the sputtering with light irradiation, respectively. **d**–**f** Cross-sectional profiles taken along *white lines* in **a**–**c**, respectively

Table 6.4. Moreover, in order to identify specific linear patterns of the scratches from the images, a Hough transform was employed, and the depths of the scratches were evaluated [43]. The results are also shown in Table 6.4. By comparing the numerical values in this table, a remarkable decrease in roughness can be seen after the sputtering with light irradiation.

6.3.3 Other Related Methods

In addition to the etching and deposition described in the previous subsections, several derivative methods have been developed:

(1) Size control of semiconductor nanoparticles: Etching and deposition using DPPs suggest the possibility of controlling the surface profiles of materials. Based on this suggestion, size control of semiconductor ZnO nanoparticles has been realized by irradiating them with light while they are grown in a solution by

Table 6.4 Depth of the scratches and surface roughness before and after the sputtering

	Before sputtering	After sputtering (without light irradiation)	After sputtering (with light irradiation)
Depth evaluated from Fig. 6.28d–f	4.0 nm	4.4 nm	1.8 nm
Surface roughness R_a	1.3 nm	1.1 nm	0.5 nm
Depths evaluated by Hough transform	3.2 nm	3.8 nm	0.8 nm

the sol–gel method [44]. A decrease in size fluctuations has been confirmed by the measurements of the particle shapes and sizes using transmission electron microscopy and photoluminescence spectroscopy. This method has also been applied to control the sizes and mole-fractional ratios of $(AgIn)_xZn_{2(1-x)}S_2$ nanoparticles [45].

(2) Homogenization of the mole-fractional ratios in composite semiconductors: Since DPPs can be localized at impurity sites, as was described in Chap. 4, spatial inhomogeneity of the mole-fractional ratio in a composite semiconductor can be improved by using DPPs. As an example, the spatial distribution of the mole fractional ratio, x, of an $In_xGa_{1-x}N$ substrate, commonly used for fabricating light emitting diodes, has been homogenized by irradiating the surface of the $In_xGa_{1-x}N$ film with light while it is grown by chemical vapor deposition [46]. The photoluminescence spectral width decreased after this irradiation, from which the homogenization was confirmed.

References

1. T. Kawazoe, Y. Yamamoto, M. Ohtsu, Appl. Phys. Lett. **79**, 1184 (2004)
2. T. Kawazoe, K. Kobayashi, S. Takubo, M. Ohtsu, J. Chem. Phys. **122**, 024715 (2005)
3. T. Kawazoe, K. Kobayashi, M. Ohtsu, Appl. Phys. B **84**, 247 (2006)
4. V. Polonski, Y. Yamamoto, M. Kourogi, H. Fukuda, M. Ohtsu, J. Microscopy **194**, 545 (1999)
5. Y. Yamamoto, M. Kourogi, M. Ohtsu, G.H. Lee, T. Kawazoe, IEICE Trans. Electron. **E85-C**, 2081 (2002)
6. T. Yatsui, T. Kawazoe, M. Ueda, Y. Yamamoto, M. Kourogi, M. Ohtsu, Appl. Phys. Lett. **81**, 3651 (2002)
7. T. Yatsui, K. Nakanishi, K. Kitamura, M. Ohtsu, Appl. Phys. B **107**, 673 (2012)
8. S. Yamazaki, T. Yatsui, M. Ohtsu, T.W. Kim, H. Fujioka, Appl. Phys. Lett. **85**, 3059 (2004)
9. H. Yonemitsu, T. Kawazoe, K. Kobayashi, M. Ohtsu, J. Photoluminescence **122–123**, 230 (2007)
10. T. Kawazoe, M. Ohtsu, in *Nanophotonics and Nanofabrication*, ed. by M. Ohtsu, Nanofabrication Principles, Practice (Wiley-VCH, Weinheim, 2009), pp. 17–34
11. T. Kawazoe, K. Kobayashi, K. Akahane, M. Naruse, N. Yamamoto, M. Ohtsu, Appl. Phys. B **84**, 243 (2006)
12. T. Kawazoe, T. Takahashi, M. Ohtsu, Appl. Phys. B **98**, 5 (2010)
13. M. Koike, S. Miyauchi, K. Sano, T. Imazono, *in Nanophotonics and Nanofabrication*, ed. by M. Ohtsu, X-ray Devices and the Possibility of Applying Nanophotonics (Wiley-VCH, Weinheim, 2009), pp. 179–192

14. Y. Inao, S. Nakasato, R. Kuroda, M. Ohtsu, Microelectron. Eng. **84**, 705 (2007)
15. R. Kuroda, Y. Inao, S. Nakazato, T. Ito, T. Yamaguchi, T. Yamada, A. Terao, N. Mizutani, *in Nanophotonics and Nanofabrication*, ed. by M. Ohtsu, Lithography by Nanophotonics (Wiley-VCH, Weinheim, 2009), pp. 131–146
16. The Energy Conservation Center (ed.), *Handbook of Energy Conservation 2010* (The Energy Conservation Center, Tokyo, 2010), pp. 150–351
17. L.M. Cook, J. Non-Cryst, Solids **120**, 152 (1990)
18. V.K. Jain, Mach. Sci. Technol. **12**, 257 (2008)
19. B. Wua, A. Kumar, J. Vac. Sci. Technol. B **25**, 1743 (2007)
20. T. Yatsui, K. Hirata, W. Nomura, Y. Tabata, M. Ohtsu, Appl. Phys. B **93**, 55 (2008)
21. T. Izawa, N. Inagaki, Proc. IEEE **68**, 1184 (1980)
22. T. Yatsui, K. Hirata, Y. Tabata, W. Nomura, T. Kawazoe, M. Naruse, M. Ohtsu, Nanotechnology **21**, 355303 (2010)
23. M. Naruse, T. Yatsui, W. Nomura, K. Hirata, Y. Tabata, M. Ohtsu, J. Appl. Phys. **105**, 063516 (2009)
24. K. Hirata, Proc. SPIE **7921**, 79210M (2011)
25. W. Nomura, T. Kawazoe, T. Yatsui, M. Naruse, N. Tabata, K. Hirata, M. Haraguchi, M. Ohtsu, Extended Abstracts (The 58th Spring Meeting, 2011) (The Jpn. Soc. Appl. Phys., Tokyo, 2011) paper number 24p-KF-10
26. T. Yatsui, *Nanophotonic Fabrication* (Springer, Berlin, 2012), p. 79
27. T. Yatsui, K. Hirata, Y. Tabata, Y. Miyake, Y. Akita, M. Yoshimoto, W. Nomura, T. Kawazoe, M. Naruse, M. Ohtsu, Appl. Phys. B **103**, 527 (2011)
28. W. Nomura, T. Yatsui, T. Kawazoe, M. Ohtsu, Extended Abstracts (The 59th Spring Meeting, 2012) (The Jpn. Soc. Appl. Phys., Tokyo, 2012) paper number 17p–B11-14
29. R. Teki, A. John Kadaksham, M. House, J. Harris-Jones, A. Ma, S.V. Babu, A. Hariprasad, P. Dumas, R. Jenkins, J. Provine, A. Richmann, J. Stowers, S. Meyers, U. Dietze, T. Kusumoto, T. Yatsui, M. Ohtsu, Proc. SPIE **8322**, 83220B (2012)
30. M. Naruse, T. Yatsui, W. Nomura, T. Kawazoe, M. Aida, M. Ohtsu, Appl. Phys. Lett. **102**, 071603 (2013)
31. T. Yatsui, W. Nomura, M. Naruse, M. Ohtsu, J. Phys. D **45**, 475302 (2012)
32. T. Morimoto, K. Hirata, N. Tabata, W. Nomura, T. Kawazoe, T. Yatsui, M. Ohtsu, Extended Abstracts (The 59th Spring Meeting, 2012) (The Jpn. Soc. Appl. Phys., Tokyo, 2012) paper number 17p–B11-13
33. W. Nomura, T. Yatsui, Y. Yanase, K. Suzuki, M. Fujita, A. Kamata, M. Naruse, M. Ohtsu, Appl. Phys. B **99**, 75 (2010)
34. A. Ikesue, I. Furusato, J. Am. Ceram. Soc. **78**, 225 (1995)
35. J. Lu, J. Son, M. Prabhu, J. Xu, K. Ueda, H. Yagi, T. Ynagitani, A. Kudryashov, Jpn. J. Appl. Phys. **39**, L1048 (2000)
36. N. Tanaka, Bull. Ceram. Soc. Jpn. **38**, 967 (2003)
37. A. Krell, P. Blank, H. Ma, T. Hutzler, J. Am. Ceram. Soc. **86**, 12 (2003)
38. N. Pavel, M. Tsunekane, T. Taira, Opt. Express **19**, 9378 (2011)
39. D. Graham-Rowe, Nat. Photon. **2**, 515 (2008)
40. F.J. Himpsel, J.E. Ortega, G.J. Mankey, R.F. Willis, Adv. Phys. **47**, 511 (1998)
41. E.J. Menke, Q. Li, R.M. Penner, Nano Lett. **4**, 2009 (2004)
42. F. Benabid, M. Notcutt, V. Loriette, L. Ju, D.G. Blair, J. Phys. D **33**, 589 (2000)
43. R.O. Duda, P.E. Hart, Commun. ACM **15**, 11 (1972)
44. Y. Liu, T. Morishima, T. Yatsui, T. Kawazoe, M. Ohtsu, Nanotechnology **22**, 215605 (2011)
45. F. Morigaki, T. Yatsui, T. Kawazoe, T. Torimoto, M. Ohtsu, Extended Abstracts (The 59th Spring Meeting, 2012) (The Jpn. Soc. Appl. Phys., Tokyo, 2012) paper number 17p–B11-10
46. T. Yatsui, S. Yamazaki, K. Ito, H. Kawamura, M. Mizumura, T. Kawazoe, M. Ohtsu, Appl. Phys. B **92**, 375 (2009)

Chapter 7
Energy Conversion Using Dressed Photons

Ab uno disce omnes.
Publius Vergilius Maro, *Aeneis,* II, 66

Chapter 4 described the localization of a dressed-photon–phonon (DPP) at the apex of a nanomaterial or at an impurity atom site. The present chapter reviews the application of this localization to energy conversion. In particular, Sects. 7.2 and 7.3 describe energy conversion devices, fabricated by using DPPs, in which the energies and spatial distribution of the DPPs are autonomously optimized to achieve the highest energy conversion efficiency. Although this autonomous DPP generation is an opposite mechanism to the autonomous DPP annihilation described in Sect. 6.3, they are both based on the principles outlined in Chap. 4.

7.1 Conversion From Optical to Optical Energy

As a first example of energy conversion, this section discusses the conversion from optical to optical energy. In this case, up-conversion of optical energy, i.e., optical frequency up-conversion, is possible. This conversion is described here by using organic dye particles as an example specimen.

In this process, short-wavelength propagating light is generated by irradiating the organic dye particles with long-wavelength propagating light. Although the fluorescence from organic dye particles is known to involve the spontaneous emission of propagating light, it is an optical frequency down-conversion because long-wavelength propagating light is generated by absorbing the short-wavelength propagating light. In contrast to this, optical frequency up-conversion is possible if the organic dye particles are excited by the absorption process illustrated in Fig. 4.12a and Table 4.2. Namely, by irradiating the aggregated dye particles with infrared propagating light, DPPs are generated at the apexes of their bumps, as schematically shown in Fig. 7.1. Via energy transfer to an adjacent dye particle mediated by this DPP,

M. Ohtsu, *Dressed Photons*, Nano-Optics and Nanophotonics,
DOI: 10.1007/978-3-642-39569-7_7,

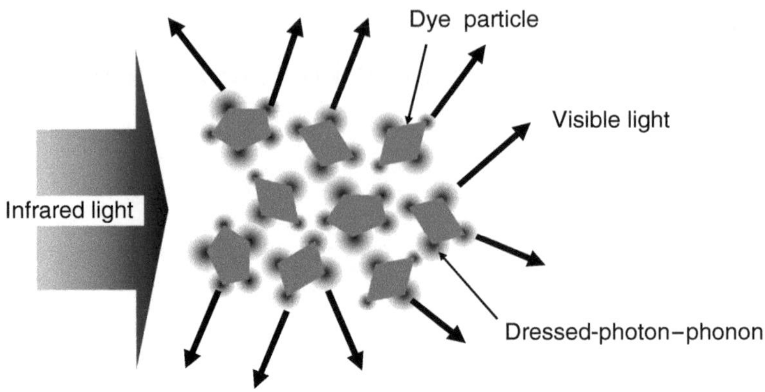

Fig. 7.1 Visible light emission from dye particles by irradiating them with infrared light

an electron in the adjacent dye molecule is excited by absorption via vibrational states, which are the modulation sidebands accompanying the electronic state. Since this excited state is an electric dipole-allowed state, spontaneous emission is possible, generating propagating light by de-excitation to the electronic ground state. As a result, propagating light is generated in the first step of the process shown in Fig. 4.12b and the right part of Table 4.3, and this light can be observed in the far field. The emission efficiency is high because this de-excitation process is equivalent to that for the above-mentioned fluorescence.

Up-converted light emission has been confirmed by irradiating a variety of dye particles with infrared light, as summarized in Table 7.1 [1, 2]. Here, the specimens were fabricated by dispersing the dye molecules in an organic solvent and coating

Table 7.1 Dye molecules used for optical frequency up-conversion

Name of dye	Coumarin 480	Coumarin 540A	DCM
Shape of the particle (Average size)	Rod (diameter, 2 μm; length, 50 μm) Fig. 7.2a	Granule (diameter, 5 μm) Fig. 7.2b	Granule (diameter, 10 nm to several μm) Fig. 7.2c
Thickness of specimen	1 mm	1 mm	100 μm
Measured photon energy of emitted light$^{(a)}$ (wavelength) [Photon energy of the infrared light used for excitation$^{(b)}$ (wavelength)]	2.25–2.76 eV (450–550 nm) {2.67 eV$^{(a)}$ at the emission peak at 465 nm} [1.53 eV$^{(b)}$ (808 nm)]	2.07–2.48 eV (500–600 nm) {2.36 eV$^{(a)}$ at the emission peak at 525 nm} [1.53 eV$^{(b)}$ (808 nm)]	Lower than 1.80–1.91 eV$^{(a)}$ (650 nm to >690 nm) [1.54 eV$^{(b)}$ (805 nm)]
Amount of energy up-conversion (difference between *a* and *b*)	1.14 eV	0.83 eV	0.37 eV

Fig. 7.2 Scanning electron microscope images of dye particles. **a** Coumarin 480. **b** Coumarin 540A. **c** DCM

aggregated dye particles on the inner wall of a quartz glass container. In addition to the experimental results in Table 7.1, red light emission at wavelengths of 600–680 nm from rhodamine (Rh)-6G dye particles and blue light emission at a wavelength of 460 nm from stilbene 420 dye particles have also been confirmed.

Figure 7.2 shows scanning electron microscope (SEM) images of these dye particles. The spots of red, green, and blue light from the DCM, coumarin 540A, and stilbene 420 dye particles are shown in Fig. 7.3a–c, respectively, showing that the three primary colors could be obtained from infrared light. Figure 7.4a shows an optical microscope image of the light emitted from the DCM dye particles. Many small bright spots can be seen in this figure because the light is generated only at the bumps of the dye particles. For comparison, Fig. 7.4b shows an optical microscope image of the conventional fluorescence generated when the dye particles are excited by short-wavelength light. The brightness of the image is spatially homogeneous because the fluorescence is emitted from the entirety of the dye particles.

The solid curve in Fig. 7.5a shows the spectrum of the blue light emitted from the coumarin 480 dye particles when excited by 808 nm-wavelength infrared light. The maximum emitted light intensity is at a wavelength of 465 nm. The amount of energy up-conversion is 1.14 eV, which is the difference between the photon energies of the emitted light and the infrared light used for excitation. For comparison, the broken curve represents the conventional fluorescence spectrum when the dye particles are excited by ultraviolet light. The peak wavelength here is also 465 nm. In a similar way, the solid curve in Fig. 7.5b shows the spectrum of the blue light emitted from

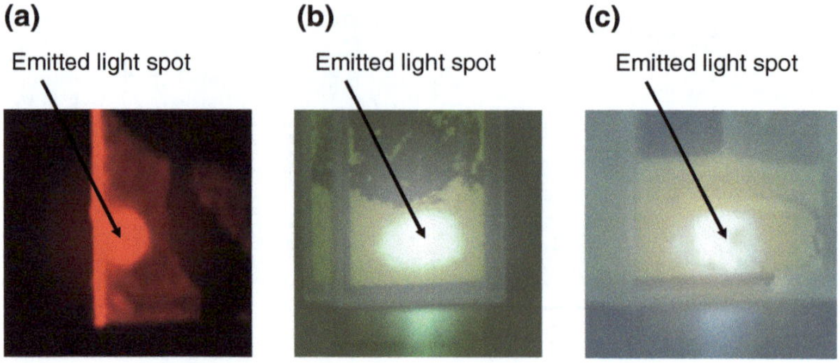

Fig. 7.3 Photographs of spots of visible light emitted from the dye particles. **a** *Red light* from DCM. **b** *Green light* from coumarine 540A. **c** *Blue light* from stilbene 420. (Images **b** and **c** are courtesy of Dr. H. Fujiwara, Hamamatsu Photonics, Co. Ltd.)

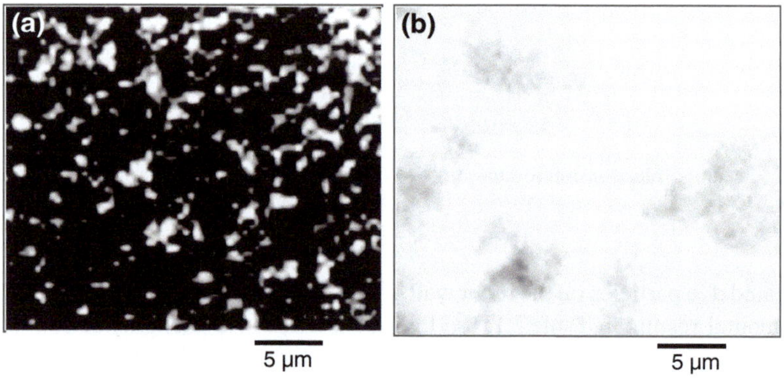

Fig. 7.4 An optical microscope images of the light emitted from the DCM dye particles. **a** Under irradiation with infrared light. **b** Fluorescence image under irradiation with short-wavelength light

the coumarin 540A dye particles. The peak emission wavelength is 525 nm, and the amount of energy up-conversion is 0.83 eV. This peak wavelength is 54 nm shorter than that of the fluorescence spectrum of the broken curve. The reason for this is still under study. Figure 7.5c shows the spectrum of the blue light emitted from the DCM dye particles. Although this figure covers the wavelength range of 650–690 nm, emission has been confirmed also in a longer wavelength range. The amount of energy up-conversion is 0.37 eV. The peak wavelength of the conventional fluorescence spectrum is 650 nm.

The experimental results reviewed above show that the peak wavelengths of the novel up-converted light emission by infrared excitation and the conventional fluorescence are equal to each other, confirming that the initial energy states of electrons for these emissions are the same. Furthermore, it has been confirmed that the up-converted emission originated from the de-excitation from the electronic excited state ($|E_{ex}; el\rangle \otimes |E_{em}; vib\rangle$ in the last row of Table 7.3 (to be described later)) to

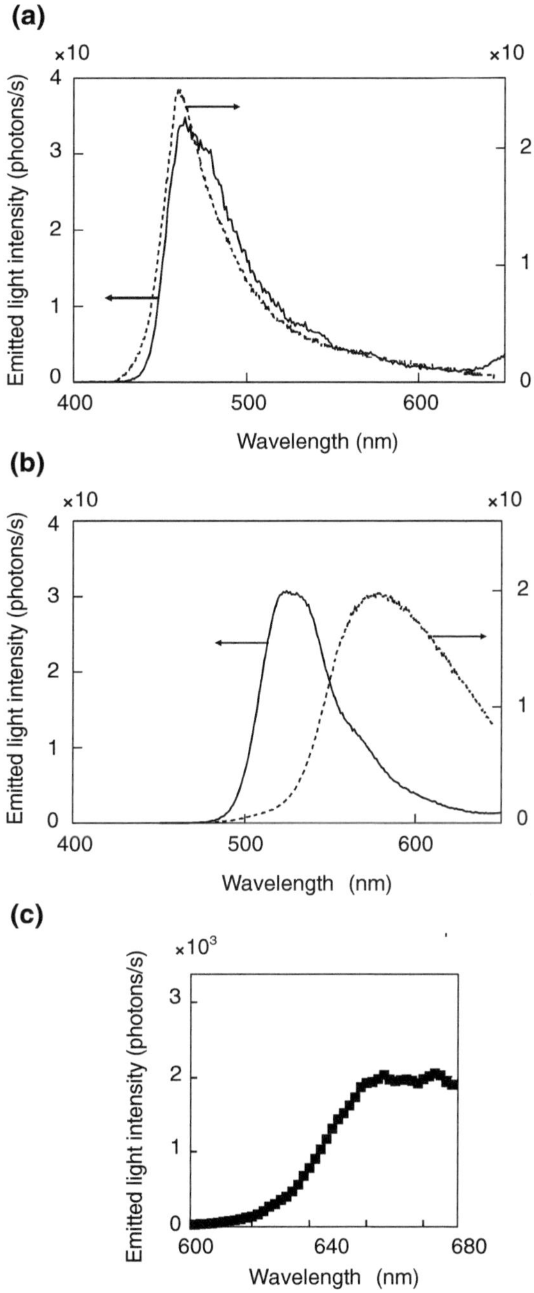

Fig. 7.5 Spectrum of *blue light* emitted from the dye particles excited by infrared light. **a** Coumarin 480. **b** Coumarin 540A. **c** DCM. The *broken curves* in **a** and **b** represent the fluorescence spectra when excited by ultraviolet light

the ground state. This is because the intensities of the up-converted emission and fluorescence decayed with two time constants, $\tau_1 = 0.45$ ns and $\tau_2 = 1.37$ ns, under pulsed light excitation [3, 4]. However, the infrared photon energy for excitation is much lower than that of the short-wavelength light used for fluorescence, which suggests that the up-conversion originated from the multi-step excitation of electrons in the dye particles, mediated by the DPPs. The amounts of up-conversion of the photon energy in the three dye particles in Table 7.1 are 0.37–1.14 eV, confirming that the up-conversion does not originate from thermal effects because these amounts correspond to thermal energies at temperatures as high as 1.32×10^3 to 4.29×10^3 K.

7.1.1 Multi-Step Excitation

(a) Three-step excitation

Three-step excitation is required for the coumarin 480 and 540A because the amounts of energy up-conversion are as high as 1.14 and 0.83 eV, respectively. Figure 7.6 shows the measured relation between the excitation light intensity I_{ex} and the emitted light intensity I_{em}. In the case of coumarin 480 (closed circles in this figure), I_{em} is the intensity at a wavelength of 460 nm (photon energy, 2.70 eV). For the coumarin 540A (closed squares), I_{em} is the intensity at a wavelength of 520 nm (photon energy, 2.38 eV). In the three-step excitation, the dependence of I_{em} on I_{ex} is expressed by the third-order polynomial function

$$I_{em} = a_{\omega_p} I_{ex} + b_{\omega_p} I_{ex}^2 + c_{\omega_p} I_{ex}^3, \tag{7.1}$$

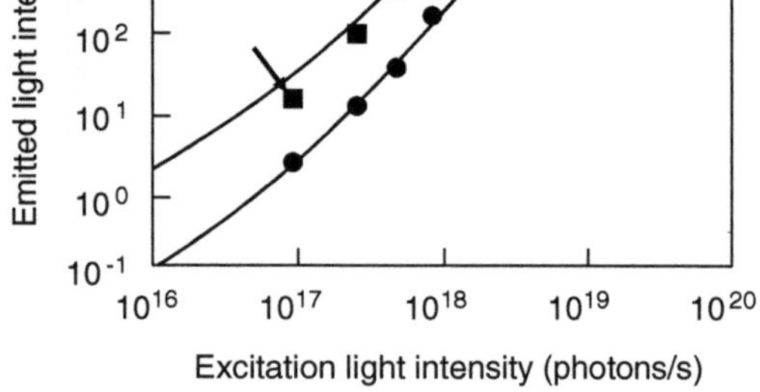

Fig. 7.6 Relation between the excitation light intensity I_{ex} and the emitted light intensity I_{em}. *Closed circles and squares* represent the results for coumarin 480 and coumarin 540A, respectively. The *arrow* represents the measured values of $I_{ex} = 9.2 \times 10^{16}$ photon/s and $I_{em} = 14.8$ photons/s

which is equivalent to Eq. (6.13) in Chap. 6. The values of the coefficients used for least-squares fitting to the measured values are

$$
\begin{aligned}
a_{2.70} &= (1.19 \pm 0.85) \times 10^{-17}, \\
b_{2.70} &= (1.71 \pm 0.10) \times 10^{-34}, \\
c_{2.70} &= (3.51 \pm 1.43) \times 10^{-54}
\end{aligned}
\tag{7.2}
$$

for coumarin 480, and

$$
\begin{aligned}
a_{2.38} &= (2.09 \pm 0.20) \times 10^{-16}, \\
b_{2.38} &= (1.42 \pm 0.02) \times 10^{-33}, \\
c_{2.38} &= (9.20 \pm 0.39) \times 10^{-53}
\end{aligned}
\tag{7.3}
$$

for coumarin 540A. Solid curves in this figure represent Eq. (7.1) drawn by using these values.

Equations (7.2) and (7.3) show that the values of the ratio $c_{\omega_p}/b_{\omega_p}$ are 2.1×10^{-20} and 6.5×10^{-20}, respectively. Theoretical expressions for these coefficients are given by Eqs. (6.14b) and (6.14c) in Chap. 6, and their ratio is given by Eq. (6.17). Here, the measured value of the coefficient a_{ω_p} to be inserted into the right-hand side of Eq. (6.17) must be replaced with the ratio I_{em}/I_{ex} by noting that I_{em} and I_{ex} are both represented by the photon number flux incident on the dye particles. Thus, the ratio is expressed as

$$
\frac{c_{\omega_p}}{b_{\omega_p}} = \frac{\hbar}{2\pi}\left(\frac{p^{el}}{p^{vib}}\right)^2 \frac{1}{\gamma_{vib}^2}\frac{I_{em}}{I_{ex}}. \tag{7.4}
$$

In order to compare with the experimental results for coumarin 540A, the measured values I_{ex}= 9.2×10^{16} s^{-1} and I_{em}= 14.8 s^{-1} (identified by an arrow in Fig. 7.6) are inserted into Eq. (7.4). Other values are similar to those for Eq. (6.17), i.e., p^{el}/p^{vib}=1×10^{-4} and γ_{vib}=0.1 eV. As a result, one obtains $c_{\omega_p}/b_{\omega_p}$=$1.1\times10^{-20}$, which agrees with the measured values given above. One may find that the values of a_{ω_p} in Eqs. (7.2) and (7.3) are larger than the theoretical value, which is due to the low accuracy of measuring the low emitted light intensity $a_{\omega_p} I_{ex}$.

Figure 7.7 shows the three-step process of exciting an electron in the dye particle to the excited state for light emission, which corresponds to Fig. 4.12a in Chap. 4. Here, $|E_\alpha; el\rangle$ and $|E_\beta; vib\rangle$ are the electronic state and vibrational state of the dye molecule, respectively. E_α is the eigenenergy of the electronic state, where $\alpha = g$ and $\alpha = ex$ respectively represent the ground and excited states. E_β is the eigenenergy of the vibrational state, where β specifies the relevant vibrational level (i, a, b, c, em). Since Fig. 4.12 in Chap. 4 illustrates a semiconductor material, the state $|E_\beta; phonon\rangle$ of the phonon in Fig. 4.12 corresponds to the vibrational state $|E_\beta; vib\rangle$ of the organic molecules in Fig. 7.7. Transitions and relaxations in each step are summarized in Table 7.2, in which the relaxations to the thermal equilibrium state after the first- and

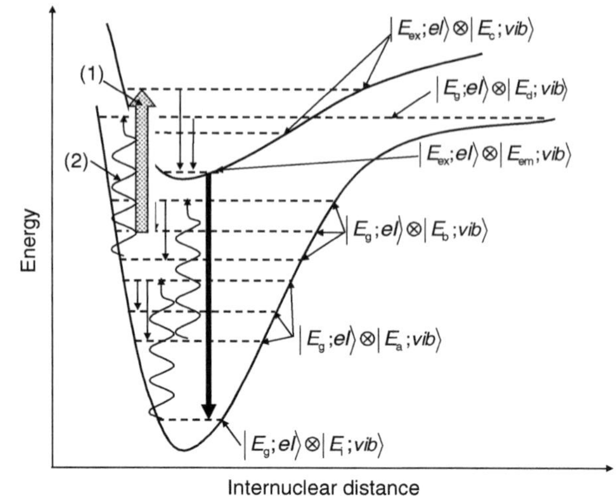

Fig. 7.7 *Three-step* process of excitation. *Upward thick arrow (1)* and *wavy arrow (2)* represent excitation by electric dipole-allowed and -forbidden transitions, respectively. The *downward thin and thick arrows* respectively represent the relaxation and light emission

second-step excitations are also represented. Although Table 4.2 of Chap. 4 did not represent these relaxations, it has been confirmed by pump–probe spectroscopy of the DCM dye particles that the ratio of the probabilities of the next-step excitation after relaxation and without relaxation was 1:1 [5].[1] Based on this result, it is appropriate to insert the relaxation process in this table. Here, it should be noted that there are two route s in the third step.

After the electron reaches the excited state $|E_{ex}; el\rangle \otimes |E_{em}; vib\rangle$ by the three-step excitation of Table 7.2, propagating light is emitted as a result of the electric dipole-allowed transition (identified by a downward thick arrow in Fig. 7.7) to the ground state $|E_g; el\rangle \otimes |E_i; vib\rangle$. This spontaneous emission process is equivalent to the emission of fluorescence.

Route 1 in the third step is an electric dipole-allowed transition whose transition probability is 10^6 times greater than that of the electric dipole-forbidden transition [1, 6]. Therefore, the emission probability by the excitation through Route 1 is governed by the probability of the electric dipole-forbidden transition, and thus, I_{em} is proportional to I_{em}^2. On the other hand, the probability of the electric dipole-forbidden transition of Route 2 is equal to those of the electric dipole-forbidden transitions in the first and second steps. As a result, I_{em} by Route 2 is proportional to I_{ex}^3. Based on the discussion given above, it is appropriate to express I_{em} by Eq. (7.1).

(b) Two-step excitation

[1] As will be given by Eqs. (7.11) and (7.12) later, the temporal evolution of the emitted light intensity has two components with different lifetimes. Among them, the shorter lifetime corresponds to that of the intermediate state. The longer one is the lifetime of the state through which the molecule passes in the process of relaxation to the thermal equilibrium state. The energy of this state is sufficiently high to excite the electron to the excited state by the second-step excitation. By comparing the emitted light intensity components of longer and shorter lifetimes, the probability of the next-step excitation after and without relaxation can be estimated. The ratio of the probabilities given in the text was obtained by this comparison.

Table 7.2 Routes of the three-step excitation

First step	Initial state: Ground state $\lvert E_g; el\rangle \otimes \lvert E_i; vib\rangle$ ⇓ Electric dipole-forbidden transition (upward wavy arrow in Fig. 7.7) Vibrational excited state $\lvert E_g; el\rangle \otimes \lvert E_a; vib\rangle$	
	⇓ Relaxation to the thermal equilibrium state[1] (downward thin arrows in Fig. 7.7) Intermediate state: Vibrational excited state whose eigenenergy is slightly lower than that of $\lvert E_g; el\rangle \otimes \lvert E_a; vib\rangle$	
Second step	⇓ Electric dipole-forbidden transition (upward wavy arrow in Fig. 7.7) Higher vibrational excited state $\lvert E_g; el\rangle \otimes \lvert E_b; vib\rangle$	
	⇓ Relaxation to the thermal equilibrium state (downward thin arrow in Fig. 7.7) Intermediate state: Vibrational excited state whose eigenenergy is slightly lower than that of $\lvert E_g; el\rangle \otimes \lvert E_b; vib\rangle$	
Third step[2]	Route 1	Route 2
	⇓ Electric dipole-forbidden transition (upward thick arrow (1) in Fig. 7.7) Electronic excited state $\lvert E_g; el\rangle \otimes \lvert E_c; vib\rangle$	⇓ Electric dipole-forbidden transition (upward wavy arrow (2) in Fig. 7.7) Higher excited state $\lvert E_g; el\rangle \otimes \lvert E_d; vib\rangle$
	⇓ Relaxation by cooling towards the thermal equilibrium state or by electron–phonon coupling[3] (downward thin arrow in Fig. 7.7) Final state: The initial state $\lvert E_{ex}; el\rangle \otimes \lvert E_{em}; vib\rangle$ for light emission	

a The experimentally measured rate of this relaxation is 100 meV/ps [7].
b Routes 1 and 2 are both possible in the third-step excitation if the eigenenergy of the vibrational level is larger than the amount of energy up-conversion in Table 7.1, even after relaxation to the thermal equilibrium state.
c Refer to references [8, 9]

It is reasonable to assume that the light emission from DCM particles originates from the two-step excitation because the amount of energy up-conversion in Table 7.1 (0.37 eV; the difference between the photon energies of the emitted green light and the excitation light) is much smaller than those of coumarin 480 and 540A. Closed squares and circles in Fig. 7.8 represent the measured emitted light intensity at a wavelength of 650 nm (1.91 eV) and 690 nm (1.80 eV), respectively, as a function of the excitation light intensity with a wavelength of 805 nm (1.54 eV). These measured values are fitted by the quadratic function

$$I_{em} = a_{\omega_p} I_{ex} + b_{\omega_p} I_{ex}^2, \tag{7.5}$$

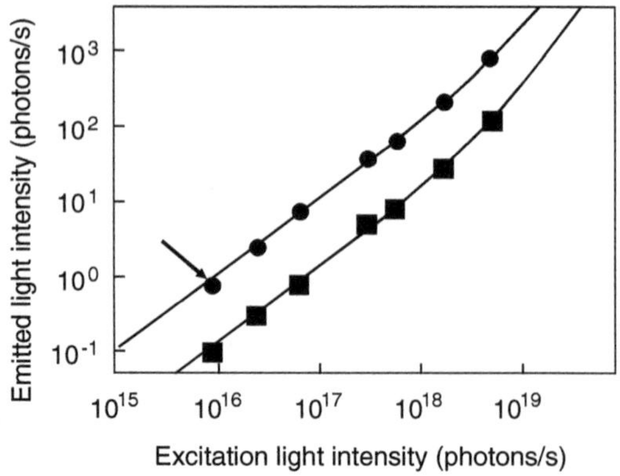

Fig. 7.8 Relation between the excitation light intensity I_{ex} and the emitted light intensity I_{em} in the case of DCM. The *closed squares and circles* represent the emitted light intensity at wavelengths of 650 and 690 nm, respectively. The *arrow* represents the measured values of $I_{ex(1.80)} = 8.9\times10^{15}$ photon/s and $I_{em(1.80)} = 0.75$ photons/s

in which the values of the coefficients used for fitting are

$$a_{1.91} = (1.37 \pm 0.33) \times 10^{-17}, b_{1.91} = (2.61 \pm 0.92) \times 10^{-36}, \tag{7.6a}$$

$$a_{1.80} = (1.12 \pm 0.09) \times 10^{-16}, b_{1.80} = (1.17 \pm 0.25) \times 10^{-35}. \tag{7.6b}$$

Solid curves in this figure represent the fitted results.

From Eqs. (7.6a) and (7.6b), the values of the ratio are 1.9×10^{-19} and 1.0×10^{-19}, respectively. On the other hand, these theoretical coefficients have been given by Eqs. (6.14a) and (6.14b) in Chap. 6, respectively, and their ratio is given by Eq. (6.17). Based on the same consideration as given in relation to Eq. (7.4), the ratio is finally expressed as

$$\frac{b_{\omega_p}}{a_{\omega_p}} = \frac{\hbar}{2\pi}\left(\frac{p^{el}}{p^{vib}}\right)^2 \frac{1}{\gamma_{vib}^2}\frac{I_{em}}{I_{ex}}. \tag{7.7}$$

In order to compare with the measured value, the values $I_{ex(1.80)}$=8.9×10^{15} s^{-1} and $I_{em(1.80)}$=0.75 s^{-1} (identified by the arrow in Fig. 7.8) are inserted into Eq. (7.7), while other values are set to p^{el}/p^{vib}=1×10^{-4} and γ_{vib}=0.1 eV. As a result, one obtains $b_{\omega_p}/a_{\omega_p}$=$0.5\times10^{-19}$, which agrees with the measured values given above.

Figure 7.9 and Table 7.3 summarize the origin of the visible light emission by the two-step excitation , including the relaxation to the thermal equilibrium state after the first-step excitation. Although this relaxation is not seen in the first step in Table 4.2 in Chap. 4, it is inserted in Fig. 7.9 and Table 7.3 for the same reason as in the case of Table 7.2. It should be noted that the second-step excitation has two routes.

After this two-step excitation, propagating light is emitted by the electric dipole-allowed transition (downward thick arrow in Fig. 7.9) from the electronic excited state $|E_{ex}; el\rangle \otimes |E_{em}; vib\rangle$ to the ground state $\left|E_g; el\right\rangle \otimes |E_i; vib\rangle$. This spontaneous emission process is equivalent to the emission of fluorescence.

The reason why I_{em} in Fig. 7.8 has components that are proportional to I_{ex} and I_{ex}^2 is that Routes 1 and 2 of the second-step excitation are electric dipole-allowed

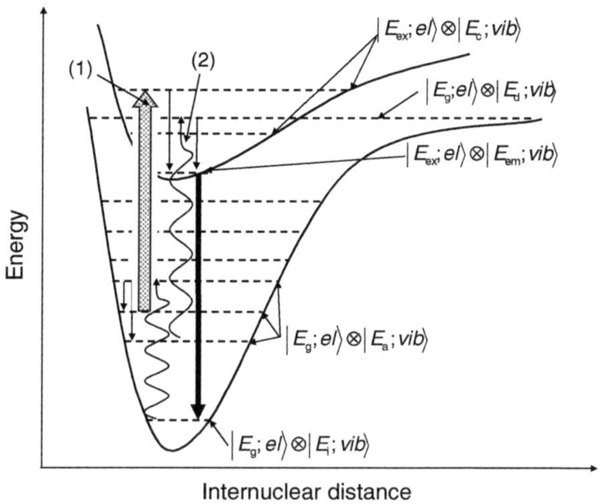

Fig. 7.9 *Two-step* process of excitation. *Upward thick arrow* (*1*) and *wavy arrow* (*2*) represent excitation by electric dipole-allowed and -forbidden transitions, respectively. The *downward thin and thick arrows* respectively represent the relaxation and light emission

and -forbidden transitions, respectively, as in the case of the above-described three-step excitation. Recent pump–probe spectroscopy experiments have succeeded in observing the emission spectra originating from the phonon energy levels involved in the multi-step excitation described here [10].

(c) Intermediate states and their lifetimes

The intermediate state in Fig. 7.7 is the vibrational excited state of the dye particle. Its lifetime represents the time constant required to relax to the thermal equilibrium state, which has been measured by pump–probe spectroscopy using a pulsed light source (pulse width, 100 fs) [2]. The lifetimes of coumarin 480 and 540A were measured to be 1.9 ps and 1.1 ps, respectively. On the other hand, by comparing the intensity of light emitted from the coumarin 480 with that emitted from the coumarin 540A under CW light excitation, their ratio was 1×10^5 even though the ratio between the excitation light intensities was as low as 3×10^3. Such a large ratio between the emitted light intensities confirms that the intermediate state is a real energy state and its lifetime is longer than the pulse width of the light source used for pump–probe spectroscopy.

Furthermore, blue light (wavelength, 460 nm) was emitted from the stilbene 420 particles by infrared excitation, and the lifetime of the intermediate state was estimated to be 2.5 ps by pump–probe spectroscopy. These measured lifetimes of the three kinds of dye particles agree with the time constants (several fs to 10 ps) of relaxation from the vibrational excited state to the thermal equilibrium state in other organic dye molecules and the semiconductor GaAs [7, 11, 12]. Therefore, the measured values presented above are confirmed to be typical lifetimes of the intermediate states involved in the up-conversion discussed here.

Since the efficiency of generating phonons, and thus, the efficiency of converting the incident photon to a DPP, is inversely proportional to the lifetime of the intermediate state, the efficiency of generating the frequency up-converted light is lower

Table 7.3 Routes of the two-step excitation

First step	Initial state: Ground state $\lvert E_g; el\rangle \otimes \lvert E_i; vib\rangle$ $\Downarrow$ Electric dipole-forbidden transition (upward wavy arrow in Fig. 7.9) Vibrational excited state $\lvert E_g; el\rangle \otimes \lvert E_a; vib\rangle$	
	$\Downarrow$ Relaxation to the thermal equilibrium state(1) (downward thin arrows in Fig. 7.9) Intermediate state: Vibrational excited state whose eigenenergy is slightly lower than that of $\lvert E_g; el\rangle \otimes \lvert E_a; vib\rangle$	
Second step(2)	Route 1	Route 2
	$\Downarrow$ Electric dipole-allowed transition (upward thick arrow (1) in Fig. 7.9) Electronic excited state $\lvert E_g; el\rangle \otimes \lvert E_c; vib\rangle$	$\Downarrow$ Electric dipole-forbidden transition (upward wavy arrow (2) in Fig. 7.9) Higher excited state $\lvert E_g; el\rangle \otimes \lvert E_d; vib\rangle$
	$\Downarrow$ Relaxation by cooling to the thermal equilibrium state or by phonon–electron coupling(3) (downward thin arrow in Fig. 7.9) Final state: The initial state for light emission $\lvert E_{ex}; el\rangle \otimes \lvert E_{em}; vib\rangle$	

a The experimentally measured rate of this relaxation is 100 meV/ps [7].
b Routes 1 and 2 are both possible in the second-step excitation if the eigenenergy of the vibrational level is larger than the amount of energy up-conversion in Table 7.1, even after relaxation to the thermal equilibrium state.
c Refer to references [8, 9]

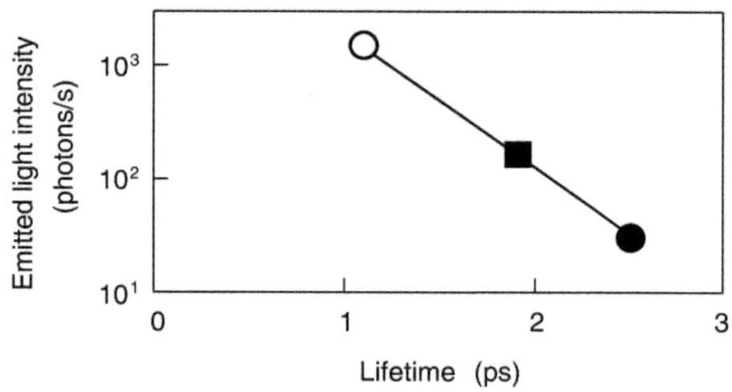

Fig. 7.10 Relation between the lifetime of the intermediate state and the emitted light intensity. *Closed circles, closed squares, and open circles* are for stilbene 420, coumarin 480, and coumarin 540A, respectively

when the lifetime of the intermediate state is longer. Figure 7.10 shows the measured results of such dependence of the generation efficiency on the lifetime of the intermediate state [2]. This figure shows the measured efficiencies of the emissions from stilbene 420, coumarin 480, and 540A under the same excitation conditions, clearly showing that the efficiency is lower for longer lifetimes.

(d) Conversion efficiency

In order to demonstrate that the efficiency of this optical frequency up-conversion is fairly high, the conversion efficiency for DCM particles, C_{DCM}, is defined as

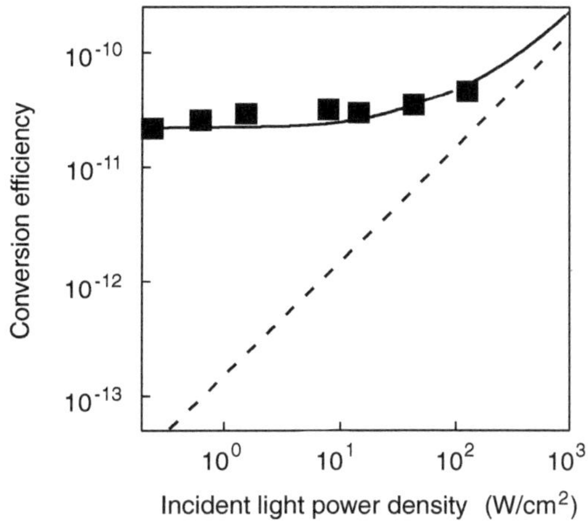

Fig. 7.11 Relation between the incident light power density and the conversion efficiency for DCM. *Broken line* represents the theoretical efficiency of second harmonic generation from a KDP crystal

the ratio between the power density P_{em} (W/cm^2) of the emitted visible light (wavelength, 660–690 nm) and the power density (W/cm^2) of the infrared light (wavelength, 805 nm) used for excitation. Closed squares in Fig. 7.11 represent the measured values of C_{DCM}. The solid curve is the theoretical value fitted by using Eq. (7.5) to the experimental values, which is expressed as

$$C_{DCM} = (2.77 \pm 0.22) \times 10^{-11} + (2.03 \pm 0.43) \times 10^{-13} P_{ex} \tag{7.8}$$

For comparison, the broken line in this figure represents the theoretical efficiency C_{KDP} of conventional second harmonic generation (SHG) by a KDP crystal having the same thickness (100 μm) as the DCM specimen [13]. This line is expressed as

$$C_{KDP} = 1.50 \times 10^{-13} P_{ex}. \tag{7.9}$$

By comparing these values, it is confirmed that C_{DCM} is larger than C_{KDP} for the whole range of the horizontal axis in Fig. 7.11. In particular, it is much larger than C_{KDP} for $P_{ex} < 100$ W/cm^2, which is due to the contribution of the coefficient a_{ω_p} in Eq. (7.5). This means that the efficiency of up-conversion is fairly high even under weak excitation.

(e) Possible applications

Although SHG has been conventionally used for optical frequency up-conversion of infrared light, a high-power coherent light source and an optical built-up cavity are required to realize sufficiently high conversion efficiency. Phosphorescence based on a multi-step transition has also been used for up-conversion [14]. However, the converted light intensity easily saturates. In order to avoid this saturation, the electron must be excited to the triplet state, requiring an additional high-power ultraviolet light source because this excitation is electric dipole-forbidden. A further problem is that the optical response is slow, which causes temporal fluctuations in the up-converted light intensity.

The present method using DPPs can solve these problems because of its very high efficiency. It should also be pointed out that the average lifetime of the intermediate state in the excited dye molecule is as short as 1.1–1.9 ps, which allows the operating wavelength regions of high-speed photodetection, imaging, and sensing devices to be expanded. Furthermore, applications to novel display systems are expected because three primary colors can be generated without using ultraviolet light or high-peak-intensity pulsed light sources [15].

7.1.2 Non-Degenerate Excitation and Applications

(a) Non-degenerate excitation and emission

This part reviews non-degenerate excitation by using DCM particles as a specimen. This excitation occurs even though the wavelengths of the propagating light for the first- and second-step excitation in Table 7.3 are different from each other.

First, in order to study the optical frequency up-conversion due to this non-degenerate excitation, the following two CW propagating light beams are used [5]:

- Signal light (wavelength $\lambda_1 = 1150\,\text{nm}$ (1.08 eV), intensity[2] $I_{ex(1.08)} = 0.55$–2.28 W/cm^2, linearly polarized): Infrared light whose wavelength is longer than that of the excitation light in Table 7.1.
- Reference light (wavelength $\lambda_2 = 808\,\text{nm}$ (1.53 eV), intensity $I_{ex(1.53)} = 2.0$–19.9 W/cm^2, elliptically polarized): Infrared light whose wavelength is nearly equal to that of the excitation light in Table 7.1.

It has been confirmed experimentally that the light intensity $I_{1.08+1.53}(\lambda)$ emitted by making these two light beams incident simultaneously on the DCM particles is larger than the sum of $I_{1.08}(\lambda_1 = 1150)$ and $I_{1.53}(\lambda_2 = 808)$ emitted by making only the signal light or reference light incident, respectively. This means that a novel excitation process was induced by making the two light beam simultaneously incident. In order to study this process, the difference in the intensity is expressed as

$$\Delta I = I_{1.08+1.53}(\lambda = 680) - [I_{1.08}(\lambda_1 = 1150) + I_{1.53}(\lambda_2 = 808)] \quad (7.10)$$

at the wavelength $\lambda = 680\,\text{nm}$. Figure 7.12a shows the relation between the signal light intensity $I_{ex(1.08)}$ and ΔI, using the reference light intensity $I_{ex(1.53)}$ as a parameter. The solid lines were obtained by least-squares fitting of linear functions to the experimental values, and their agreement with the experimental values indicates

[2] In order to express the light intensity, several technical terms are used in this section, which depend on the measurement methods employed. In Figs. 7.5, 7.6, 7.8, and 7.10, the photon number flux (s^{-1}) was used, whereas in Fig. 7.11, the power density (W/cm^2) was used. Here, in this subsection, the light power density (W/cm^2) is used in Fig. 7.12.

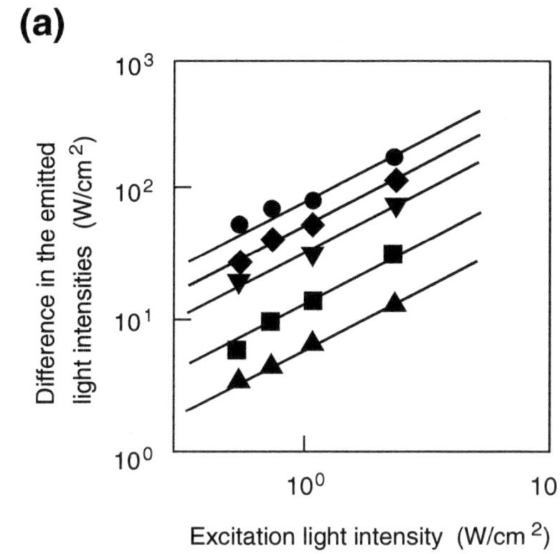

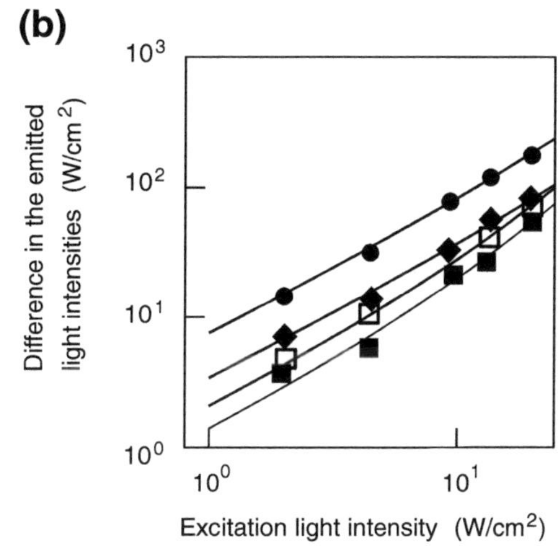

Fig. 7.12 Relation between the excitation light intensity I_{ex} and difference ΔI in the emitted light intensity in the case of DCM. **a** The signal light intensity is used for the horizontal axis. The *closed upward triangles, closed squares, closed downward triangles, closed diamonds, and closed circles* are for reference light intensities of 2.0, 4.4, 9.3, 13.6, and 19.9 mW/cm^2, respectively. **b** The reference light intensity is used for the horizontal axis. The *closed squares, open squares, closed diamonds, and closed circles* are for signal light intensities of 0.55, 0.73, 1.08, and 2.28 mW/cm^2, respectively

that one photon of the signal light contributes to the excitation. On the other hand, Fig. 7.12b shows the relation between $I_{ex(1.53)}$ and ΔI, using $I_{ex(1.08)}$ as a parameter. The solid curves were obtained by least-squares fitting of quadratic functions, as in Eq. (7.5). The results show that ΔI is approximately proportional to $I_{ex(1.53)}$ and $I^2_{ex(1.53)}$ under weak excitation and strong excitation, respectively. The components of this function that are proportional to $I_{ex(1.53)}$ and $I^2_{ex(1.53)}$ correspond to one-step and two-step electric dipole-forbidden transitions, respectively.

The excitation process of Fig. 7.12 is attributed not to the multi-photon excitation process via virtual energy states but to the multi-step excitation because the signal and reference light beams are mutually non-degenerate and also because they are low-intensity CW light. Seven types of absorption process can contribute to this

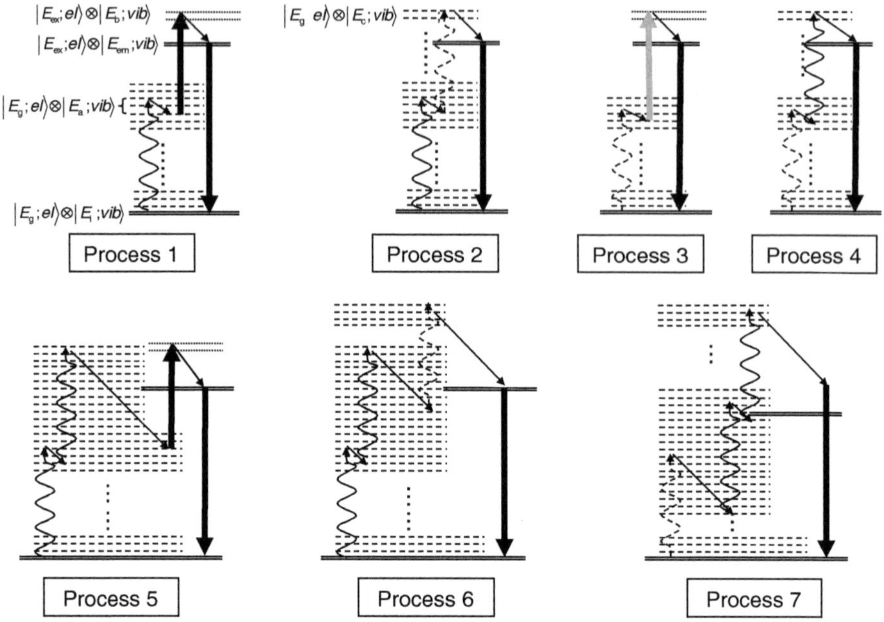

Fig. 7.13 *Seven* excitation processes. *Horizontal dotted lines and broken lines* represent the excited and ground states of the electron, respectively (compare $|E_{ex}; el\rangle \otimes |E_b; vib\rangle$ of process 1 and $|E_g; el\rangle \otimes |E_c; vib\rangle$ of process 2). *Upward solid and broken wavy arrows* respectively represent the electric dipole-forbidden excitations by the reference light (wavelength, 808 nm) and signal light (wavelength, 1150 nm). *Upward light gray and dark gray arrows* represent the electric dipole-allowed excitations by the reference light and signal light, respectively. *Downward thin and thick arrows* represent relaxation and light emission, respectively

multi-step excitation, which supports the dependence of ΔI on the excitation light intensities $I_{ex(1.53)}$ and $I_{ex(1.08)}$. They are two-step excitation s (processes 1–4) and three-step excitations (processes 5–7), as summarized in Fig. 7.13 and Table 7.4.

The value of ΔI in Fig. 7.12b is proportional to $I_{ex(1.53)}$ when the reference light intensity $I_{ex(1.53)}$ is low. This means that the emission originates from the two-step excitation processes 1, 2, and 4 in Table 7.4 because these processes include one electric dipole-forbidden transition that is excited by the reference light. By increasing the reference light intensity, the three-step excitation processes 5–7, which include two electric dipole-forbidden transitions, become dominant. As a result, a component proportional to $I^2_{ex(1.53)}$ appears. On the other hand, the value of ΔI in Fig. 7.12a is proportional to $I_{ex(1.08)}$. This means that the emission originates from the two- or three-step excitation processes 3, 4, 6, and 7 because these processes include only one electric dipole-forbidden transition that is excited by the signal light.

It has been confirmed that the value of ΔI does not depend on the difference $\Delta\theta$ in the polarization angles of the signal and reference light because of the random orientations of the DCM particles. This means that the optical frequency up-conversion reviewed in the present section does not require any polarization control of the incident light. This is a great advantage over conventional methods of optical sum–frequency conversion which are strongly influenced by the polarization of the incident light [16].

Next, the temporal evolution of the optical pulse emitted from the DCM particles due to excitation with two non-degenerate optical pulses is discussed. The widths of the optical pulses of the signal light (wavelength λ_{ex1}=1250 nm (0.99 eV), light intensity I_{ex1}=1.3 W/cm^2) and the reference light (wavelength λ_{ex2}=750 nm (1.65 eV), light intensity I_{ex2}=3.2 W/cm^2) are 0.2 and 0.1 ps, respectively.

Figure 7.14 shows the measured temporal evolution of the emitted optical pulse intensity $I_{em(1.82)}$ (Δt) at a wavelength of 680 nm (1.82 eV). Here, is the delay time of the reference light incident on the DCM particles with respect to the incident signal light. The full width at half-maximum of the emitted optical pulse is 0.8 ps. The physical origin of the light emission can be summarized as follows, depending on whether Δt <0 or Δt >0:

Δt <0: Since the reference light arrives at the DCM particle earlier than the signal light, the DCM particle emits light by processes 1, 2, 5, and 6 in Fig. 7.13 and Table 7.4.

Table 7.4 Seven excitation processes induced by two non-degenerate light beams

	Number of steps, kind of excitation light, electric dipole-forbidden or -allowed
Process 1	[First step] Reference light Forbidden: excitation to $\lvert E_g; el\rangle \otimes \lvert E_a; vib\rangle$ and subsequent relaxation to a slightly lower vibrational excitation state. [Second step] Signal light Allowed: excitation to $\lvert E_{ex}; el\rangle \otimes \lvert E_b; vib\rangle$ and subsequent relaxation to $\lvert E_{ex}; el\rangle \otimes \lvert E_{em}; vib\rangle$.
Process 2	[First step] Reference light Forbidden: excitation to $\lvert E_g; el\rangle \otimes \lvert E_a; vib\rangle$ and subsequent relaxation to a slightly lower vibrational excitation state. [Second step] Signal light Forbidden: excitation to $\lvert E_g; el\rangle \otimes \lvert E_c; vib\rangle$ and subsequent relaxation to $\lvert E_{ex}; el\rangle \otimes \lvert E_{em}; vib\rangle$.
Process 3	[First step] Signal light Forbidden: excitation to $\lvert E_g; el\rangle \otimes \lvert E_d; vib\rangle$ and subsequent relaxation to a slightly lower vibrational excitation state. [Second step] Reference light Allowed: excitation to $\lvert E_{ex}; el\rangle \otimes \lvert E_e; vib\rangle$ and subsequent relaxation to $\lvert E_{ex}; el\rangle \otimes \lvert E_{em}; vib\rangle$.
Process 4	[First step] Signal light Forbidden: excitation to $\lvert E_g; el\rangle \otimes \lvert E_d; vib\rangle$ and subsequent relaxation to a slightly lower vibrational excitation state. [Second step] Reference light Forbidden: excitation to $\lvert E_g; el\rangle \otimes \lvert E_f; vib\rangle$ and subsequent relaxation to $\lvert E_{ex}; el\rangle \otimes \lvert E_{em}; vib\rangle$.

(continued)

Table 7.4 (continued)

Process 5	[First step] Reference light Forbidden: excitation to $\left|E_g; el\right\rangle \otimes \left|E_h; vib\right\rangle$ and subsequent relaxation to a slightly lower vibrational excitation state. [Second step] Reference light Forbidden: excitation to $\left|E_g; el\right\rangle \otimes \left|E_i; vib\right\rangle$ and subsequent relaxation to a slightly lower vibrational excitation state. [Third step] Signal light Allowed: excitation to $\left|E_{ex}; el\right\rangle \otimes \left|E_j; vib\right\rangle$ and subsequent relaxation to $\left|E_{ex}; el\right\rangle \otimes \left|E_{em}; vib\right\rangle$.
Process 6	[First step] Reference light Forbidden: excitation to $\left|E_g; el\right\rangle \otimes \left|E_h; vib\right\rangle$ and subsequent relaxation to a slightly lower vibrational excitation state. [Second step] Reference light Forbidden: excitation to $\left|E_g; el\right\rangle \otimes \left|E_i; vib\right\rangle$ and subsequent relaxation to a slightly lower vibrational excitation state. [Third step] Signal light Forbidden: excitation to $\left|E_g; el\right\rangle \otimes \left|E_k; vib\right\rangle$ and subsequent relaxation to $\left|E_{ex}; el\right\rangle \otimes \left|E_{em}; vib\right\rangle$.
Process 7	[First step] Signal light Forbidden: excitation to $\left|E_g; el\right\rangle \otimes \left|E_l; vib\right\rangle$ and subsequent relaxation to a slightly lower vibrational excitation state. [Second step] Reference light Forbidden: excitation to $\left|E_g; el\right\rangle \otimes \left|E_m; vib\right\rangle$ and subsequent relaxation to a slightly lower vibrational excitation state. [Third step] Reference light Forbidden: excitation to $\left|E_g; el\right\rangle \otimes \left|E_n; vib\right\rangle$ and subsequent relaxation to $\left|E_{ex}; el\right\rangle \otimes \left|E_{em}; vib\right\rangle$.

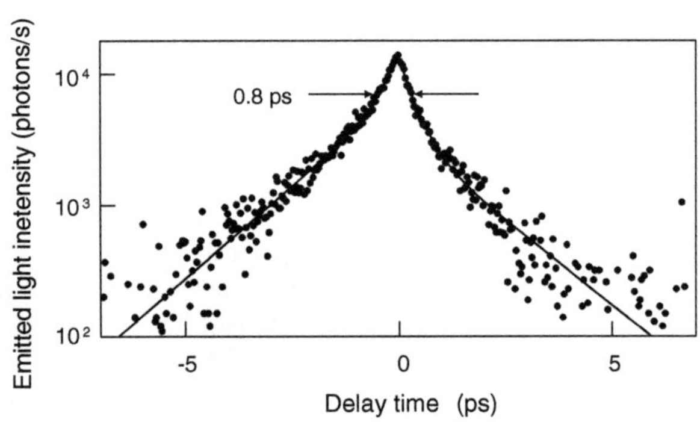

Fig. 7.14 Relation between the delay time and the emitted optical pulse intensity

$\Delta t > 0$: Since the reference light arrives at the DCM particle later than the signal light, the DCM particle emits light by processes 3, 4, and 7 in Fig. 7.13 and Table 7.4.

By least-squares fitting of exponential functions to the measured values of $I_{em(1.82)}(\Delta t)$ in Fig. 7.14, one derives

$$\Delta t < 0: \ I_{em(1.82)}(\Delta t) = A_{-}\exp\left(\Delta t/\tau_{fast}\right) + B_{-}\exp\left(\Delta t/\tau_{slow}\right)$$
$$\tau_{fast} = 0.3\text{ps}, \quad \tau_{slow} = 1.6\text{ps}, \quad B_{-}/A_{-} = 1, \tag{7.11}$$

$$\Delta t > 0: \ I_{em(1.83)}(\Delta t) = A_{+}\exp\left(-\Delta t/\tau_{fast}\right) + B_{+}\exp\left(-\Delta t/\tau_{slow}\right)$$
$$\tau_{fast} = 0.35\text{ps}, \quad \tau_{slow} = 1.7\text{ps}, \quad B_{+}/A_{+} = 1/4, \tag{7.12}$$

which are shown by the solid curves in Fig. 7.14. The two decay time constants (τ_{fast}, τ_{slow}) in these equations are due to the fact that the reference light is involved in the two-step excitation (the first step of processes 1 and 2 for $\Delta t<0$, and the second step of processes 3 and 4 for $\Delta t>0$) and the three-step excitation (the first and second steps of processes 5 and 6 for $\Delta t<0$, and the third step of process 7 for $\Delta t>0$). Since the values of the time constants τ_{fast} and τ_{slow} ($\cong$0.3–1.7 ps) are nearly equal to that of the measured lifetime of the intermediate state for the emission from coumarin 480 and coumarin 540A under excitation by degenerate optical pulses [2], these time constants are identified as the lifetimes of the intermediate states. The full width at half-maximum (0.8 ps) of the emitted optical pulse depends on these lifetimes.

The optical pulse shape shown in Fig. 7.14 is asymmetric. This is because the DCM particles accept different amounts of excitation energy depending on which light arrives earlier than the other. It is also because the lifetimes of the intermediate states are different from each other: In the case of $\Delta t>0$, the DCM particles are excited to the intermediate state by the longer-wavelength signal light, whose photon energy is lower than that of the reference light. Therefore, the lifetime of the intermediate state is shorter, and thus, the slope of the solid curve at $\Delta t>0$ is steeper than that at $\Delta t<0$.

As an additional experiment, by increasing the wavelength λ_{ex1} of the signal light to 1350 nm (0.92 eV) and the wavelength λ_{ex2} of the reference light to 775 nm (1.60 eV) (light intensities I_1=1.3 W/cm^2 and I_2=5.1 W/cm^2), the full width at half-maximum (FWHM) of the optical pulse was measured to be 1.1 ps. Least-squares fitting of Eqs. (7.11) and (7.12) to these experimental values yields

$$\Delta t < 0: \quad \tau_{fast} = 0.69\text{ps}, \quad \tau_{slow} = 3.0\text{ps}, \tag{7.13}$$

$$\Delta t > 0: \quad \tau_{fast} = 0.62\text{ps}, \quad \tau_{slow} = 5.2\text{ps}. \tag{7.14}$$

Furthermore, it was confirmed that the optical pulse shape was asymmetric and its slope was steeper at $\Delta t>0$, as was the case in Fig. 7.14.

(b) Possible applications

Non-degenerate excitation can be applied to pulse shape measurements for evaluating the temporal evolution of the signal light pulse, for which the reference light pulse is used as a time standard. The resolution of this measurement technique is as short as the FWHM (0.8–1.1 ps) of the emitted optical pulse shape, which depends on the lifetime of the intermediate state of the dye particle. Furthermore, since the longest measurable wavelength is 1250–1350 nm, this measurement can be used for diagnos-

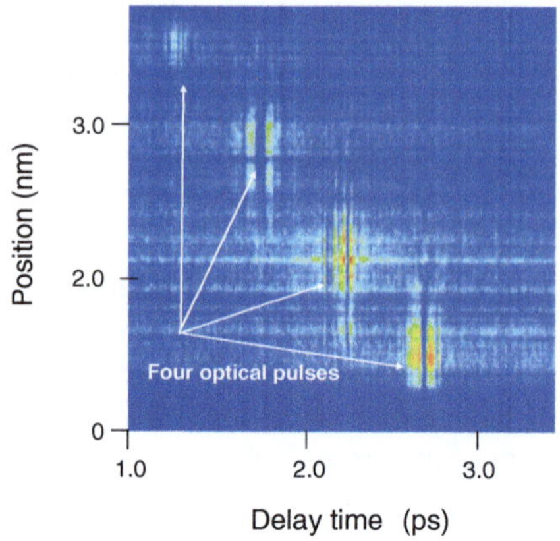

Fig. 7.15 Measured optical pulse shapes using coumarin 540A particles. (By the courtesy of Dr. H. Fujiwara, Hamamatsu Photonics, Co. Ltd.)

ing the optical pulse shapes in optical fiber communication systems. No polarizers or other additional optical elements are required for this measurement because of the lack of dependence on the polarizations of the two light beams. Furthermore, measurement is possible in a wide wavelength range because the incidence angle of the light does not depend on the wavelength.

Although an electro-optical streak camera has been conventionally used for optical pulse shape measurement [17], it has several technical problems: its energy consumption is large due to its high-voltage power supply, and the signal-to-noise ratio of the measurement is limited by the noise generated in the electrical circuits in the camera. Moreover, the sensitivity of the measurement in the wavelength range of 1250–1350 nm is low because of the narrow wavelength range of the photodetector. In contrast, the all-optical method presented here can solve these problems inherent to the streak camera method.

Figure 7.15 shows measured optical pulse shapes using coumarin 540A particles. The wavelengths of the light source and the light emitted from the coumarin 540A particles are 808 and 520 nm, respectively. In this figure, the series of four pulses arriving at the coumarin 540A particle with 0.5 ps-time difference is clearly resolved, confirming that the temporal resolution of the measurements was shorter than 0.5 ps.

7.2 Conversion From Optical to Electrical Energy

As the second example of energy conversion, this section discusses the conversion from optical to electrical energy. Representative examples of the devices for this conversion include photovoltaic devices, which have been used for photodetectors, solar cells, and so on. A semiconductor is typically used as the material for these devices, as already examined in Sect. 4.2 (refer to Fig. 4.12 and Tables 4.2–4.4).

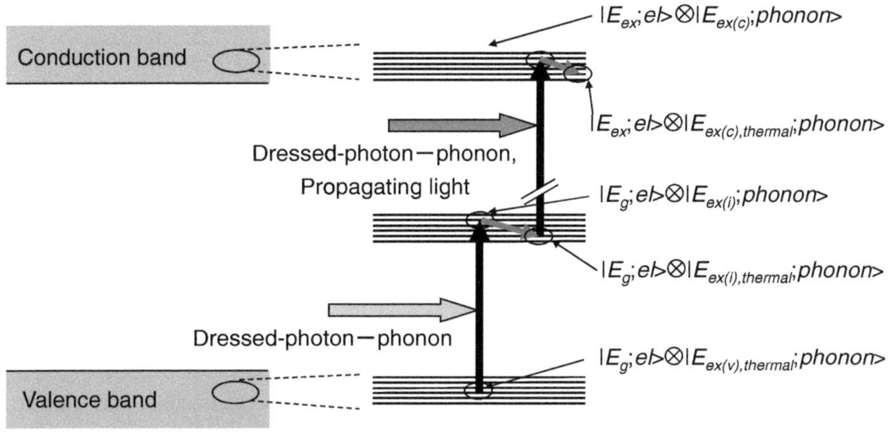

Fig. 7.16 *Two*-step excitation process

7.2.1 Multi-Step Excitation and Autonomous Fabrication

In a photovoltaic device, the converted optical energy is governed by the bandgap energy E_g of the semiconductor material used. That is, light with a photon energy lower than E_g, i.e., with a wavelength longer than $\lambda_c = E_g/hc$, is not absorbed by the device, and thus, its energy is not converted to electrical energy. Here, λ_c is called the cut-off wavelength, whose values are 390 nm, 1.11 μm, and 3.0 μm for GaN, Si, and InGaAs, respectively. In order to convert the optical energy of long-wavelength light, conventional technology has explored novel semiconductors with smaller and smaller E_g based on advances in materials science.

This section reviews a novel method of converting the optical energy of light with a wavelength longer than λ_c. This method does not rely on advances in materials science but by exploiting DPPs while using the same material: Namely, by converting the incident propagating light to DPPs, electron–hole pairs are created using the phonon energy state (the modulation sideband), even though the photon energy of the incident light is lower than E_g. Therefore, this conversion is nothing more than an energy up-conversion, which is possible by the two-step excitation described above (refer to Fig. 7.16 and is similar to that explained by Fig. 4.12a and Table 4.2. Table 7.5 summarizes this excitation process, which includes relaxation to a thermal equilibrium state after the first-step excitation, for the same reason as given in Tables 7.2 and 7.3. It should be noted that there are two route s in the second step. Since electrons and holes are respectively excited to the conduction and valence bands by this excitation process, electron–hole pairs are created, allowing the electrical energy to be extracted from the photovoltaic device to an external circuit.

Although inorganic semiconductor materials such as GaN, Si, and InGaAs have been popularly used as conventional photovoltaic materials, organic semiconductor materials are also being used recently. Sections 6.1, 6.2, and 7.1 discussed organic molecular gas, organic polymers, and organic dye particles, respectively, and this

Table 7.5 Routes of the two-step excitation

First step	Initial state: $\lvert E_g; el\rangle \otimes \lvert E_{ex(v),thermal}; phonon\rangle$ in the valence band$^{(1)}$ ⇓ Electric dipole-forbidden transition $\lvert E_g; el\rangle \otimes \lvert E_{ex(i)}; phonon\rangle^{(2)}$	
	⇓ Relaxation Intermediate state: $\lvert E_g; el\rangle \otimes \lvert E_{ex(i),thermal}; phonon\rangle^{(3)}$	
Second step	Route 1 (by the DPP)	Route 2 (by the propagating light)
	⇓ Electric dipole-allowed transition $\lvert E_g; el\rangle \otimes \lvert E_{ex(c)}; phonon\rangle$ in the conduction band$^{(4)}$	
	⇓ Relaxation Final state: $\lvert E_{ex}; el\rangle \otimes \lvert E_{ex(c),thermal}; phonon\rangle^{(5)}$	

a $\lvert E_g; el\rangle$: ground state of an electron. $\lvert E_{ex(v),thermal}; phonon\rangle$: thermal equilibrium state of a phonon.
b $\lvert E_{ex(i)}; phonon\rangle$: the state of a phonon whose eigenenergy depends on the energy of the DPP.
c $\lvert E_{ex(i),thermal}; phonon\rangle$: thermal equilibrium state of a phonon.
d $\lvert E_g; el\rangle$: excited state of an electron. $\lvert E_{ex(c)}; phonon\rangle$: excited state of a phonon whose energy depends on the energy of the propagating light or the DPP.
e $\lvert E_{ex(c),thermal}; phonon\rangle$: thermal equilibrium state of a phonon

section also examines the use of organic P3HT molecules (poly(3-hexylthiophene)) to fabricate a thin-film photovoltaic device [18].

Here, the DPPs are utilized not only to increase the conversion efficiency but also to fabricate the device. The process of autonomous DPP generation is used in this fabrication, which is an opposite process to the autonomous annihilation described in Sect. 6.3. That is, the fabrication finishes autonomously once the spatial distribution of the DPPs, generated by propagating light illumination, reaches a stationary state. A photovoltaic device fabricated by this method is expected to exhibit not only energy up-conversion but also energy selectivity in the conversion process. In other words, this device should exhibit selectively large energy conversion efficiency for incident light whose photon energy is the same as that used for fabrication.

Figure 7.17 schematically explains the autonomous fabrication of the device. A P3HT film is used as a p-type semiconductor whose E_g is 2.18 eV (λ_c=570 nm) [19], whereas a ZnO film is used as an n-type semiconductor (E_g=3.37 eV, λ_c=367 nm) [20]. A transparent ITO film and an Ag film are used as two electrodes. The principal features of this photovoltaic device originate from the P3HT, because a depletion layer of the pn-junction is formed inside the P3HT. The films of ITO, ZnO, P3HT, and Ag are deposited successively on a sapphire substrate to thicknesses of 200 nm, 100 nm, 50 nm, and several nm, respectively.

After the photovoltaic device shown in Fig. 7.17a is preliminarily formed by this deposition, Ag particles are deposited on the Ag film by the method shown in Fig. 7.17b in order to generate the DPPs: While Ag particles are being deposited by RF-sputtering, the surface of the Ag film is illuminated by the propagating light, and a reverse bias voltage V_b is applied to the pn-junction. Here, the wavelength λ_0 of the propagating light is longer than the cutoff wavelength λ_c. As an example, λ_0

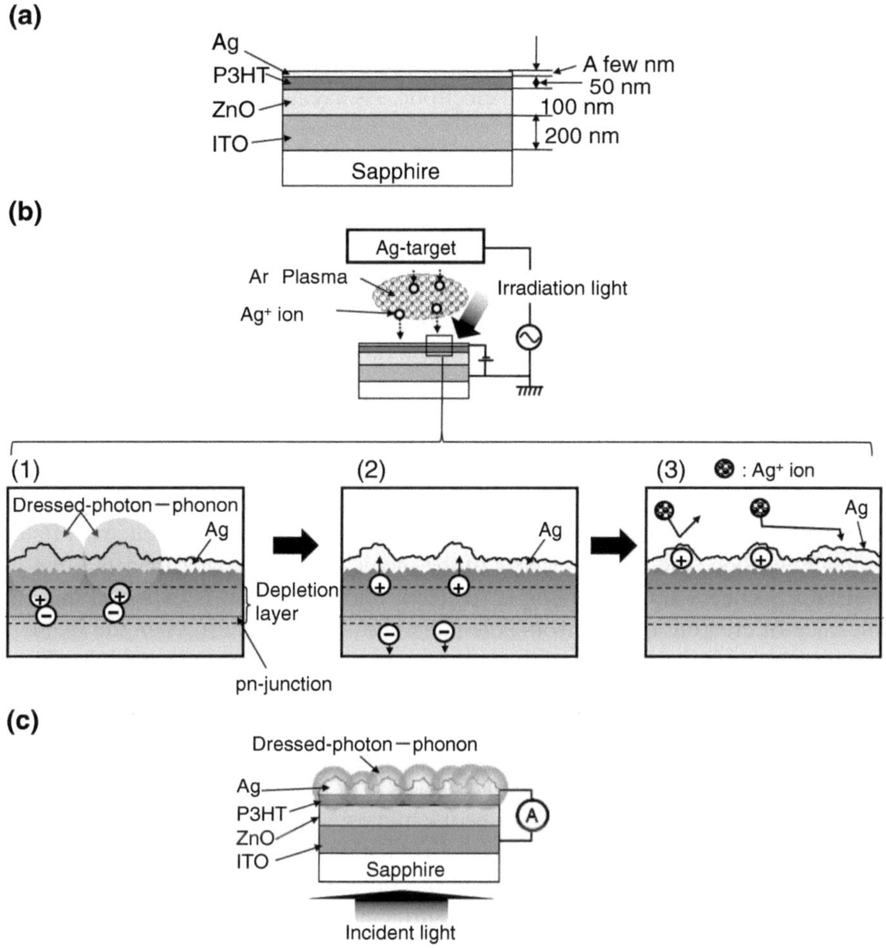

Fig. 7.17 Autonomous fabrication of photovoltaic device using organic P3HT molecules. **a** Preliminarily formed photovoltaic device by using an Ag film as an electrode. **b** Deposition of Ag particles by RF-sputtering under light illumination. (*1*), (*2*), and (*3*) represent the generation of electron–hole pairs by the dressed-photon–phonon, charging of the Ag film, and autonomous control of the Ag particle deposition, respectively. **c** Structure of the fabricated photovoltaic device

is fixed to 660 nm, and the value of V_b is -1.5 V. The fabrication principle is that the DPPs and the reverse bias voltage control the amount of Ag particles that flow into and out of the Ag film surface. This is summarized as follows:

(1) Generation of electron–hole pairs by the DPPs (Part (1) in Fig. 7.17b): The DPPs are generated at bumps in the Ag film surface by the irradiation light. If the field of the DPPs extends to the pn-junction, the two-step excitation takes places, by which electrons are excited to create electron–hole pairs even though the photon energy of the irradiation light is lower than E_g.

(2) Charging the Ag film (Part (2) in Fig. 7.17b): The created electron–hole pairs are annihilated by the electric field produced by the reverse bias voltage, and the holes are attracted to the Ag film electrode. As a result, the Ag film is positively charged.
(3) Autonomous control of Ag particle deposition (Part (3) in Fig. 7.17b): The Ag particles flying into the Ag film surface are positively charged because they pass though the Ar plasma for RF-sputtering (or because the Ar ions in the plasma collide with the Ag target material) [21]. Therefore, these Ag particles are repulsed from the area of the Ag film surface that is locally positively charged as a result of efficient generation of DPPs by processes (1) and (2). As a result, the repulsed Ag particles are deposited on other areas of the Ag film surface.

By processes (1)–(3), a unique surface morphology is formed on the Ag film, which is governed by the spatial distribution of the DPPs. By using the resultant Ag film as the electrode for the photovoltaic device and by irradiating propagating light from the rear surface of the sapphire substrate (refer to Fig. 7.17c), the DPPs are generated efficiently on the Ag electrode surface, creating electron–hole pairs in the pn-junction. Since the photon energy of the incident light is lower than E_g, optical to electrical energy up-conversion is realized. Furthermore, it is expected that the efficiency of creating the electron–hole pairs will be the highest when the incident light wavelength is equal to λ_0 of the light irradiated in the fabrication process. If the incident light wavelength is different from λ_0, the spatial distribution of the generated DPP field must be different from that of the DPPs generated in the fabrication process. Therefore, the efficiency of creating the electron–hole pairs must be lower, which means that this device shows wavelength selectivity in generating the photocurrent, and the efficiency of generation must be the highest at the wavelength λ_0.

Table 7.6 summarizes the values of the irradiation light power P and the reverse bias voltage V_b used for device fabrication. Figure 7.18 shows scanning electron microscope (SEM) images of the Ag film surfaces of devices A–C in Table 7.6. The figure shows that the surfaces of devices B and C (Fig. 7.18b, c) are rougher than that of device A (Fig. 7.18a) due to the larger grains grown on the surface. The lower parts in Fig. 7.18b, c show histograms of the distribution of grain diameters, assuming spherical grains, where the solid curves are lognormal functions fitted to these histograms. In Fig. 7.18b, the average and the standard deviation of the diameter are 90 and 64 nm, respectively, whereas they are respectively 86 and 32 nm in Fig. 7.18c. By comparing these values, it is found that the standard deviation decreases with increasing irradiation light power P, which means that a surface morphology with unique-sized grains is autonomously formed by the high irradiation power.

Since the spatial distributions of the DPPs on the Ag grains in Fig. 7.18b, c depend on the grain sizes (the average diameters of the Ag grains are 90 and 86 nm, respectively), the DPP fields of devices B and C extend to the pn-junction because the sum of the thicknesses of the Ag film and the P3HT is smaller than 70 nm. As a result, electron–hole pairs are created by these DPPs when light is incident for device operation. In the case of device A, on the other hand, the field of the DPPs does not

Table 7.6 The values of the irradiation light power P and the reverse bias voltage V_b for device fabrication

Name of the device	Irradiation light power P	Reverse bias voltage V_b
A[a]	0	0
B	50 mW	−1.5 V
C	70 mW	−1.5 V

[a] A is a reference device that was fabricated to compare its performance with that of devices B and C

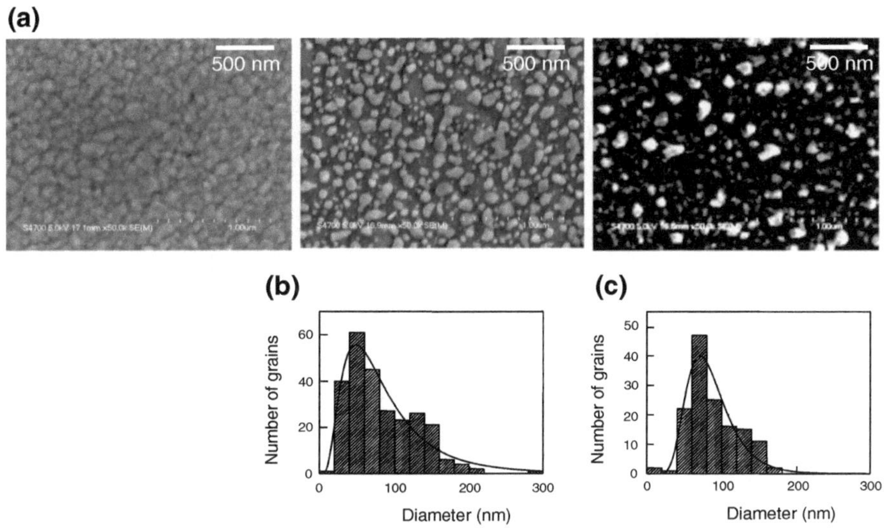

Fig. 7.18 Scanning electron microscope images of Ag film surfaces of fabricated photovoltaic devices. **a, b,** and **c** show images of devices A, B, and C, respectively. *Lower parts* of **b** and **c** show histograms for the distribution of the Ag grain diameters

reach the pn-junction because the Ag film is quite thick at 800 nm, and therefore, electron–hole pairs are not created by the DPPs.

7.2.2 Wavelength Selectivity and Light Emission

(a) Wavelength selectivity

The dependences of the generated photocurrent density on the incident light wavelength can be analyzed by using a wavelength-tunable laser as a light source. The power density is decreased to 125 mW/cm^2, and the tuned wavelength range is 580–670 nm, which is longer than the cutoff wavelength λ_c (= 570 nm) of the P3HT. It is expected that the generated photocurrent density will be proportional to the power density at this low incident light power because the rate of creating electron–hole pairs is governed by the electric dipole-forbidden transition of the first-step excitation in Fig. 7.16 and Table 7.5.

Figure 7.19 shows the measured dependence, in which curve A is the very small photocurrent density generated from device A, shown as a reference. Curves B and C are for devices B and C, respectively. They show that photocurrents are generated for incident light wavelengths longer than λ_c, from which the energy up-conversion is confirmed. Curve C has a peak at 620 nm, clearly confirming the wavelength selectivity. The quantum efficiency at the peak of this curve (wavelength, 620 nm) is 0.24 %, which is as high as the value reported for a hetero-structured P3HT photovoltaic device [22]. However, it should be noted that this high efficiency was realized at a wavelength longer than λ_c as a result of the energy up-conversion.

The peak wavelength (620 nm) of curve C is 40 nm shorter than that λ_0 (=660 nm) of the irradiation light used in the fabrication process. This difference in the wavelength originates from the DC Stark effect induced by the reverse bias voltage V_b applied to control the morphology: From the value of the thickness (about 10 nm) of the depletion layer of the P3HT/ZnO pn-junction [23] and the value of the relative dielectric constant (3.0) of the P3HT [24], the DC electric field applied to the depletion layer is estimated to be -1.0×10^6 V/m when V_b is -1.5 V. Furthermore, by assuming that the reduced mass of the electron–hole pair at the pn-junction is equal to the electron mass in vacuum, the shift of the cutoff wavelength λ_c induced by the DC electric field is estimated using the formula for the optical absorption coefficient to be 40 nm [25]. This estimated value agrees with the measured value given above.

The wavelength selectivity of curve B (device B) is not so clear as that of curve C, because of the lower light power (50 mW) irradiated in the fabrication process. The clear wavelength selectivity of curve C is due to efficient generation of DPPs by the higher irradiation power (70 mW).

(b) Light emission

It was confirmed in supplementary experiments that curve B in Fig. 7.19 maintains a sufficiently large value even at a wavelength shorter than λ_c. In contrast, curve C rapidly decreased with decreasing wavelength, resulting in a low photocurrent for incident light having a wavelength shorter than λ_c. This decrease originates from

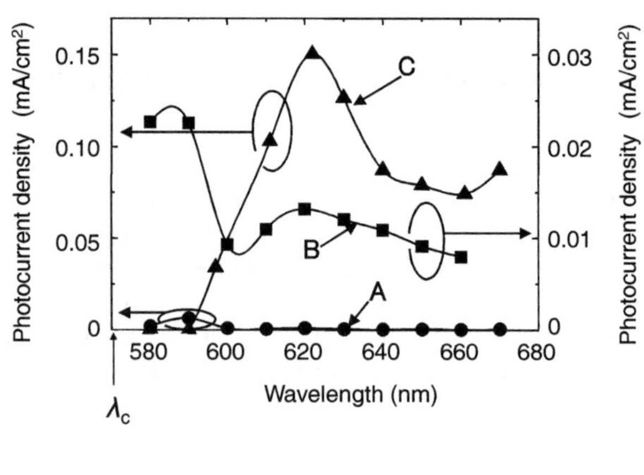

Fig. 7.19 Relation between the incident light wavelength and the photocurrent density. *Curves A, B,* and *C* represent the measured results for devices A, B, and C, respectively

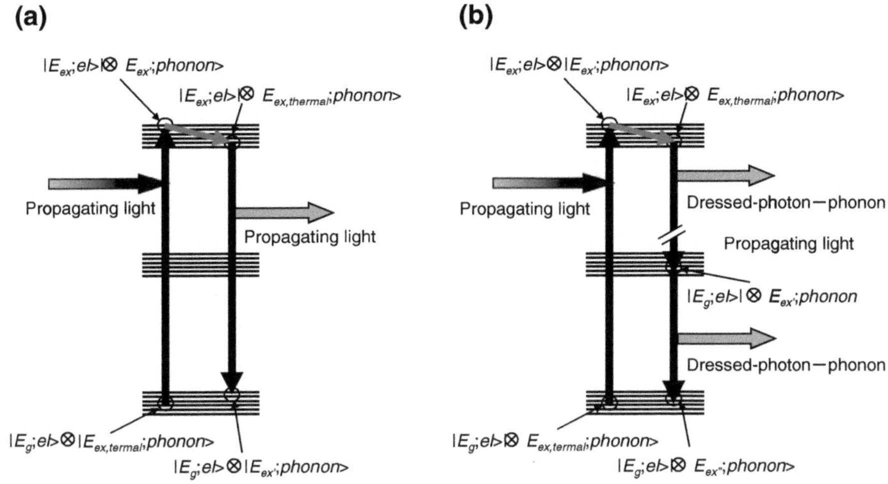

Fig. 7.20 Excitation and de-excitation processes. **a** Device A. **b** Devices B and C

the light emission from the device C due to the de-excitation induced by the DPPs, which is summarized as follows:

(1) Device A: Because the photon energy of the incident light is higher than E_g of the P3HT, electrons in device A are excited from the valence band to the conduction band (from HOMO to LUMO in the present case of the organic P3HT semiconductor) by the conventional electric dipole-allowed transition due to the light absorption. This transition is summarized by Fig. 7.20a and Table 7.7. Since the de-excitation in Table 7.7 causes spontaneous light emission, the photocurrent decreases if this emission rate is high. These excitation and de-excitation processes are due to the conventional absorption and spontaneous emission of propagating light, respectively, which are different from those of Fig. 4.12a–c and Tables 4.2, 4.3 and 4.4.
(2) Devices B and C: In the case of the devices B and C, not only spontaneous emission but also stimulated emission takes place after excitation by light absorption, which is due to DPPs (refer to Fig. 7.20b and Table 7.8). Thus, further decreases in the photocurrent are possible.

Figure 7.21a shows the spectrum of the propagating light emitted from device A as a result of the de-excitation in Fig. 7.20a and Table 7.7. Its peak wavelength is 585 nm, which is 15 nm longer than λ_c because of the conventional Stokes shift induced by the collision between the electron and phonon in P3HT [26]. The full width at half-maximum (FWHM) is 90 nm.

On the other hand, curve A of Fig. 7.21b shows the emission spectrum of device C. Its peak wavelength is 620 nm, which is 50 nm longer than λ_c. This curve A is the spectrum of the propagating light generated as a result of the de-excitation in the first step of Fig. 7.20b and Table 7.8. Its FHWM is as large as 150 nm, which means that the spectrum contains several components that originate from the densely distributed

Table 7.7 Routes of the two-step excitation

<table>
<tr><td>Excitation</td><td colspan="2">$|E_g; el\rangle \otimes |E_{ex,thermal}; phonon\rangle$ in HOMO
⇓ Electric dipole-allowed transition
$|E_{ex}; el\rangle \otimes |E_{ex'}; phonon\rangle$ in LUMO
⇓ Relaxation
$|E_{ex}; el\rangle \otimes |E_{ex,thermal}; phonon\rangle$ in LUMO</td></tr>
<tr><td>De-excitation</td><td>Route 1
(Photocurrent generation)</td><td>Route 2
(Propagating light emission)
⇓ Electric dipole-allowed transition
$|E_g; el\rangle \otimes |E_{ex'}; phonon\rangle$ in HOMO</td></tr>
</table>

Table 7.8 Routes of the two-step excitation

<table>
<tr><td colspan="2">Excitation</td><td colspan="2">$|E_g; el\rangle \otimes |E_{ex,thermal}; phonon\rangle$ in HOMO
⇓ Electric dipole-allowed transition
$|E_{ex}; el\rangle \otimes |E_{ex'}; phonon\rangle$ in LUMO
⇓ Relaxation
$|E_{ex}; el\rangle \otimes |E_{ex,thermal}; phonon\rangle$ in LUMO</td></tr>
<tr><td rowspan="2">De-excitation</td><td>First step</td><td colspan="2">Route 1 (DPP generation) | Route 2 (Propagating light generation)
⇓ Electric dipole-allowed transition
Intermediate state: $|E_g; el\rangle \otimes |E_{ex'}; phonon\rangle$</td></tr>
<tr><td>Second step</td><td colspan="2">⇓ Electric dipole-forbidden transition
(DPP generation)
$|E_g; el\rangle \otimes |E_{ex''}; phonon\rangle$ in HOMO</td></tr>
</table>

energy levels of phonons. The curve A has a kink at the position identified by an upward arrow. This means that this curve contains a propagating light component, which is equivalent to the emission spectrum of device A. This component is extracted from curve A and represented by the curve B, by referring to the spectral profile of Fig. 7.21a. The difference between curves A and B is represented by curve C. As a result of this extraction, the ratio between the emitted light intensities due to the conventional de-excitation in device A (Fig. 7.20a) and the first step de-excitation in device C (Fig. 7.20b) is derived by calculating the ratio of the areas under curves B and C. As a result, this ratio was derived to be 1:4, by which it is confirmed that the de-excitation in device C is mostly due to the DPP-mediated process (Fig. 7.20b).

(c) Possible applications

Since the energy up-conversion reviewed in the present section can be realized merely by controlling the morphology of the electrode surface, it can be applied not only to P3HT but also to a variety of organic and inorganic semiconductors. Higher efficiency is expected by adjusting the power and wavelength of the irradiation light and the reverse bias voltages used for controlling the morphology. In particular, the peak wavelength of the selective energy conversion depends not on E_g but on the wavelength of the irradiation light. Using this dependence, novel photovoltaic devices are expected, even when using conventional semiconductor materials, resulting in

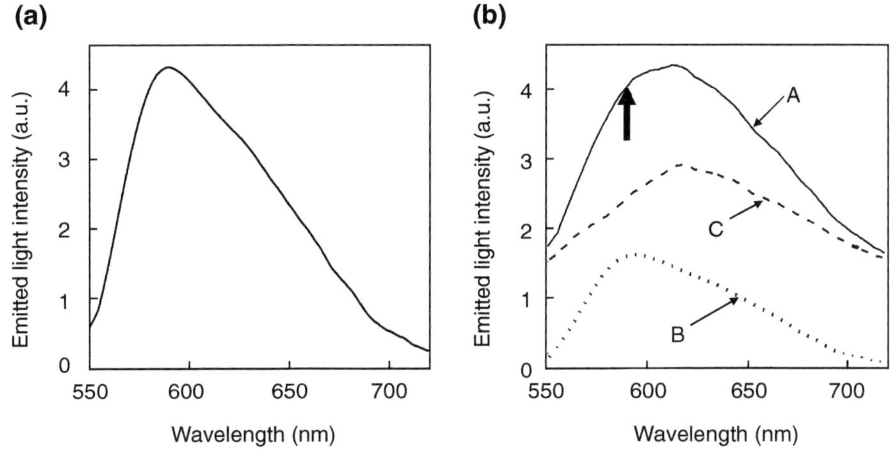

Fig. 7.21 Emission spectra. **a** Device A. **b** *Curve A* is for Device C. *Upward arrow* represents the position of a kink in *curve A*. *Curve B* represents the propagating light component. *Curve C* represents the difference between *curves A* and *B*

energy up-conversion and wavelength selectivity even for incident light whose photon energy is lower than E_g.

When this device is used as a photodetector for an optical sensing system, it is usually illuminated by monochromatic light of a specific wavelength. In order to realize high conversion efficiency for monochromatic incident light, it is advantageous to use this same monochromatic light in the fabrication process for controlling the morphology. On the other hand, in the case where this device is used as a solar cell, since the solar spectrum covers a wide range from the ultraviolet to the infrared, in the fabrication process, it is advantageous to simultaneously irradiate several light beams whose wavelengths are longer than λ_c.

The optical energy density $\rho(\nu)$ (J/Hz·m^3) of sunlight is expressed by Planck's formula for thermal radiation:

$$\rho(\nu) = \frac{8\pi h\nu^3}{c^3} \frac{1}{\exp(h\nu/k_B T) - 1}, \tag{7.15}$$

where ν, c, h, k_B, and T are the optical frequency, the speed of light, Planck's constant, Boltzmann's constant, and the temperature of the sun, respectively [27]. However, it should be noted that the photon number density $n(\nu)$ is more important than $\rho(\nu)$ in discussing the quantum efficiency of the energy conversion because this efficiency is defined as the number of electrons generated by one incident photon. From the relation $n(\nu) = \rho(\nu)/h\nu$, the photon number density $n(\lambda)$ $[= n(\nu)\, d\nu/d\lambda]$ per unit wavelength width is expressed as

$$n(\lambda) = \frac{8\pi}{\lambda^4} \frac{1}{\exp(hc/\lambda k_B T) - 1}. \tag{7.16}$$

Curves A and B in Fig. 7.22 represent the calculated values of $\rho(\lambda)$ and (λ), respectively, for the case where $T = 5800$ K [28]. Since curve B shows that sunlight contains a large number of photons in the infrared region, it is advantageous for increasing the conversion efficiency of the solar cell in the infrared region. However, the problem is that light having a wavelength longer than λ_c cannot be absorbed by the solar cell. Even if a novel material with sufficiently low E_g were to be found, the additional problem is that the low E_g results in a low open-circuit voltage, which is disadvantageous for extracting large electrical energy from the solar cell. However, the energy up-conversion reviewed in this section can solve these problems.

The energy up-conversion described here has been used for other applications, including electrolysis of water by irradiating TiO_2 nanorods with visible light [29]. Similarly, electrolysis of water [30] and activation of charge transfer in water solution [31] by irradiating ZnO nanorods with visible light have also been reported.

7.3 Conversion From Electrical to Optical Energy

As a third example of energy conversion, this section reviews the conversion from electrical to optical energy by taking a light emitting diode (LED) as an example. The discussion focuses on the possibility of using an indirect transition-type semiconductor to construct an LED (and also a laser, to be reviewed in Sect. 7.3.3), which is not possible by using conventional methods of materials science and technology. The principle of the device operation is equivalent to those in Sects. 7.1 and 7.2.

As was the case in Sect. 7.2, the wavelength of the light emitted from a conventional LED is governed by the bandgap energy E_g of the semiconductor material used. Although there is a Stokes wavelength shift [32], its magnitude is negligibly small. Therefore, the value of E_g must be decreased for longer wavelength light emission, which has been realized by exploring novel semiconductor materials.

Direct transition-type semiconductors have been used for conventional LEDs. Among the materials commonly used, InGaAsP has been used for optical fiber com-

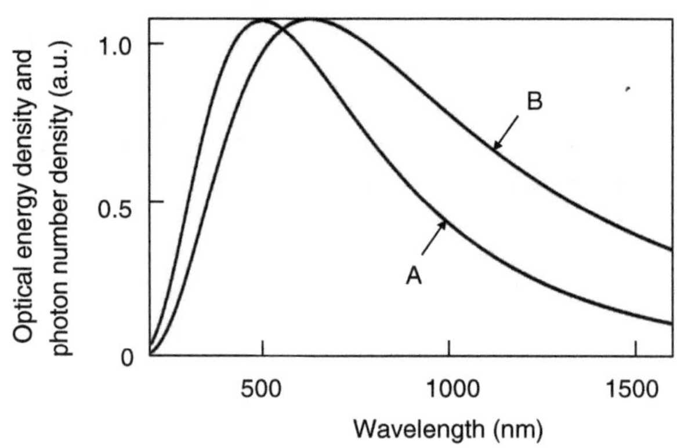

Fig. 7.22 Calculated values of the optical energy density (*curve A*) and the photon number density (*curve B*) of sunlight at a black-body radiation temperature of 5800 K

munication systems because its cutoff wavelength λ_c is as long as 1.00–1.70 μm (E_g=0.73–1.24 eV) [33, 34]. As another example, GaN, which has λ_c as short as 365 nm ($E_g = 3.40$ eV) has been used for visible LEDs.

In contrast to these examples, the LEDs reviewed in this section can use indirect transition-type semiconductors: Although silicon (Si) has been popularly used for electronic devices, there is a long-held belief in materials science and technology that Si is not suitable for use in LEDs because its light emitting efficiency is very low. However, it is possible to realize high efficiency with Si if DPPs could be used, instead of exploring other semiconductor materials [35].

Electrons have to transition from the conduction band to the valence band to spontaneously emit light. However, in the case of an indirect transition-type semiconductor, the wave-numbers (momenta) of the electron at the bottom of the conduction band and at the top of the valence band are different between each other. Therefore, for electron–hole recombination, a phonon is required to satisfy the momentum conservation law. In other words, electron–phonon interaction is required. However, the probability of this interaction is low, resulting in a low interband transition probability.

However, since the DPPs are accompanied by phonons with sufficiently large wave-number for satisfying the momentum conservation law, the probability of spontaneously emitting DPPs can become high.[3] Since some of the emitted DPPs are converted to propagating light, a highly efficient LED can be realized even by using an indirect transition-type semiconductor.

7.3.1 Autonomous Device Fabrication

DPPs can be used in two ways for realizing an indirect transition-type semiconductor LED. They are:

(1) In order to emit light spontaneously for device operation.
(2) In order to fabricate the device, especially, to autonomously control the spatial distribution of the doped boron (B) concentration.

Among them, this subsection reviews (2), and the next subsection reviews (1).

Here, an n-type Si crystal with low arsenic (As) concentration is used. By doping the crystal with B, the Si crystal surface is transformed to the p-type, forming a homostructured pn-junction. An ITO film and an Al film are coated on opposite surfaces of the Si crystal to be used as positive and negative electrodes, respectively. A forward

---

[3] This high probability of spontaneous emission can be alternatively explained as follows: Let's start from the uncertainty principle $\Delta k \cdot \Delta x \geq 1$ between the uncertainty Δk of the wave-number k (proportional to the momentum) and the uncertainty Δk of the position x (refer to Sect. 1.1 of Chap. 1). Since the relation $\Delta x < \lambda$ holds because the size of DPP is smaller than the wavelength λ of light, the uncertainty principle leads to the relation $\Delta k > k$, which means that the wave-number and momentum are uncertain. This uncertainty suggests that the momentum conservation law is invalid, which means that the distinction between direct and indirect transitions is also invalid.

bias voltage of 16 V is applied to inject the current (current density, 4.2 A/cm^2) in order to generate Joule-heat for performing annealing, causing the B to be diffused and varying the spatial distribution of its concentration. During the annealing, the Si crystal is irradiated, through the ITO electrode, with laser light (light power density, 10 W/cm^2) whose photon energy $h\nu_{anneal}$ (= 0.95 eV, wavelength; 1.30 μm) is lower than the band-gap energy of Si (E_g = 1.12 eV, λ_c = 1.11 μm) [36]. This light irradiation generates DPPs at the domain boundaries of the inhomogeneous distribution of B at the pn-junction[4] [37]. These DPPs induce the stimulated emission in Fig. 4.12c simultaneously with the absorption in Fig. 4.12a.[5] Then, a part of the electrical energy applied to the Si crystal by the current injection is converted to optical energy for the stimulated emission, which is dissipated out from the Si crystal. This dissipation affects the B diffusion and forms characteristic minute inhomogeneous domain boundaries autonomously. This autonomous formation occurs via the following three processes:

(1) Since the energy of the electrons driven by the forward bias voltage (16 V) is higher than E_g (=1.12 eV), the difference $E_{Fc} - E_{Fv}$ between the quasi Fermi energies in the conduction band E_{Fc} and the valence band E_{Fv} is larger than E_g. Therefore, the Benard–Duraffourg inversion condition is satisfied [38]. Furthermore, since the photon energy $h\nu_{anneal}$ of the irradiation light is lower than E_g, this light propagates through the Si crystal without any attenuation and reaches the pn-junction. As a result, it generates DPPs efficiently around the domain boundaries of the inhomogeneous distribution of B. Since the energy of the generated DPPs is sufficiently high to induce stimulated emission even though $h\nu_{anneal} < E_g$, the electrons generate photons by the stimulated emission and transition from the conduction band to the valence band via the phonon energy level.
(2) The annealing rate decreases because a part of the electrical energy for generating the Joule-heat is spent for the stimulated emission of photons. As a result, at the sites where the DPPs in (1) are easily generated, the shape and dimensions of the B inhomogeneous domain boundaries become more difficult to change.[6]

---

[4] This annealing utilizes DPPs localized at the impurity sites, as was described in Chap. 4, in contrast to the process described in Sects. 7.1 and 7.2, which utilized the DPPs localized at the apexes of the one-dimensional nanomaterial.

[5] This optical absorption corresponds to the two-step excitation , and thus, it corresponds to the energy up-conversion reviewed in Sects. 7.1 and 7.2.

[6] If the shape of the domain boundary is assumed to be a sphere of radius r that is smaller than the thickness of the depletion layer, the probability of generating stimulated emission in one domain boundary is proportional to the product of the number of photons incident on the domain boundary, the transition probability, and the volume of the DPP. Since these are respectively proportional to r^2, r^{-2}, and r^3 [6], the probability of stimulated emission, i.e., the annealing inhibition rate, is proportional to r^3. On the other hand, since the amount of generated Joule heating is proportional to the electric current passing through the surface of the domain boundary, the annealing rate is proportional to r^2. Therefore, the temporal evolution of the radius r is expressed as $dr/dt = ar^2 - br^3$, where a and b are constants depending on the forward bias current and the irradiation light power density, respectively. Thus, in the stationary state ($dr/dt = 0$), r takes the constant

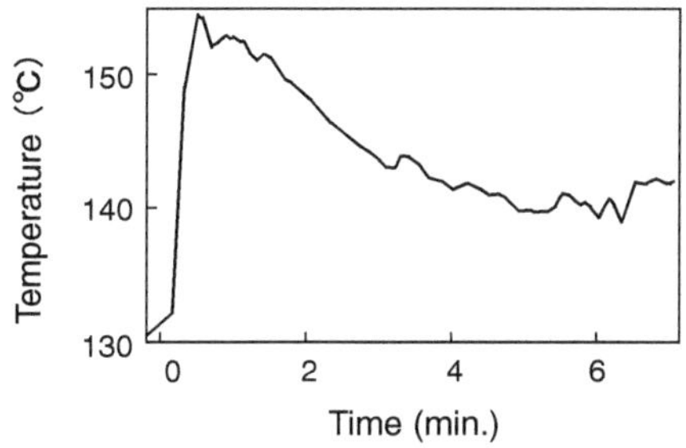

Fig. 7.23 Temporal evolution of the temperature of the device surface

(3) Spontaneous emission occurs more efficiently at the areas in which the DPPs are generated easily because the probability of spontaneous emissions is proportional to that of stimulated emission [39]. Furthermore, with temporal evolution of process (2), the light from stimulated and spontaneous emission spreads through the whole Si crystal, and as a result, process (2) takes place autonomously in the entire volume of the device.

It is expected that this method of annealing will form the optimum spatial distribution of the B concentration for efficient generation of DPPs, resulting in efficient device operation. Figure 7.23 shows the temporal evolution of the temperature of the device surface as annealing progresses. After the temperature rapidly rises to 154 °C, it falls and asymptotically approaches a constant value (140 °C) after 6 min., at which the temperature inside the device is estimated to be about 300 °C. The feature of this temporal evolution are consistent with that of the principle of annealing under light irradiation described above: The temperature rises due to the Joule heating by the applied electrical energy. However, the temperature gradually falls because stimulated emission is induced by the DPPs generated at the domain boundary of the inhomogeneous distribution of the B concentration. Finally, the system reaches the stationary state.

7.3.2 Device Operation

Figure 7.24 shows the measured relation between the forward bias voltage V and the current I injected into a fabricated large-area device (about 10 mm^2). The curve in this figure represents the negative resistance feature at I >50 mA, where the breakover voltage V_b at the bending point on this curve is 73 V. It should be pointed out that small-area devices do not show such a negative resistance feature. This feature is due to the filament current originating from the spatially inhomogeneous current density [40]. In other words, the current is concentrated in a domain boundary of the

value a/b. It should be pointed out that the mathematical scientific model to be reviewed in Sect. 8.4 of Chap. 8 is more effective in describing the details of this temporal evolution because the shape of the domain boundary in the actual system is much more complicated than a simple sphere.

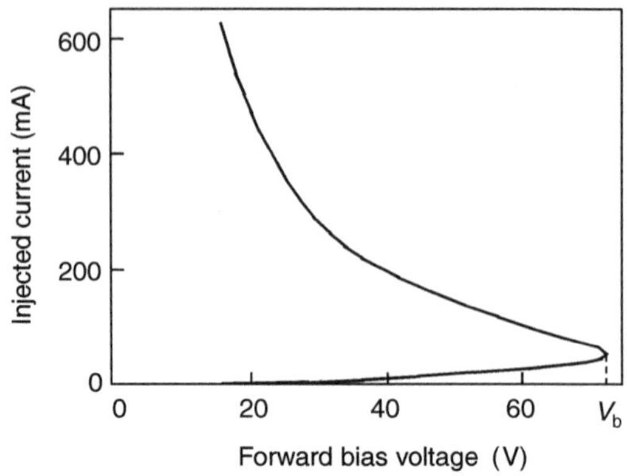

Fig. 7.24 Relation between the forward bias voltage and the injected current. V_b represents the break-over voltage

B distribution. In this boundary region, a center of localization is easily formed, to which the electrons are easily bound, and a DPP is efficiently generated. Thus, the negative resistance feature supports the principle of device fabrication reviewed in Sect. 7.3.1. The value of V_b given above is higher than the built-in potential of the pn-junction of Si, which is mainly due to the high total resistance of the thick Si crystal and the high contact resistance between the electrode and the crystal surface. Although the spatially averaged temperature inside the device may not be sufficiently high for efficiently diffusing the B, the device must be locally heated by the filament current described above, which enables efficient diffusion. High-resolution, three-dimensional atom-probe imaging has been used to confirm the inhomogeneous spatial distribution of B in the Si crystal after the annealing.

Figure 7.25a, b are photographs showing the device without and with current injection (current density, 4.2 A/cm^2), respectively, at room temperature, which were taken by an infrared CCD camera (photosensitive bandwidth, 1.73–1.38 eV; wavelengths, 0.90–1.70 μm) under fluorescent lamp illumination. Figure 7.25b shows a bright spot of light with a power as high as 1.1 W, which was emitted by applying 11 W of electrical power. Figure 7.25c shows a prototype packaged commercial Si LED, to which a convex lens is attached for light beam collimation.

It should be pointed out that even a conventional Si photodiode can emit light even though its efficiency is extremely low. Figure 7.26a shows the emission spectrum of a commercial photodiode (Hamamatsu Photonics, L10823) at an injection current density of 0.2 A/cm^2. Higher current injection damaged the photodiode. The main part of the emission spectrum in this figure is located at energies higher than E_g (=1.12 eV), which is due to the indirect transition caused by phonon scattering[7] [34].

Figure 7.26b shows the emission spectra of the devices at an injected current density of 1.5 A/cm^2, which were fabricated by the method described in Sect. 7.3.1.

[7] The band-edge emission spectrum is generally located in the lower energy region due to the Stokes shift. It is shifted to higher energy with increasing current, which is a common phenomenon observed when the device temperature is maintained constant. However, in the case of the indirect transition-type semiconductors, the main part of the emission spectrum is located at energies higher than E_g because of the increasing carrier density due to the increasing current.

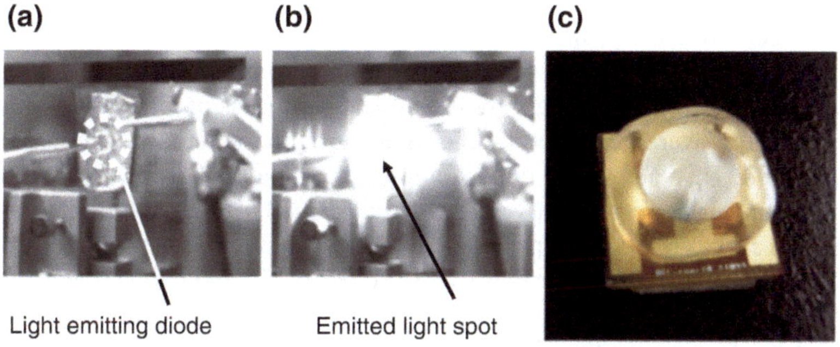

Fig. 7.25 External appearance of the device. **a** and **b** are without and with current injection, respectively. **c** A prototype packaged commercial Si LED, to which a convex lens is attached for light beam collimation

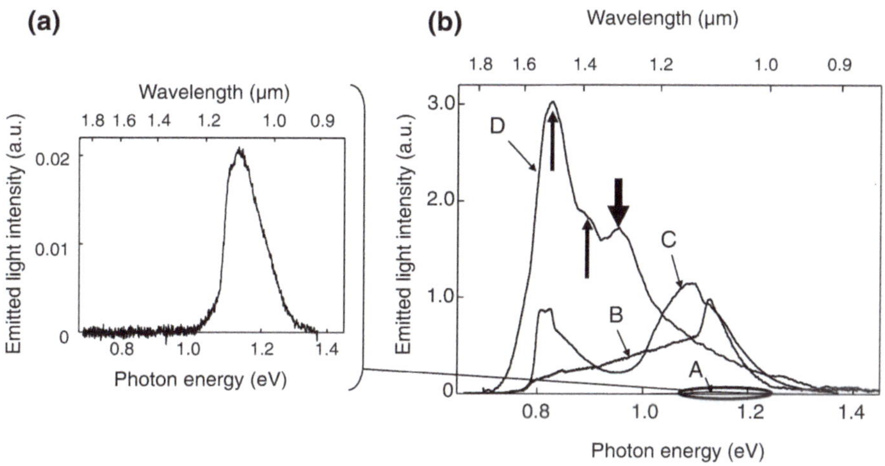

Fig. 7.26 Emission spectra. **a** A commercial photodiode (Hamamatsu Photonics, L108023). **b** Fabricated devices. *Curve A* is identical to the spectrum in (**a**). *Curves B, C,* and *D* are spectra of the devices fabricated by the annealing for 1, 7, and 30 min, respectively. *Downward thick arrow* on *curve D* represents the peak at the energy that corresponds to the photon energy of the light irradiated in the annealing process. *Two upward arrows* represent the contributions of phonons

Curve A in this figure is identical to the curve in Fig. 7.26a for comparison. Curves B–D are the spectra of the devices fabricated by annealing for 1, 7, and 30 min., respectively. Their profiles are significantly different from that of curve A, and the spectra are located at lower energies than E_g. The light intensity values of the curves B–D look low at energies lower than 0.8 eV, which is due to the low sensitivity of the photodetector used for the measurement. Although the emission spectrum of the device fabricated by 1 min. of annealing (curve B) still has a clear peak at around E_g, the spectrum broadened and reached an energy of 0.75 eV (wavelength, 1.65 μm). That of the device annealed for 7 min (curve C) shows a new peak at around 0.83 eV (wavelength, 1.49 μm). In the case of the device annealed for 30 min. (curve D), no peaks are seen around E_g. Instead, a new peak appears, identified by a downward thick arrow, at an energy that corresponds to the photon energy $h\nu_{anneal}$ (0.95 eV; wavelength, 1.30 μm) of the light irradiated in the process of annealing. This peak means that DPPs were generated by the light illumination and that the B diffusion was controlled. The value of the emission intensity at the highest peak (identified by the left thin upward arrow) of curve D is 14-times and 3.4-times higher than those of the peaks on curves B and C, respectively. Here, the separations between the energies identified by two upward thin arrows (0.83 and 0.89 eV) , and by the downward thin arrow (0.95 eV) are respectively 0.06 eV, which is equal to the energy of an optical phonon in Si. This means that the two upward thin arrows show that the DPP at an energy of 0.95 eV is converted to a free photon after emitting one and two optical phonons.

These conversion processes demonstrate that the light emission described here uses the phonon energy levels as an intermediate state. The spectrum of curve D extends over the energies 0.73–1.24 eV (wavelengths, 1.00–1.70 μm), which covers the wavelength band of optical fiber communication systems. The spectral width of curve D is 0.51 eV, which is more than 4-times greater than that (0.12 eV) of a conventional commercial InGaAs LED with a wavelength of 1.6 μm.

The solid curve in Fig. 7.27 represents the relation between the applied electric power and the emitted light power for the device annealed for 30 min. The slope of the broken curve corresponds to the differential external power conversion efficiency. Figure 7.27a is the measured optical power at photon energies higher than 0.73 eV (wavelength, 1.70 μm). By applying 11 W of electrical power, the external power conversion efficiency and the differential external power conversion efficiency reach as high as 1.3 and 5.0 %, respectively. This device operated in a stable manner by applying higher electrical power without inducing any damage. Figure 7.27b shows the relation for the emitted photon energy of 0.11–4.96 eV (wavelengths, 0.25–11.0 μm). The external power conversion efficiency and the differential external power conversion efficiency are as high as 10 and 25 %, respectively. For Fig. 7.27b, more detailed evaluations are required because infrared radiation may be included in the emitted power due to the current-induced temperature rise in the device.

In order to estimate the quantum efficiency, the relation between the injected current density I_d and the emitted light power density P_d was measured for the device annealed for 30 min., and the results are shown by Fig. 7.28. Figure 7.28a shows the values of P_d at photon energies higher than 0.73 eV (wavelength, 1.70 μm). The solid curve in this figure represents a quadratic function fitted to the measured values identified by the closed squares. This fitting means that P_d is proportional to ${I_d}^2$, which is because the two-step spontaneous emission process is dominant, i.e., one electron is converted to two photons. For comparison, it is proportional to I_d in the conventional LED device. Furthermore, the external quantum efficiency is estimated to be 15 % at $I_d = 4.0\,\mathrm{A/cm^2}$, and the differential external quantum efficiency is 40 % at $I_d = 3.0$–$4.0\,\mathrm{A/cm^2}$. Figure 7.28b shows the relation between I_d and P_d at photon energies of 0.11–4.96 eV (wavelengths, 0.25–11.0 μm). The external quantum efficiency is as high as 150 %. The reason why this value is higher

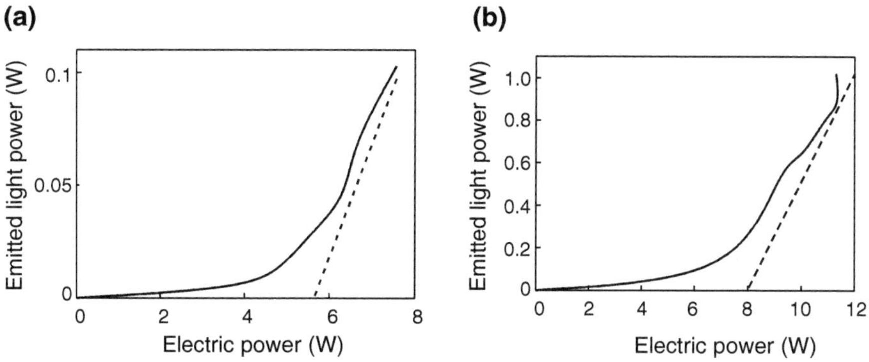

Fig. 7.27 Relation between the applied electric power and the emitted light power for the device annealed for 30 min. **a** and **b** represent the measured optical powers at energies higher than 0.73 eV (wavelength, 1.70 μm) and between 0.11 and 4.96 eV (wavelengths, 0.25–11.0 μm), respectively

Fig. 7.28 Relation between the injected current density and the emitted light power density for the device annealed for 30 min. **a** and **b** represent the measured optical powers at energies higher than 0.73 eV (wavelength, 1.70 μm) and between 0.11 and 4.96 eV (wavelengths, 0.25–11.0 μm), respectively

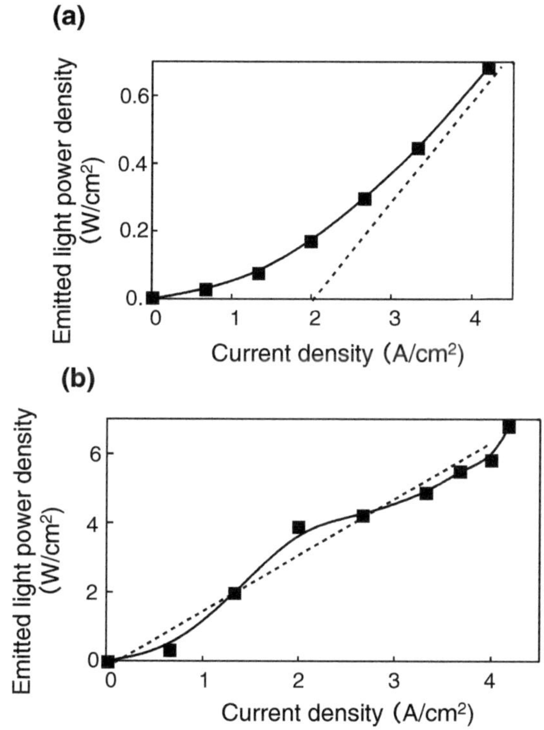

than 100 % is that the two-step spontaneous emission process converts one electron to two photons.

In the case of fabricating a highly efficient infrared light emitting device using a conventional direct transition-type InGaAsP semiconductor, one has to use a double hetero-structure composed of an InGaAsP active layer and an InP carrier confinement layer. The problems are the complexity of its structure and the high toxicity of the element As [41]. It should also be noted that the element In is a rare metal. On the other hand, in the case of composite semiconductors for emitting visible light, such as AlGaInP and InGaN, they have a green gap[8] around the wavelength of 550 nm ($E_g = 2.25\,\mathrm{eV}$), at which the emission efficiency is extremely low [42]. Although this efficiency has been increasing recently by improving the dopant materials and fabrication methods, there still exist several technical problems because highly toxic materials or rare materials are required, which increases the cost of fabrication.

In order to solve these problems, several methods have been recently proposed using Si. For example, porous Si [43], a super-lattice structure of Si and SiO_2 [44, 45], and Si nano-precipitates in SiO_2 [46] have been used to emit visible light. To emit infrared light, Er-doped Si [47] and Si-Ge [48] have been employed. However, since Si still works as an indirect transition-type semiconductor in these materials, the emission efficiency is still very low. In contrast to them, the present method has realized an extremely high-efficiency device by using a simple homo-structured pn-junction in a bulk Si crystal. Further increases in the efficiency and extension of the wavelength region are expected by decreasing the thickness of the Si crystal and increasing the efficiency of extracting the light generated in the crystal to the outside.

7.3.3 Applications to Other Related Devices

Novel devices have been realized by applying the autonomous fabrication method reviewed in Sect. 7.3.1. They are reviewed in the following:

(1) Visible LED using Si: By annealing a Si crystal by Joule heating while irradiating it with visible light, it is possible to develop novel LEDs that emit red, green, and blue light at wavelengths shorter than λ_c, as shown by Fig. 7.29a–c, respectively [49, 50]. A blue light emitting device is also possible by further modification of the fabrication method.

(2) Optical and electrical relaxation oscillator by using a Si LED: By using the above-described negative resistance feature of a large-area LED, it has been possible to develop a relaxation oscillator whose emitted light power and terminal voltage vary periodically [51]. This was realized by connecting the LED in parallel with a capacitor and driving it with a DC current supply. As shown in Fig. 7.30, the output light power and the terminal voltage vary synchronously. These oscillatory varying features have been theoretically reproduced by rate equations for the

[8] A green light emitting device may be fabricated by the following two methods: (1) Using blue light emitting InGaN, the mole-fractional ratio of In is increased in order to increase the emission wavelength to the green region. (2) Using red light emitting AlGaInP, the mole-fractional ratio of Al is increased in order to decrease the emission wavelength to the green region. However, in method (1), the emission efficiency decreases due to the increase of the internal electric field. In method (2), the efficiency also decreases due to the deterioration of the crystal quality. For these reasons, the emission efficiency is extremely low in the green spectral region, a phenomenon called the green gap.

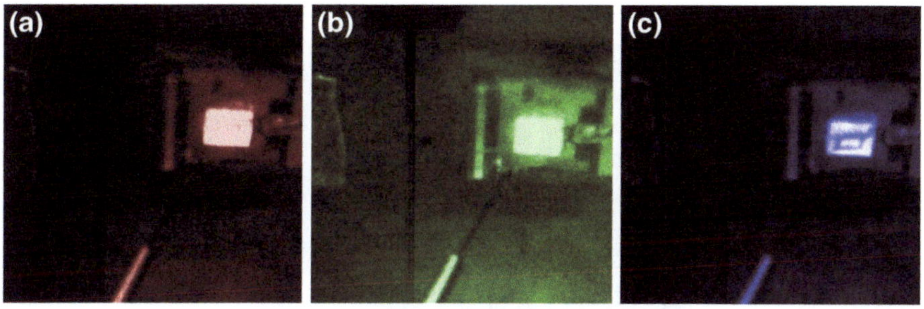

Fig. 7.29 *Red* (**a**), *green* (**b**), and *blue* (**c**) light spots emitted from Si LEDs. The center wavelengths of their emission spectra are 640, 530, and 430 nm, respectively

Fig. 7.30 Temporal evolution of the output signals from the Si optical and electrical relaxation oscillator. *Curves A* and *B* represent the values of the output light power and the terminal voltage, respectively

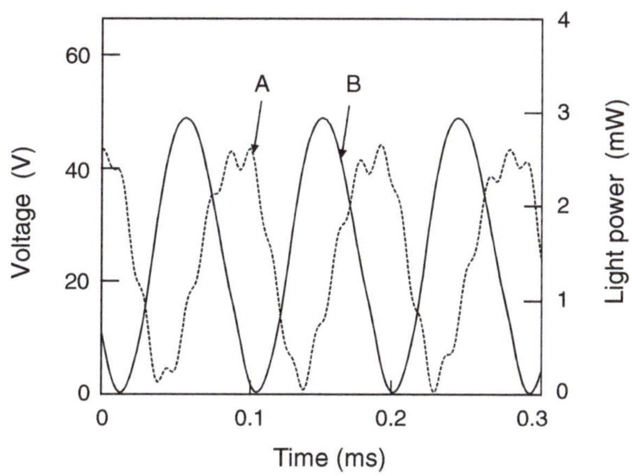

numbers of photons and electrons. In a practical system, parasitic capacitance of the LED can contribute to the relaxation oscillation even when no capacitors are connected. Although directly modulated semiconductor lasers and mode-locked lasers have been used for generating optical pulse trains in conventional optical technology, the present method provides a much simpler method for generating optical pulse trains.

(3) Laser using Si: A ridge waveguide-type optical cavity (refer to Fig. 7.31a) was employed to confine stimulated emission light in a Si crystal, resulting in the world?fs first laser device that emits CW coherent light at room temperature [52]. When the injection current is below the oscillation threshold, a broadband and low-intensity emission spectrum is observed, as shown in Fig. 7.31b. Above the threshold, the spectrum is remarkably narrowed and the emitted light power increases, resulting in a very sharp oscillation spectrum, as shown in Fig. 7.31c). The oscillation wavelength is about 1.3 μm, which corresponds to that employed for optical fiber communication systems. In addition to the narrow oscillation spectrum of Fig. 7.31c, sharp directivity of the output light beam (refer to Fig. 7.31a), TE mode-selective emission, and a rapid increase in the

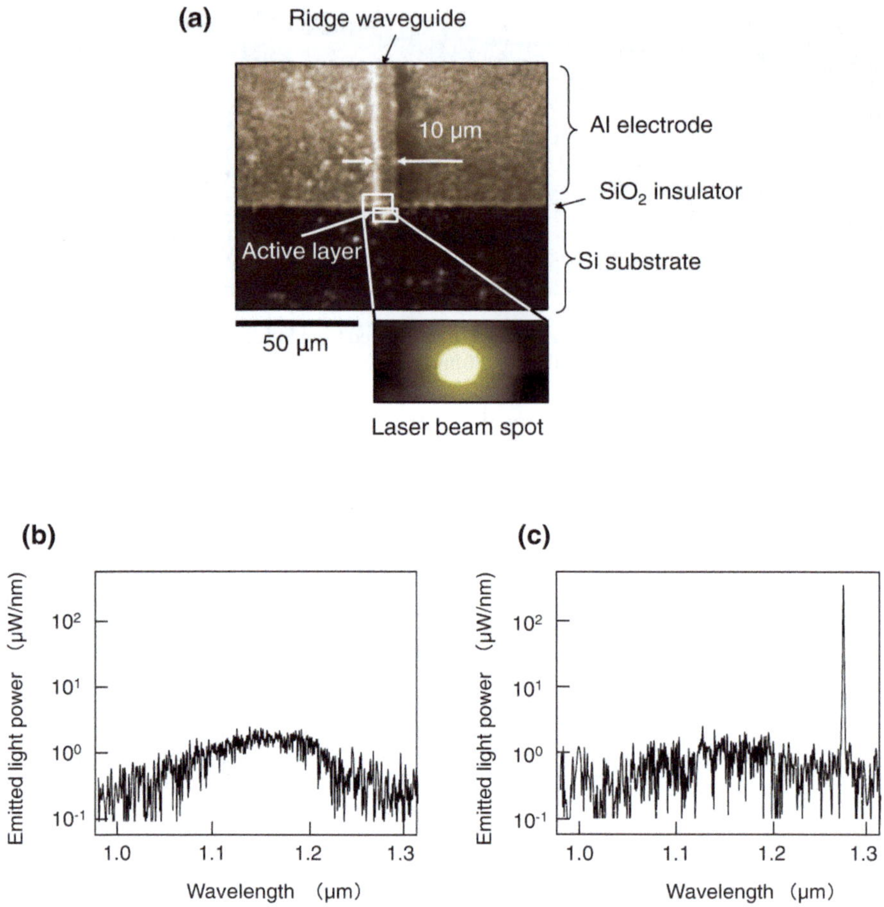

Fig. 7.31 A Si laser. **a** A scanning electron microscope image of a ridge waveguide used for the optical cavity. An image of the emitted laser beam spot is also shown. **b** and **c** represent the measured spectra below and above the oscillation threshold, respectively

output light power above the threshold have been observed, confirming laser oscillation.

(4) A Si photodiode with optical gain: A novel photodiode has been realized, which generates a photocurrent even when the photon energy of the incident light is lower than E_g of Si [53]. The device is fabricated by annealing a Si crystal by Joule heating while irradiating it with infrared light at a wavelength of 1.32 μm. As represented by curve A in Fig. 7.32, the detection sensitivity at wavelengths longer than λ_c is higher than that of a conventional photodiode (curve B). Furthermore, by injecting current into this device, the incident light power was amplified due to stimulated emission. As a result, the closed triangle (current density, 9 A/cm^2) in this figure shows that the detection sensitivity at a wavelength of 1.32 μm is as high as that at wavelengths shorter than λ_c.

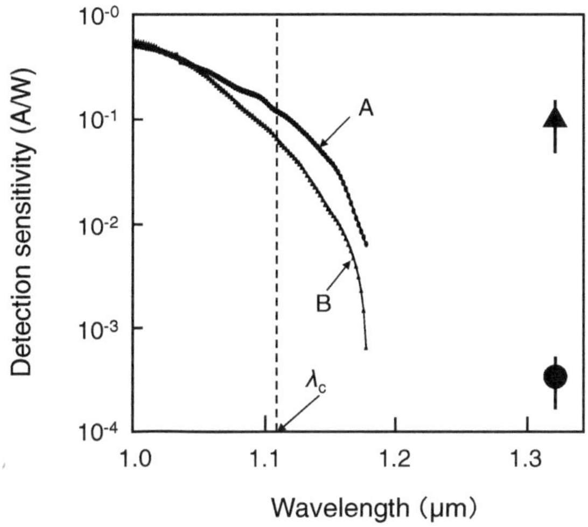

Fig. 7.32 Wavelength dependences of the detection sensitivities of Si photodiodes. *Curve A* is for a device fabricated by the present method. For comparison, *curve B* is for a commercially available Si photodiode (Hamamatsu Photonics, S3590). The *closed triangle and circle* represent the measured sensitivities at injected current densities of 9 A/cm^2 and 60 mA/cm^2, respectively

In this case, the small-signal gain coefficient and the gain-saturation power of the light amplification are estimated to be 2.2×10^{-2} and 710 mW, respectively. The closed circle represents the measured result at an injected current density of 60 mA/cm^2, at which the small-signal gain coefficient and the gain-saturation power are respectively estimated to be 3.2×10^{-4} and 17 mW. Application of this device fabrication method is expected to realize a high-efficiency Si solar cell device that can convert solar energy whose wavelength is longer than λ_c. It should be noted that this device uses the DPPs generated at the domain boundary of B inside the device, in contrast to the organic thin-film photovoltaic device of Sect. 7.2, which uses DPPs generated on the electrode surface.

As exemplified in (1)–(4) above, novel active and passive devices, such as an LED, a relaxation oscillator, a laser, and a photodiode, have been realized by using Si. This means that the principal functions for optical devices can be realized by using Si only, which can be integrated with conventional Si electronic devices, resulting in low-power-consumption optoelectronic integrated circuit systems.

The autonomous fabrication method described in Sect. 7.3.1 has been applied not only to Si but also to other materials, allowing other types of novel LEDs to be realized. These include:

(5) LEDs using GaP and SiC: Since GaP and SiC are also indirect transition-type semiconductors, it has been difficult to use them in LEDs based on conventional materials science and technology. However, by modifying the method of Sect. 7.3.1, a novel GaP LED has been realized by annealing while irradiating it with visible light. This LED emits yellow light with wavelengths shorter than λ_c, as shown by Fig. 7.33a [54]. Also, by using SiC, it has been possible to realize novel LEDs that emit green, blue, and violet light, as shown by Figs. 7.33b–d, respectively [55].

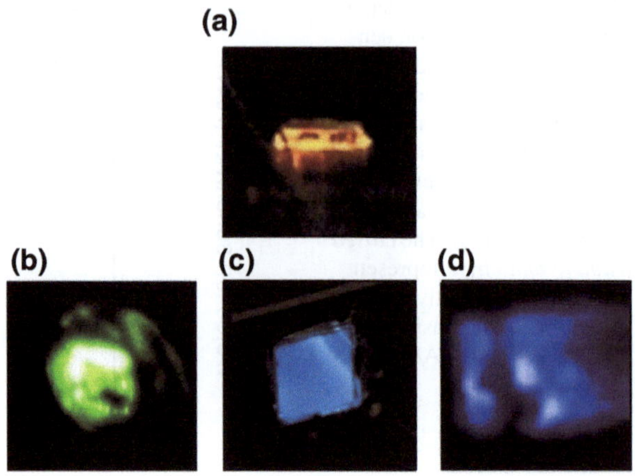

Fig. 7.33 Light spots emitted from GaP and SiC LEDs. **a** *Yellow light spot* from a GaP LED. **b–d** *Green, blue,* and violet*light spots* from SiC LEDs, respectively

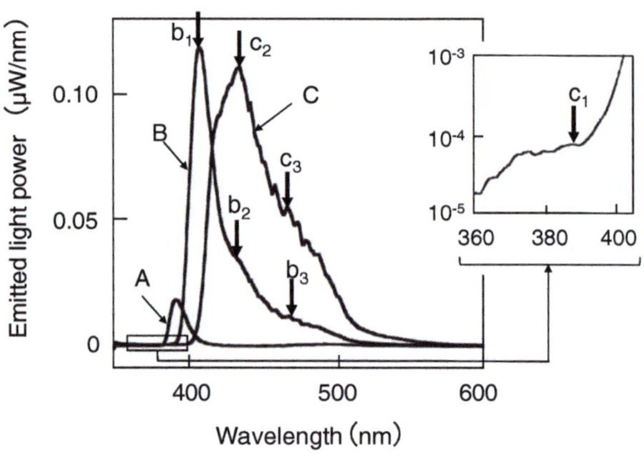

Fig. 7.34 Emission spectra of a ZnO LED. *Curves A, B,* and *C* represent the measured results for injected currents of 10, 15, and 20 mA, respectively. The *arrow* b_1 at the peak of *curve B* corresponds to the wavelength of the light irradiated during the annealing process. The *arrows* b_2, b_3, and $c_1 - c_3$ represent the positions of sidebands originating from the phonon energy levels

(6) LED using ZnO: Although ZnO is a direct transition-type semiconductor, its use in LEDs has been difficult because this p-type semiconductor is difficult to grow based on conventional materials science and technology [56]. However, a blue-violet LED has been realized by annealing while irradiating it with visible light (wavelength, 407 nm), based on the method described in Sect. 7.3.1 [57]. Curves A, B, and C in Fig. 7.34 represent the emission spectra at injected currents of 10, 15, and 20 mA, respectively. The emission peak on curve B (identified by the downward arrow b_1 at a wavelength of 407 nm) corresponds to the wavelength of the light irradiated in the process of annealing. Furthermore, this curve has kinks at the downward arrows b_2 and b_3, which represent sidebands originating from the phonon energy levels involved in spontaneous emission. On curve C, these sidebands (identified by the downward arrows $c_1 - c_3$) are also clearly seen.

As was demonstrated in this section, novel optical devices have been realized by indirect transition-type semiconductors. Their emission wavelengths are not only in the infrared but also cover the whole visible region. It is expected that the conventional materials used for optical devices will be replaced with these materials as DPP technology advances.

References

1. T. Kawazoe, H. Fujiwara, K. Kobayashi, M. Ohtsu, IEEE J. Select. Top. Quantum Electron. **15**, 1380 (2009)
2. H. Fujiwara, T. Kawazoe, M. Ohtsu, Appl. Phys. B **98**, 283 (2010)
3. W. Cao, P. Palffy-Muhoray, B. Taheri, A. Marino, G. Abbate, Mol. Cryst. Liq. Cryst. **429**, 101 (2005)
4. S.K. Pal, D. Sulul, D. Mandal, S. Sen, K. Bhattacharyya, Chem. Phys. Lett. **327**, 9 (2000)
5. H. Fujiwara, T. Kawazoe, M. Ohtsu, Appl. Phys. B **100**, 85 (2010)
6. T. Kawazoe, K. Kobayashi, S. Takubo, M. Ohtsu, J. Chem. Phys. **122**, 024715 (2005)
7. F. Laermer, T. Elsaesser, W. Kaizer, Chem. Phys. Lett. **156**, 381 (1989)
8. H. Sumi, Phys. Rev. B **29**, 4616 (1984)
9. M.S. Miao, S. Limpijumnong, W.R.L. Lambrecht, Appl. Phys. Lett. **79**, 4360 (2001)
10. T. Kawazoe, H. Fujiwara, M. Niigaki, M. Ohtsu, *Proceedings of the 31st International Congress on Applications of Lasers and Electro-Optics*, (Laser Institute of America, Orlando FL., 2012) pp. 947–948
11. H.J. Polland, W.W. Ruhle, K. Ploog, C.W. Tu, Phys. Rev. B **36**, 7722 (1987)
12. Y. Rosenwaks, M.C. Hanna, D.H. Levi, D.M. Szmyd, R.K. Ahrenkei, A.J. Nozik, Phys. Rev. B **48**, 14675 (1993)
13. A. Yariv, *Introduction to Optical Electronics*, 1st edn. (Rinehert and Winston, New York, 1985), pp. 177–221
14. P.W. Atkins, *Physical Chemistry*, 6th edn. (Oxford University Press, Oxford, 1998)
15. G.S. He, R. Signorini, P.N. Pasad, Appl. Opt. **37**, 5720 (1998)
16. K. Kato, IEEE J. Quantum Electron. **QE-22**, 1013 (1986)
17. E.M. Conwell, IEEE J. Quantum Electron. **QE-9**, 867(1973)
18. S. Yukutake, T. Kawazoe, T. Yatsui, W. Nomura, K. Kitamura, M. Ohtsu, Appl. Phys. B **99**, 415 (2010)
19. M. Bredol, K. Matras, A. Szatkowski, J. Sanetra, A. Prodi-Schwab, Sol. Energy Mater. Sol. Cells **93**, 662 (2009)
20. D.C. Reynolds, D.C. Look, B. Jogai, Phys. Rev. B **60**, 2340 (1999)
21. J. Joo, J. Vac. Sci. Technol. **A18**, 23 (2000)
22. S. Guenes, H. Neugebaur, S. Sariciftci, Chem. Rev. **107**, 1324 (2007)
23. M. Onoda, K. Tada, H. Nakayama, J. Appl. Phys. **86**, 2110 (1999)
24. A. Khaliq, F. Xue, K. Varahramyan, Microelectron. Eng. **86**, 2312 (2009)
25. J. Callaway, Phys. Rev. **130**, 549 (1963)
26. M. Wang, X. Wang, Sol. Energy Mater. Sol. Cells **91**, 1782 (2007)
27. M. Planck, Ann. Phys. **4**, 553 (1901)
28. P. Würfel, *Physics of Solar Cells*, 2nd edn. (Updated and Expanded) (Wiley-VCH, Weinheim, 2009), p. 25
29. T.H.H. Le, K. Mawatari, Y. Pihosh, T. Kawazoe, T. Yatsui, M. Ohtsu, M. Tosa, T. Kitamori, Appl. Phys. Lett. **99**, 213105 (2011)
30. T. Mochizuki, K. Kitamura, T. Yatsui, M. Ohtsu, *Schedule and Abstract of the XIV International Conference on Phonon Scattering in Condensed Matter* (American Institute of Physics, Melville, NY, 2012), pp. 236–237
31. T. Yatsui, K. Iijima, K. Kitamura, M. Ohtsu, *Schedule and Abstract of the XIV International Conference on Phonon Scattering in Condensed Matter* (American Institute of Physics, Melville, NY, 2012), pp. 234–235

32. F. Yang, M. Willkinson, E.J. Austin, K.P. OfDonnell. Phys. Rev. Lett. **70**, 323 (1993)
33. T.P. Lee, C.A. Burus, A.G. Dentai, IEEE J. Quantum Electron. **17**, 232 (1981)
34. R.A. Milano, P.D. Dapkus, G.E. Stillman, IEEE Tran. Electron Devices **29**, 266 (1982)
35. T. Kawazoe, M.A. Mueed, M. Ohtsu, Appl. Phys. B **104**, 747 (2011)
36. R.J. Van Overstraeten, P. Mertens, Sold-State Electron. **30**, 1077 (1987)
37. J.A. Van den Berg, D.G. Armour, S. Zhang, S. Whelam, H. Ohno, T.-S. Wang, A.G. Cullis, E.H.J. Collart, R.D. Goldberg, P. Bailey, T.C.O. Noakes, J. Vac. Sci. Technol. B **20**, 974 (2002)
38. M.G.A. Bernard, G. Duraffourg, Phys. Status Solidi **1**, 699 (1961)
39. A. Einstein, P. Ehrenfest, Z. Phys. **19**, 301 (1923)
40. E. Scholl, *Nonequilibrium Phase Transition in Semiconductors* (Springer-Verag, Berlin, 1987), pp. 5–6
41. U.S. Department of Health and Human Services, Public Health Service, National Inst. Health, National toxicology program (ed.), *NTP technical report on the toxicology and carcinogenesis studies of indium phosphide* (U.S. Department of Health and Human Services, Washington, DC., 2012) NTP TR 499.
42. K.T. Delaney, P. Rinke, C.G. Van de Walle, Appl. Phys. Lett. **94**, 191109 (2009)
43. K.D. Hirschman, L. Tysbekov, S.P. Duttagupta, P.M. Fauchet, Nature **384**, 338 (1996)
44. Z.H. Lu, D.J. Lockwood, J.-M. Baribeau, Nature **378**, 258 (1995)
45. L. Dal Negro, R. Li, J. Warga, S.N. Beasu, Appl. Phys. Lett. **92**, 181105 (2008)
46. T. Komoda, Nucl. Instrum. Methods Phys. Res. Sect. B, Beam Interact. Mater. Atoms **96**, 387 (1995)
47. S. Yerci, R. Li, L. Dal Negro, Appl. Phys. Lett. **97**, 081109 (2010)
48. S.K. Ray, S. Das, R.K. Singha, S. Manna, A. Dhar, Nanoscale Res. Lett. **6**, 224 (2011)
49. T. Kawazoe, M. Ohtsu, *Extended Abstracts* (The 59th Spring Meeting, 2012) (The Japan Society of Applied Physics, Tokyo, 2012), paper number 17p–B11-1
50. M.-A. Tran, T. Kawazoe, M. Ohtsu, Appl. Phys. A, doi:10.1007/s00339-013-7907-9
51. N. Wada, T. Kawazoe, M. Ohtsu, Appl. Phys. B **108**, 25 (2012)
52. T. Kawazoe, M. Ohtsu, K. Akahane, N. Yamamoto, Appl. Phys. B **107**, 569 (2012)
53. H. Tanaka, T. Kawazoe, M. Ohtsu, Appl. Phys. B **108**, 51 (2012)
54. T. Hayashi, T. Kawazoe, M. Ohtsu, *Extended Abstracts* (The 73rd Autumn Meeting, 2012) (The Japan Society of Applied Physics, Tokyo, 2012), paper number 13p–F8-11
55. T. Kawazoe, M. Ohtsu, Appl. Phys. A, doi:10.1007/s00339-013-7930-x
56. D. Seghier, H.P. Gislason, J. Mater. Sci., Mater. Electron. **19**, 687 (2008)
57. K. Kitamura, T. Kawazoe, M. Ohtsu, Appl. Phys. B **107**, 293 (2012)

Chapter 8
Spatial Features of the Dressed Photon and its Mathematical Scientific Model

Ars longa, vita brevis.
Lucius Annaeus Seneca, *De Brevitate Vitae,* 1.1

After Sect. 2.2 in Chap. 2 reviewed the spatial features of the dressed photon (DP), Chap. 3–7 reviewed its temporal features, which enabled analysis of the DP-mediated energy transfer. In the present chapter, the spatial features of the DP are discussed again in order to demonstrate some novel applications. Furthermore, relevant mathematical scientific models are described, and these are effectively used for analyzing the spatial features of the autonomous annihilation and creation of dressed-photon–phonons (DPPs), described in Sects. 6.3, 7.2, and 7.3.

8.1 Hierarchy

Section 2.2.2 reviewed size-dependent resonance and hierarchy, which are typical spatial features of the DP. Size-dependent resonance has been observed in the molecular dissociation described in Sect. 6.1.2 Specifically, in the deposition of Zn particles on a sapphire substrate after dissociating DEZn molecules by DPPs at the apex of a probe, the deposition rate reaches a maximum when the size of the deposited Zn particles reaches the diameter of curvature (9 nm) of the probe apex, as shown in Fig. 8.1 [1].

Furthermore, the left parts of Fig. 8.2a and b show near field optical microscope (NOM) images of several flagella of salmonellae [2], where the probe–flagellum separations are 15 and 65 nm, respectively. It can be seen that the diameters of the filament-like structures in Fig. 8.2a are smaller than those in b. In order to explain the origin of the difference in diameter, the right parts of Fig. 8.2a and b schematically show the modeled setups used to obtain these images. The probe is approximated as two spheres, where the smaller sphere represents the apex of the probe, and the larger sphere is placed on the smaller sphere. The right part of Fig. 8.2a shows that the DP

M. Ohtsu, *Dressed Photons*, Nano-Optics and Nanophotonics,
DOI: 10.1007/978-3-642-39569-7_8,

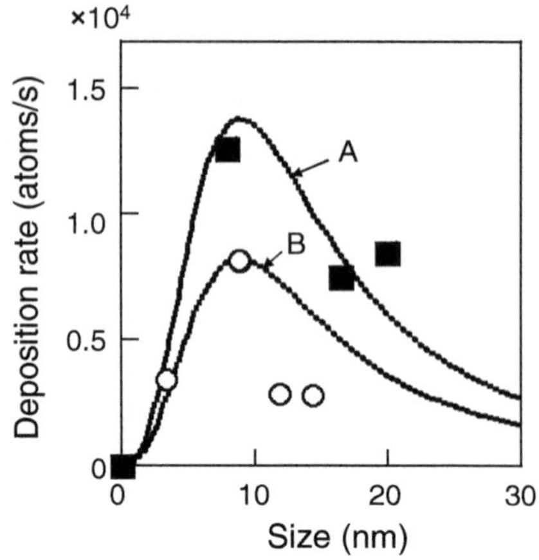

Fig. 8.1 Relation between the size of a ZnO particle and the deposition rate. The diameter of curvature of the probe apex is 9 nm. Closed squares and open circles represent the experimental results using light powers of 10 and 5 μW, respectively. The curves A and B represent Eq. 2.80 fitted to the experimental values

fields generated on the two spheres extend to the flagellum when the probe–flagellum separation is small. This means that a high-spatial-resolution image can be acquired by the DP field on the smaller sphere due to the size-dependent resonance between the flagella and the smaller sphere of the probe, even though it is partly superposed with the low-resolution-image obtained by the DP field on the larger sphere. On the other hand, as shown in the right part of Fig. 8.2b, the DP field on the smaller sphere does not extend to the flagella when the probe–flagellum separation is large. Only the DP on the larger sphere is involved in the imaging due to the size-dependent resonance between the flagella and the larger sphere of the probe, by which a low-resolution image is acquired. Furthermore, this separation-dependent feature of the resolution is evidence for the manifestation of hierarchy. The following subsections review novel applications of this phenomenon, mainly information security based on this hierarchy.

8.1.1 Hierarchical Memory

Following upon the massive increases in information storage density and capacity enabled by technological advancements, recently there has been a demand for novel functions in information security, and these demands can be met by utilizing the hierarchical feature of DPs. More concretely, abstract data, meta data, or tag data can be recorded in a different hierarchical layer, in addition to the original raw data. For example, low density, rough information is read out at a coarser scale (refer to Fig. 8.3a), whereas high-density, detailed information is read out at a finer scale (refer to Fig. 8.3b). A novel optical memory based on this read-out method has been called a "hierarchical memory" [3].

As an example of a hierarchical memory, consider a maximum of N nanoparticles distributed on the circumference of a circle with sub-wavelength diameter (400 nm),

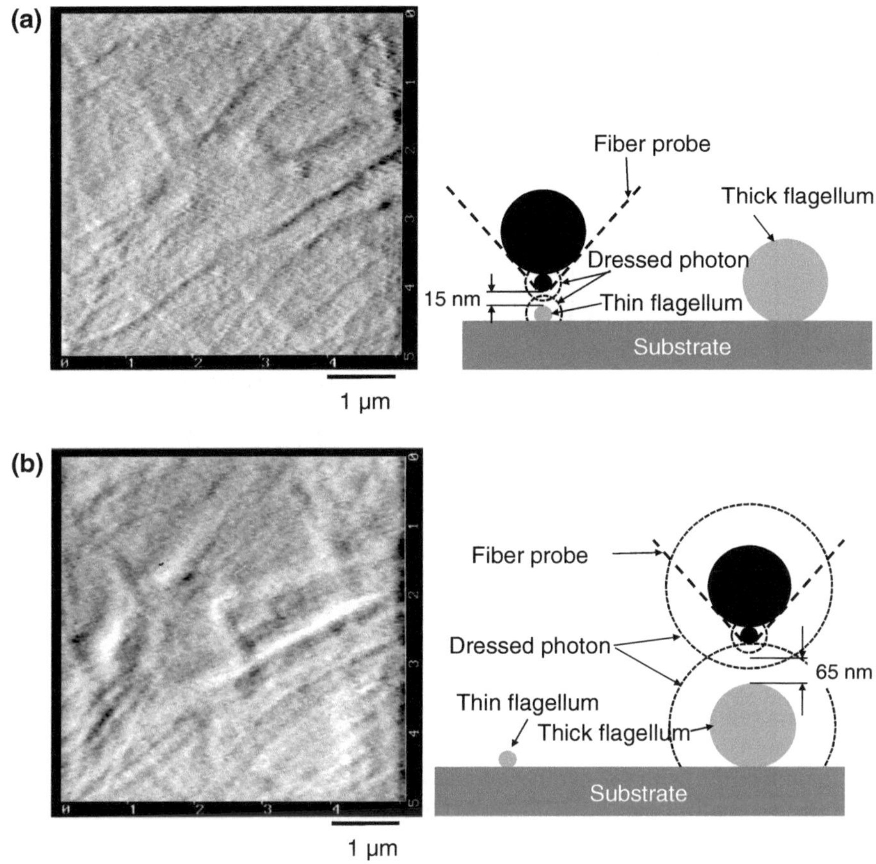

Fig. 8.2 Near-field optical microscope images of the flagella of salmonellae and schematic explanation of the setups used to obtain them. **a** and **b** represent cases with probe–flagellum separations of 15 and 65 nm, respectively

as shown in the left part of Fig. 8.4a [3]. These nanoparticles can be resolved by a NOM if the apex size of the probe is comparable to the size of individual nanoparticles. In this way, the first-layer information associated with the spatial distribution of the nanoparticles is retrievable, corresponding to 2^N different codes. On the other hand, by using a probe with a larger diameter, mean-field features with a low resolution comparable to the apex size of the probe can extracted. Namely, the number of particles within an area comparable to the apex size can be read out. Thus, the second-layer information associated with the number of particles, corresponding to $N+1$ different signal levels, is retrievable. Therefore, one can access different sets of signals, numbering 2^N or $N+1$, depending on the scale of observation using the probe. This leads to the possibility of hierarchical memory retrieval by associating this information hierarchy with the distribution and the number of nanoparticles

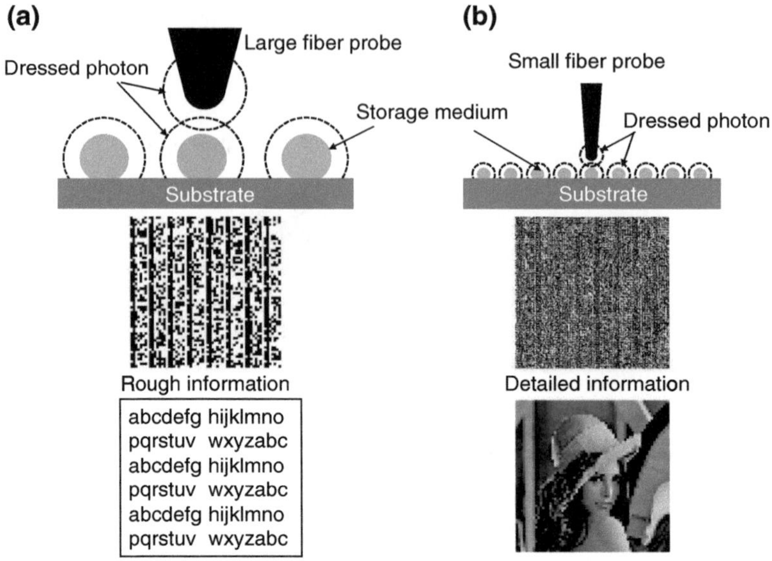

Fig. 8.3 Structure of the hierarchical memory. **a** and **b** represent the read-out of rough and detailed information, respectively

using an appropriate coding strategy. For example, in encoding N-bit information, (N−1)-bit signals can be encoded by distributions of nanoparticles, while associating the remaining 1 bit with the number of nanoparticles.

In order to confirm the above principle, simulations have been carried out assuming ideal isotropic metallic nanoparticles to see how the second-layer signal varies depending on the number of nanoparticles. Au nanoparticles with a diameter of 80 nm are distributed on the circumference of a 400 nm-diameter circle on a SiO_2 substrate, as shown by the scanning electron microscope (SEM) image in the right part of Fig. 8.4a. The closed squares and the solid curve in Fig. 8.4b show the calculated scattered light intensity as a function of the number of nanoparticles. This function can be approximated as a straight line, proportional to the number of nanoparticles, which supports the simple physical model described above.

In order to experimentally demonstrate this principle, groups of Au nanoparticles, each nanoparticle having a diameter of 80 nm, were distributed on 400 nm-diameter circles on a SiO_2 substrate. The spacing between each group of Au nanoparticles was 2 μm. An SEM image is shown in the upper part of Fig. 8.4c. The lower part shows an intensity image acquired by the NOM, from which the second-layer information is retrieved. The closed circles in Fig. 8.4b indicate the peak intensity of each spot in the lower part of Fig. 8.4c, which also increases linearly and is consistent with the simulated results. These results demonstrate hierarchical memory retrieval from nano-structures.

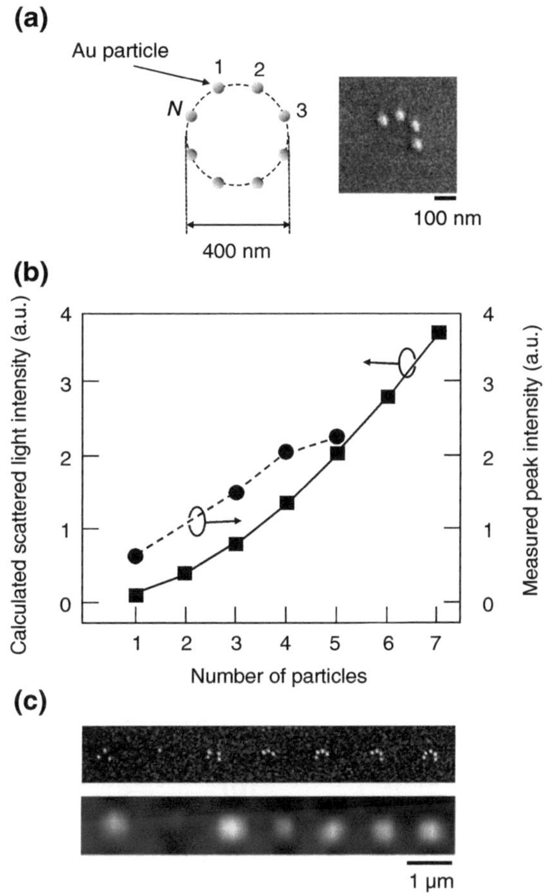

Fig. 8.4 Experimental results of hierarchical memory. **a** Au nanoparticles on the circumference of a circle. **b** Relation between the number of Au nanoparticles on the circumference and the scattered light intensity. Closed squares and circles represent the calculated and measured values, respectively. **c** Scanning electron microscope image (*upper*) and near-field optical microscope image (*lower*) of the seven circles

8.1.2 Hierarchy Based on the Constituents of Nanomaterials

The length of the effective interaction between nanomaterials is given by Eq. 2.78b, which depends not only on the size a_α of the nanomaterial but also on the effective mass m_α of the exciton. This means that the spatial extent of the DP field depends not only on the size of the nanomaterial but also on its constituents, which suggests that the hierarchical feature also depends on the constituents.

This dependency has been confirmed by experiments using a cubic core—shell structured nanomaterial [4]. As is confirmed by the SEM image in Fig. 8.5a, its size is $150 \times 150 \times 50$ nm. Its inner core and outer shell are composed of Au and Ag, respectively, as can be seen in the TEM image in Fig. 8.5b, which was acquired in the growth process. Nanomaterials 1 and 2 used as measurement specimens had Au fractional ratios of 33 and 43 %, respectively. That is, the constituents of the materials were different from each other, even though their sizes were equal. NOM images

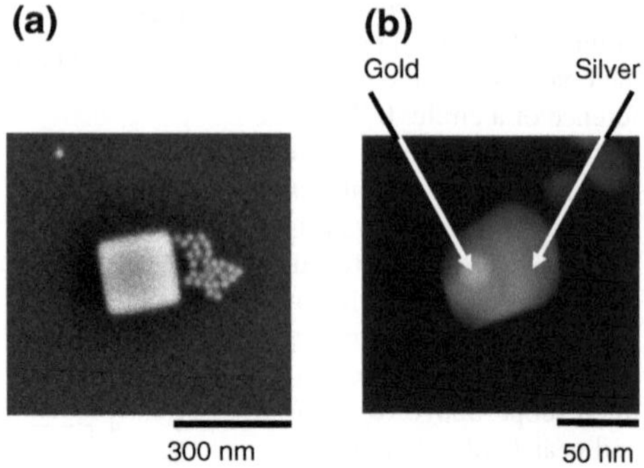

Fig. 8.5 A core–shell structured nanomaterial. **a** Scanning electron microscope image. **b** Transmission electron microscope image

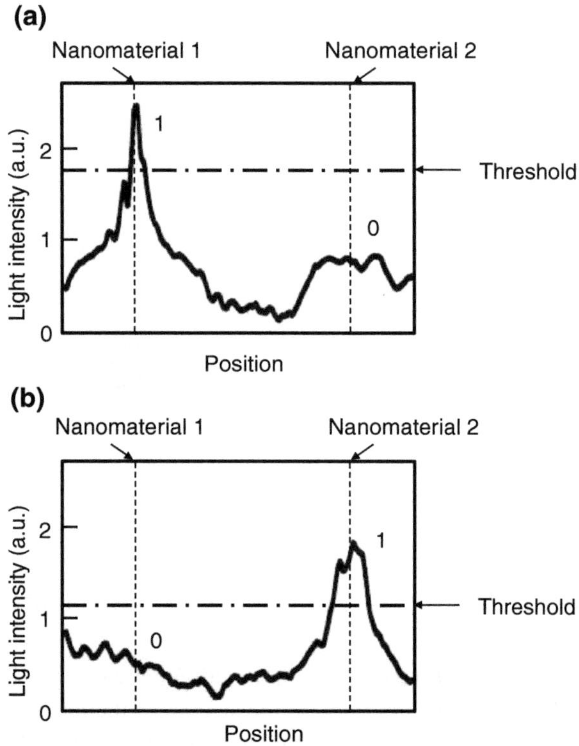

Fig. 8.6 Cross-sectional profiles of near-field optical microscope images of nanomaterials 1 and 2. **a** and **b** represent the image acquired when the probe–specimen separations are large and small, respectively

were acquired by scanning the probe two-dimensionally above the specimens. The wavelength of the light source used for imaging was 785 nm. The curves in Figs. 8.6a and b represent the cross-sectional profiles of the measured light intensities for large and small separations between the probe and the specimen, respectively. It is found from Fig. 8.6a that the measured light intensity for Nanomaterial 1 is larger than that for Nanomaterial 2. On the contrary, in Fig. 8.6b, the measured light intensity for Nanomaterial 1 is smaller than that for Nanomaterial 2. Such an inverse relation between Fig. 8.6a and b is due to the difference in the interaction lengths, given by Eq. 2.78b, for Nanomaterials 1 and 2, which means that the hierarchical feature depends on the constituents of the nanomaterials (specifically, the mole fraction of Au).

From the viewpoint of digital information retrieval, one can retrieve a logical-one level from Nanomaterial 1 in the distant layer, while at the same time retrieving a logical-zero level from the same material in the closest layer by digitizing the light intensity with an appropriate threshold level indicated by the horizontal chain lines in Figs. 8.6a and b. On the other hand, from Nanomaterial 2, one can retrieve the logical-zero level in the distant layer and the logical-one level in the closest layer. This means that logically inverted combinations are retrievable, which suggests the possibility of retrieving arbitrary information by properly designing the constituents of the nanomaterials.

8.1.3 Hierarchy and Local Energy Dissipation

By combining the hierarchical feature of Sects. 8.1.1 and 8.1.2 with light—matter interactions in nanometric space, further functions can be generated. As an example, this subsection reviews the combination of the interaction in one layer and local energy dissipation. This enables a novel traceable optical memory that can record memory accesses to each bit [5]. This memory uses different interactions between the two layers, allowing digital information to be read out. Here, the memory access events are recorded by combining the interaction in the first layer and the energy dissipation. The recorded memory access can be read out by utilizing two local energy dissipation phenomena, which are:

(1) A photochemical reaction of the material around a metallic nanomaterial: This is possible by using DPs generated by an electric charge concentrated in the local area of the metallic nanomaterial under light illumination. For example, if two metallic nanomaterials with right-triangular shapes are provided, DPs are generated at the apexes of the triangles by light illumination. By depositing a thin film of, e.g., silver oxide, around the apexes, the DPs induce a photochemical reaction in the film to record the light illumination event.
(2) Energy dissipation in a large QD after energy transfer from the small QD to the large QD (refer to Chap. 3): Optical memories with much higher storage density are expected by exploiting this method because the sizes of the QDs are much smaller than those of the metallic nanomaterials in (1) above.

This subsection reviews the application of the first phenomenon (1) by focusing on the hierarchy and local energy dissipation [6]. The application of (2) will be reviewed in Sect. 8.3.2. Here, two right-triangular metallic nanomaterials shown in Fig. 8.7a (Shapes 1 and 2) are used for the traceable optical memory; the two triangles are arranged to face in the same direction in Shape 1, whereas they face each other

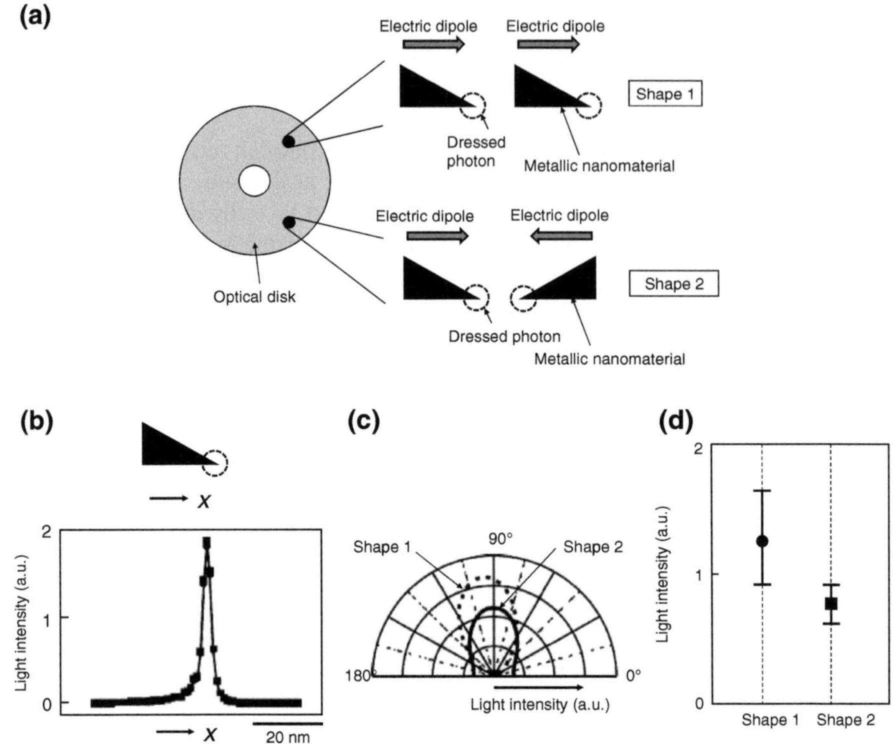

Fig. 8.7 Traceable optical memory. **a** Structure. **b** Calculated light intensities 1 nm above the apex of the right triangle of the metallic nanomaterial, i.e., in the first layer. **c** Calculated scattered light intensities in the second layer. **d** Scattered light intensities measured in the second layer

in Shape 2. The first-layer interaction occurs in the small area at the apexes of the right triangles, whereas the second-layer interaction occurs in the large-scale area covering the two right triangles. They correspond to the detailed and rough information discussed in Sect. 8.1.1, respectively.

Figure 8.7b shows the calculated light intensity at a position 1 nm above the apex of one of the right triangles of the metallic nanomaterial. Here, the apex angle, the length of the base, the height, and the thickness of the right triangle are 30°, 173, 100, and 30 nm, respectively. The separation between the two triangles is 50 nm. The wavelength of the incident light is 680 nm. This figure shows that the light intensity at the apex is more than 1000 times greater than that in the areas surrounding the apex, irrespective of whether the shape is Shape 1 or 2. With such a highly concentrated light energy at the apex, its dissipation can induce a photochemical reaction in the silver oxide film to record an access to the nanometric memory.

On the other hand, Fig. 8.7c shows the calculated scattered light intensities in the second layer for Shapes 1 and 2. It is found from this figure that the intensity for Shape 1 is larger than that for Shape 2; i.e., the two shapes show different optical responses in the second layer[1] Based on this difference, digital information read-out is possible in the second layer by allotting the logical-one and -zero levels to Shapes 1 and 2, respectively. In order to confirm this function, metallic Au nanomaterials of Shapes 1 and 2 were deposited on a SiO_2 substrate. The wavelength of the incident light was 690 nm. Figure 8.7d shows the scattered light intensities measured in the second layer, from which the read-out signal intensity from Shape 1 was confirmed to be larger than that from Shape 2.

The novel functions reviewed above are expected to find applications in the use and control of digital information, protecting personal information by managing recorded memory accesses of electronic tags, and so on.

8.1.4 Applications Exploiting the Differences Between Propagating Light and Dressed Photons

The simplest feature of the hierarchy is based on the difference in the spatial features between propagating light and DPs. Since very-low-intensity propagating light can pass through the small apex of the probe in imaging systems or the photomask aperture in fabrication systems, it is necessary to consider the difference between the spatial features of propagating light and DPs. This requirement has been met by using the phonon-assisted process in the case of energy transfer, as described in Chaps. 4 and 6. In Chaps. 3 and 5, it was also met by using the difference in energy transfer between QDs, i.e., the electric dipole-forbidden transition for propagating light was allowed in the case of DPs.

The present subsection reviews novel information security methods exploiting the difference in spatial features between propagating light and DPs. First, it should be noted that conventional optical devices exploit responses to propagating light. These responses include diffraction in the case of a hologram and a diffraction grating, and reflection and transmission in the case of a mirror and a lens. For these optical devices, an additional response can be provided by using the above-mentioned difference between propagating light and DPs, while still retaining the conventional responses. As an example, a novel optical device called a "hierarchical hologram" has been proposed. Holograms are representative examples of overt security because

---

[1] These optical responses can be understood by expressing the spatial distribution of the electronic charges as electric dipoles induced at the apexes of the right triangles under light illumination, as will also be shown in Sect. 8.2. That is, since the two triangles are arranged to face in the same direction in Shape 1, two mutually parallel electric dipoles are induced, as shown in the upper part of Fig. 8.7a. From this pair of parallel electric dipole moments, a large electric field is generated, which is easily detected in the second layer. Thus, parallel electric dipoles correspond to the bright state in Sect. 3.1 of Chap. 3. On the other hand, in Shape 2, the triangles are opposed to each other. Thus, the two induced electric dipoles are in an anti-parallel alignment, as shown in the lower part of Fig. 8.7a, forming an electric quadrupole. Since the electric fields generated from these anti-parallel electric dipoles cancel each other out, they cannot be detected in the second layer, corresponding to the dark state in Sect. 3.1.

they offer the advantage of allowing visual confirmation. Although holograms have been widely used in the anti-counterfeiting of bank notes, credit cards, and so on, conventional anti-counterfeiting methods based on the physical appearance of holograms are less than 100 % secure. It is quite difficult to add other security functions without degrading the optical response of the hologram.

In order to overcome this difficulty, the hierarchical feature of DPs has been applied to realize hierarchical holograms [7, 8]. A nano-structure is added to the hologram, and the DPs generated at the surface of the nano-structure are detected by two-dimensionally scanning a probe, producing an NOM image. This allows an additional layer of information to be read out using only the DPs. This can provide covert security, that is to say, an invisible information security technology. In the case of the hologram with an additional nano-structure, the information is read out through the interaction between the probe and the nano-structure. This achieves reliable anti-counterfeiting because of the technological difficulties that would be involved in duplicating the device and reading out the information.

Thus, the hierarchical hologram is a novel optical device in which overt and covert security features coexist, based on the difference in the spatial features of propagating light and DPs.

There are two methods for realizing a hierarchical hologram:

(1) Fabricating a nano-structure on the surface of a conventional hologram: A nano-structure is fabricated to store the additional information, while maintaining the conventional optical response of the hologram, and this additional information is read out using only the DPs [8]. Figure 8.8a shows an example, in which small circular pits of 100 nm diameter are formed on the surface of a 40 nm-thick Au film in a conventional hologram. As shown in Fig. 8.8b, the diffracted light intensity under propagating light illumination is not affected by the pits. On the other hand, by scanning the probe two-dimensionally above the surface of the specimen, specific NOM images that clearly depend on the pits are obtained, as shown in Fig. 8.8c.

(2) Fabricating a nano-structure in a hologram having a metallic grating structure [9]: As shown in the upper part of Fig. 8.9a, the fabricated hologram is composed of long linear patterns of metallic film lying in parallel along the y-axis. By fabricating a rectangular nano-structure with a side length of 80 nm in the linear patterns of the metallic film, the electric charge is concentrated at the upper and lower sides of the rectangular nano-structure when it is illuminated by light linearly polarized along the y-axis. This charge concentration is schematically explained by the lower left part of Fig. 8.9a, which represents a snap-shot of the spatial distribution of the electric charges in the temporally oscillating electric field of the incident light. When the structure is illuminated by light linearly polarized along the x-axis, the electric charges are concentrated at the left and right sides of the rectangular pattern. It should be noted that they are concentrated also at the left and right sides of the long linear patterns of the metallic film even in the absence of the rectangular nano-structure, as shown in the lower right part of Fig. 8.9a. This difference in the spatial features of the electric

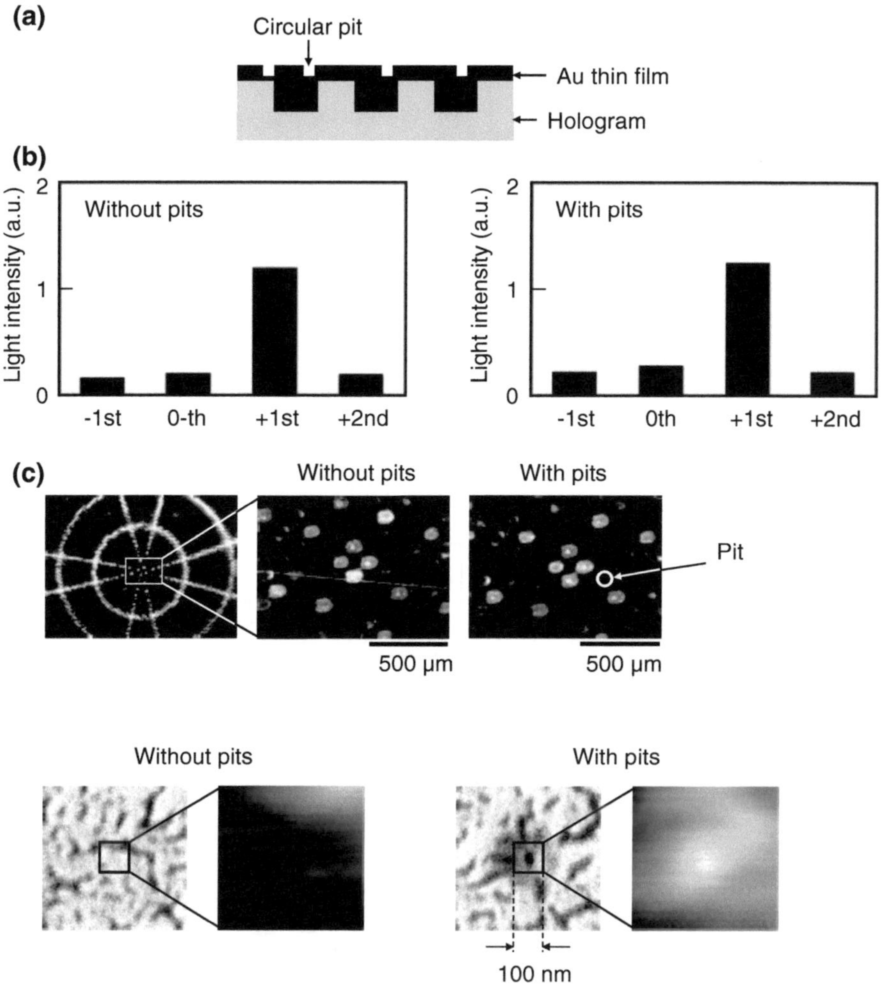

Fig. 8.8 Hierarchical hologram. **a** Structure. **b** Measured light intensity of each diffraction order. **c** Upper part shows optical images of the hologram surface and its central parts. Lower parts show scanning electron microscope images and near-field optical microscope images

charge concentration allows the information to be read out in nanometric space. Figure 8.9b shows this dependency, which has been confirmed by using a metric called recognizability, defined as the difference between the measured light intensities in the nanometric space and that averaged over the macroscopic space. Closed circles in this figure show that the recognizability due to the rectangular nano-structure fabricated on the hologram is much larger than that (open triangles) for the solitary nano-structure on a flat substrate when the incident light is linearly polarized along the y-axis (90°-polarization angle).

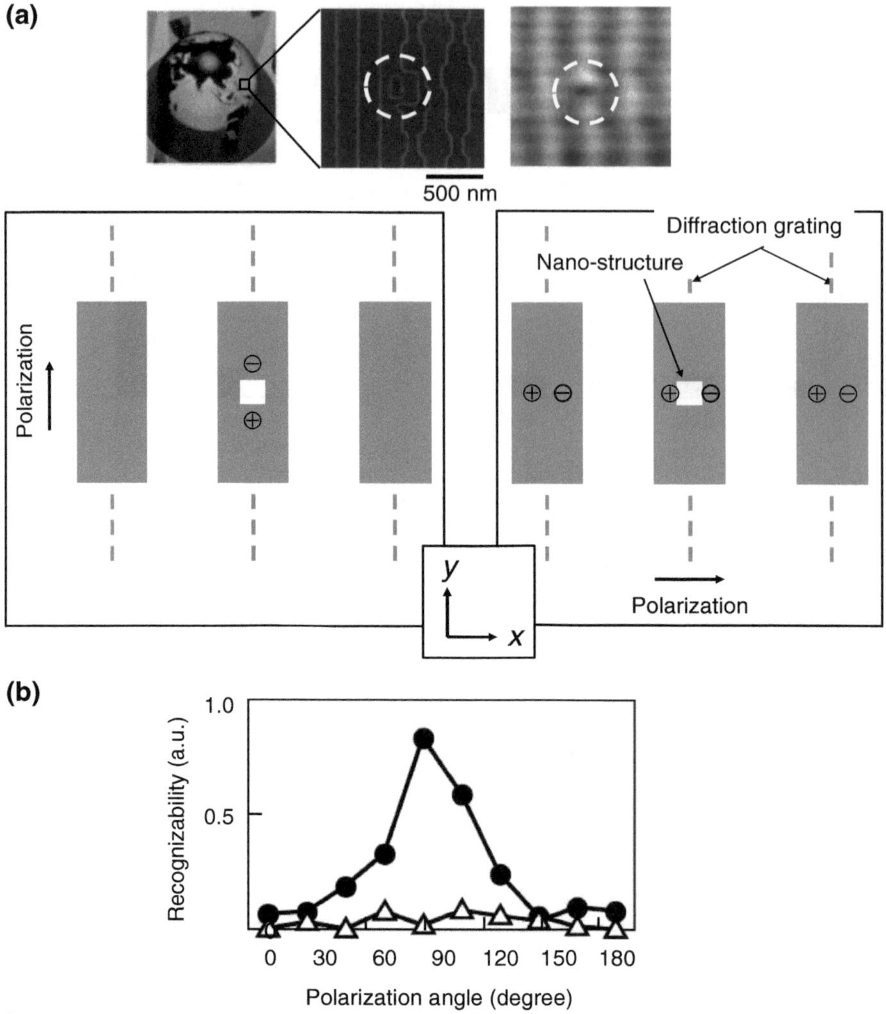

Fig. 8.9 Hologram with nano-structure. **a** *Upper left*, *center*, and *right* pictures show the surfaceimage, magnified scanning electron microscope image, and near field optical microscope image, respectively. *Lower left* and *right* figures represent the electric charges concentrated when the hologram is illuminated by y-and x-polarized light, respectively. **b** Calculated recognizability as a function of the polarization angle. The closed circles represent the values for the hologram with the nano-structure. The open triangles represent the values for a solitary nano-structure on a flat surface

To improve the anti-counterfeiting performance even further, it is also possible to fabricate a nano-structure that cannot be duplicated even by the manufacturer. This fabrication method corresponds the technology of "artifact-metrics" [10]. It is also related to the field of artificial biometrics, which uses the individuality inherent in the structures of fingerprints and veins.

8.2 Conversion From an Electric Quadrupole to an Electric Dipole

This section reviews a novel device operating principle and its application to information security based on the DP-mediated interactions between small metallic patterns placed on a large-area substrate. The principle of operation is explained from the viewpoint of conversion from an electric quadrupole to an electric dipole [11].[2] Metals have been popularly used due to the simplicity of controlling their sizes and positions in deposition on a large-area substrate. When an electron in the metal is excited by light illumination, one cannot expect any specific behavior based on the quantized discrete energy state of the electron. Although this may be possible if the size of the metal is smaller than several nanometers and at extremely low temperature, these extreme situations are very inconvenient for practical use. Based on such practical considerations, the classical model of electronic motion in the metal is used for the following discussion.

A large number of sub-wavelength metallic patterns are deposited on a transparent large-area substrate, and the array of these patterns works as a polarization controller for the incident light. First, in the case of the layout of metallic patterns having a shape like a letter "I" of sub-wavelength size, as illustrated in Fig. 8.10a, since they are aligned along the x-direction, the electric charges are concentrated at both ends of the patterns when they are illuminated by x-polarized light. The thick white arrows in this figure indicate vectors directed from the negative to the positive charges. They correspond to electric dipoles and are called intra-material polarization. On the other hand, the vectors directed from the negative to the positive electric charges in adjacent I-shaped metallic patterns are represented by thick black arrows. These electric dipoles are called inter-material polarization, which originates from the interaction between the adjacent I-shaped metallic patterns mediated by DPs. By using an appropriate layout of the I-shaped metallic patterns, the inter-material polarization can be directed along the y-axis. This figure represents a snap-shot of the temporally oscillating intra- and inter-material polarizations. It should be noted that adjacent inter-material polarizations are anti-parallel to each other in this figure, which means that the pair of adjacent polarizations corresponds to an electric quadrupole. Although the temporally oscillating polarization generates an electric field and works as the source for the light transmitted through the polarization controller, y-polarized transmitted light is not generated because the electric quadrupoles composed of the pair of adjacent inter-material polarizations are oriented along the

---

[2] This conversion has also been seen in the energy transfer from a small QD to a large QD, as was described in Chap. 3. The (1, 1, 1) and (2, 1, 1) energy levels of the cubic small and large QDs are electric dipole-allowed and -forbidden, respectively, corresponding to two electric dipoles respectively aligned in parallel and anti-parallel directions. Therefore, the energy transfer from the (1, 1, 1) energy level in the small QD to the (2, 1, 1) energy level in the large QD corresponds to the conversion from the electric dipole to the electric quadrupole. Furthermore, the subsequent relaxation from the (2, 1, 1) energy level to the (1, 1, 1) energy level in the large QD corresponds to the conversion from the electric quadrupole to the electric dipole.

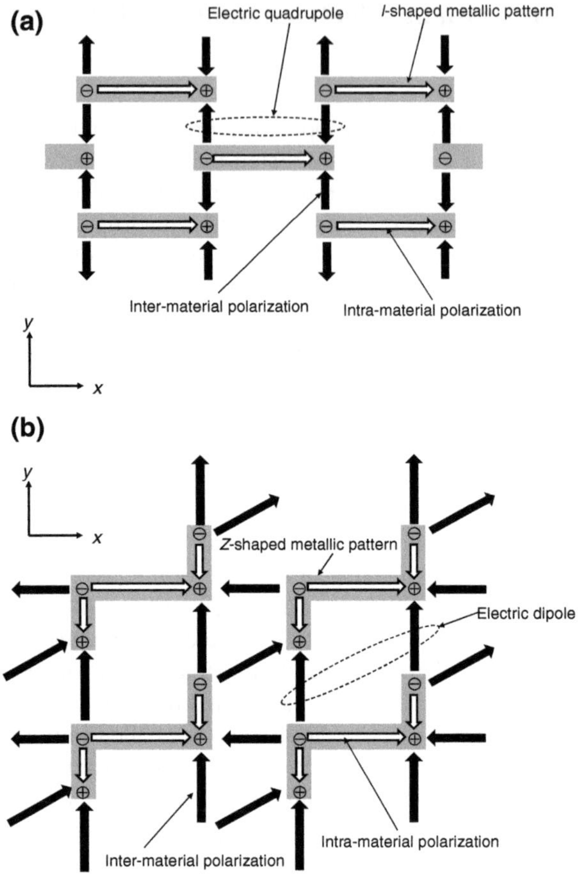

Fig. 8.10 Two-dimensional array of sub-wavelength-sized metallic patterns. **a** and **b** represent I-shaped and Z-shaped patterns, respectively

y-axis. On the other hand, x-polarized transmitted light is generated because the intra-material polarization along the x-axis works as the source. In summary, this polarization controller transmits x-polarized incident light only.

Second, Fig. 8.10b shows the layout of sub-wavelength Z-shaped metallic patterns. Since their main axes and two arms are directed along the x- and y-axis, respectively, the electric charges are concentrated on both ends of the main axis when x-polarized light is incident. They are also concentrated at the end of the two arms. Intra- and inter-material polarizations are also shown in this figure, whose directions are more complicated than those for the I-shaped patterns. However, by noting that a large number of inter-material polarizations are parallel and directed along the $+y$-axis, their sum corresponds to a large electric dipole, not an electric quadrupole. Since this electric dipole works as a source for generating y-polarized

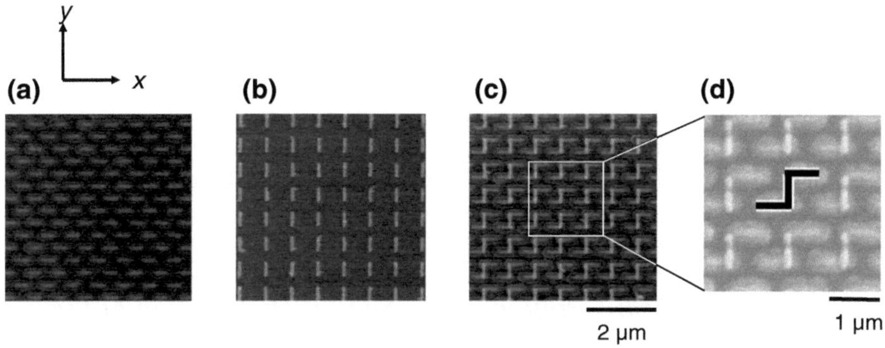

Fig. 8.11 Lock and key system using two-dimensional array of *I*-shaped metallic patterns. **a** and **b** show scanning electron microscope images of the polarization controllers 1 and 2, used for the lock and key, respectively. **c** and **d** are scanning microscope images acquired when the polarization controllers 1 and 2 are stacked

light, this polarization controller converts x-polarized incident light into y-polarized transmitted light.

As demonstrated in Fig. 8.10a and b, a two-dimensional array of sub-wavelength metallic patterns can be used as a novel polarization controller because the arrangement of the electric dipoles and quadrupoles can be controlled by controlling their shape, size, and layout. A novel "lock and key" system in which this polarization controller is applied to information security has also been proposed [12]. In this system, two I-shaped polarization controllers are used, as shown in Fig. 8.11. The polarization controller 1 (Fig. 8.11a), in which the I-shaped metallic patterns are arranged along the x-axis, is used as a lock. On the other hand, in the polarization controller 2 (Fig. 8.11b), to be used as a key, the I-shaped metallic patterns are arranged along the y-axis. First, when illuminating the polarization controller 1 with x-polarized light, no y-polarized light is transmitted. However, by stacking the polarization controller 2 on polarization controller 1, a DP-mediated interaction can be induced between the metallic patterns in the polarization controllers 1 and 2 if the gap between the two stacked controllers is narrower than the wavelength of the light. In this case, the shape of the stacked metallic patterns becomes equivalent to the Z-shaped pattern of Fig. 8.10b, as shown in Fig. 8.11c and d. Since the inter-material polarization is directed along the y-axis, y-polarized transmitted light is generated by illuminating the structure with x-polarized light. This transmitted light originates from the conversion from the electric quadrupoles in the individual polarization controllers 1 and 2 to the electric dipoles in the stacked structure. Thus, the transmission efficiency depends on the sizes, shapes, and layout of the metallic patterns in the two polarization controllers, which means that there is a correct key for each lock. When another key is stacked on the lock, the transmission efficiency is inevitably lower, making this lock and key system advantageous for information security applications.

8.3 Probe-Free Methods

The methods of the previous sections required scanning a probe or bringing two planar devices into close proximity, which may not be convenient in some practical applications. In order to overcome these issues, this section reviews methods that are free from these requirements, allowing low-resolution optical measurements in the far-field.

8.3.1 Magnified Transcription of the Spatial Distribution of the Interaction

This subsection reviews the magnified transcription of the spatial energy distribution of the interactions via DPs using a photo-induced phase transition [13]. Among several cyano-bridged metal complexes used for photo-induced phase transitions [14], this subsection examines complexes including Fe and Mn [15] as an example specimen. Figure 8.12a shows an SEM image of a single crystal of such a complex, with dimensions $1 \times 1\,\mu\text{m} \times 500\,\text{nm}$. It exhibits a phase transition between high-temperature and low-temperature phases due to the charge transfer induced by light illumination [16, 17], resulting in a structural change between a cubic and tetragonal structure caused by Jahn-Teller distortion of the Mn. It is known that the phase transition area is spontaneously expanded to the surroundings after the initial phase transition is induced at the crystal surface. Finally, the total area of the phase transition increases spontaneously to 30-times the incident light spot size [18]. Once this area is magnified via this spontaneous increase, it can be easily observed by a conventional image processing method using propagating light [19].

Pump-probe spectroscopy has been carried out to confirm this magnified transcription. Propagating light with a wavelength of 532 nm was used as pump light, which was injected into the tail of the probe to generate DPs on the apex for inducing the phase transition in the crystal. Low-power propagating light with a wavelength of 635 nm was used as probe light for confirming the phase transition.[3] This light was injected into the tail of the same probe for acquiring an NOM image of the crystal. Figure 8.12b shows a cross-sectional profile of the measured NOM image, which shows a considerable increase in the measured light intensity as a result of the pump light illumination.

Figure 8.12c shows the light power density on the crystal surface, estimated by numerical calculations. It shows that the light power density on the surface of the high-temperature phase is larger than that of the low-temperature phase, which agrees with the results of Fig. 8.12b, confirming the successful transcription of the spatial distribution of the DPs. The size of the transcripted pattern was estimated to be 1 μm, both by experiments and by numerical calculations, which is sufficiently larger than

[3] Although a probe is used here for confirming the phase transition, it is not required to read out the transcripted area after it is magnified.

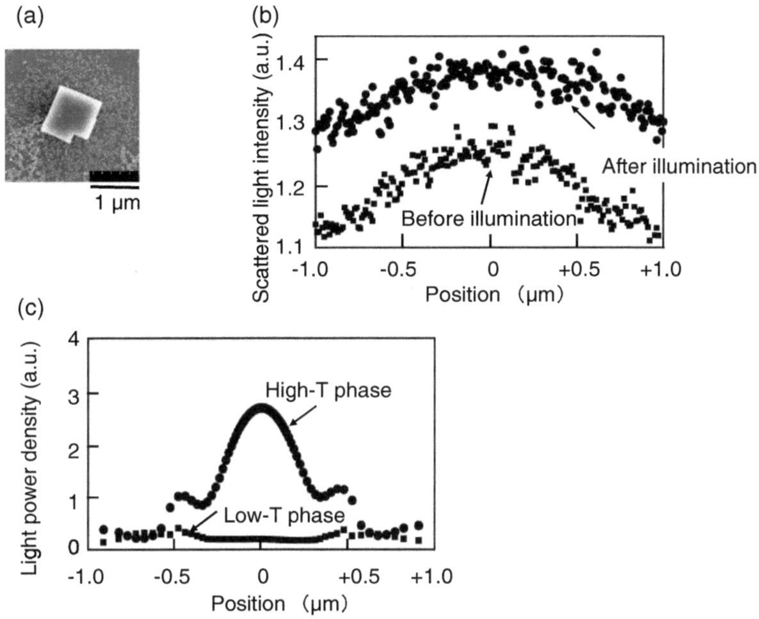

Fig. 8.12 Magnified transcription by using a photo-induced phase transition. **a** Scanning electron microscope image of a single crystal of the complex. **b** and **c** show the measured and calculated scattered light intensities, respectively

the radius of curvature (50 nm) of the probe apex. From these results, magnified transcription was confirmed, demonstrating that the magnification is sufficiently large for detection without using probes.

The present method of retrieval depends on the features of the energy transfer [20] and hierarchy [3]. It should be noted that the present method is essentially different from the conventional methods of memory retrieval such as holography using DPs [21], because only the change in the structure of the memory materials is used.

8.3.2 Spatial Modulation of the Energy Transfer Between Quantum Dots

This subsection reviews an example of applying phenomenon (2) of Sect. 8.1.3. Since the energy is transferred from a small QD to a large QD via a DP under light illumination, the light intensity emitted from the small QD decreases, whereas that from the lower energy level in the large QD increases. The wavelength of the latter is longer than that of the former because of the subsequent relaxation from the higher to the lower energy level in the large QD. Furthermore, the amount of energy transfer decreases with increasing separation between the two QDs. Therefore, when

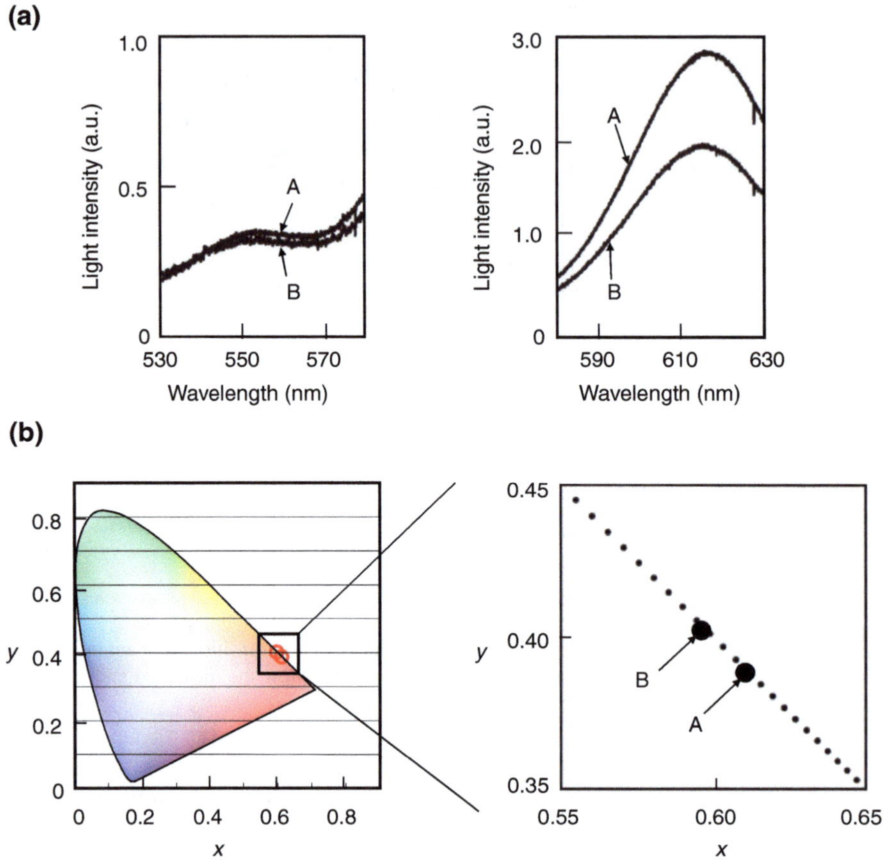

Fig. 8.13 Light emission from CdSe/ZnS core–shell structured quantum dots. **a** Spectral profiles. The curves A and B represent the measured results with and without bending the substrate, respectively. **b** The positions of the spectra of the curves A and B on a chromaticity diagram

emissions from the two QDs are simultaneously measured by a single photodetector in the far-field with a spatial resolution lower than that determined by the diffraction limit, the intensity of the longer-wavelength component of the measured spectra decreases with increasing separation. This suggests that the emission spectral profile, measured in the far-field, can be varied by varying the separation. In other words, the emission spectral profile varies depending on the radius of curvature of the bent flexible substrate on which the QDs are dispersed.

This variation has been observed by dispersing small and large QDs with CdSe/ZnS core-shell structures on a flexible, transparent polydimethilsiloxane (PFMS) substrate. Figure 8.13a shows the measured spectral profiles which vary in response to bending of the substrate [22]. Also, Fig. 8.13b shows that the color tone of the emitted light on the chromaticity diagram varies. These variations can be applied to a sensor for detecting a tiny strain and to a modulatable multi-spectral emitting device whose emission spectrum can be switched by applying mechanical modulation.

8.4 Mathematical Scientific Model

Although Sects. 6.1 and 6.2 of Chap. 6 reviewed selective molecule dissociation and nano-pattern formation by the DPPs generated on a probe and photomask, similar dissociation and nano-pattern formation are possible without using these devices. For example, Zn nanoparticles can be selectively deposited on the apex of ZnO nanorods that stand close together on a planar substrate. They are deposited by metal-organic chemical vapor deposition using DPPs generated on the apex of the ZnO nanorods by light illumination. The upper left and upper right parts of Fig. 8.14a show SEM images of the Zn nanoparticles deposited on ZnO nanorods without and with light illumination, respectively. Here, the ZnO nanorods themselves work as the probe for generating the DPPs. The SEM images were analyzed to determine the representative scales; they are highlighted as shown in the lower left and lower right parts of Fig. 8.14a, which were digitized from the two upper images, respectively. Physically, the lower left part represents the projected area of the ZnO nanorods, and the lower right part shows that of the deposited Zn nanoparticles on the top of the ZnO nanorods. The representative scale, which is denoted by the horizontal extent S of the structures, is evaluated as is schematically shown in the magnified image of the lower left part. Figure 8.14b shows the measured incidences of the scales S, where the closed squares and circles indicate the structures formed without light and with light, respectively. The structures formed without light exhibit a maximum incidence around 30 nm, which is equivalent to the representative scale S of this structure. On the other hand, the structures fabricated with light exhibit a quite different incidence pattern with smaller scales and higher incidences. In other words, smaller-scale structures were generated from larger-scale ones by the light illumination. Furthermore, it should also be noted that the histogram of the structures formed with light exhibits a power-law distribution; the closed circles were fitted by the broken line in a double logarithmic plot at scales larger than around 6 nm. This means that a fractal nature emerged from a non-fractal structure [23].

In contrast to the topics of Sects. 6.1 and 6.2 discussed above, Sect. 6.3 reviewed smoothing of a glass surface utilizing the autonomous annihilation of DPPs. Furthermore, Sect. 7.2 reviewed the fabrication of metallic electrodes of photo-voltaic devices utilizing the autonomous generation of DPPs. And Sect. 7.3 reviewed control of the spatial distribution of B by utilizing the autonomous generation of DPPs at the domain boundaries of B in an LED. It should be pointed out that, in the processes described in Sects. 6.3, 7.2, and 7.3, the autonomy governed the smoothing, device fabrication, and device operation.

On the other hand, conventional methods for smoothing, device fabrication, and device operation have explored novel materials based on advances made in materials science and technology. However, rare materials or toxic materials have to be used

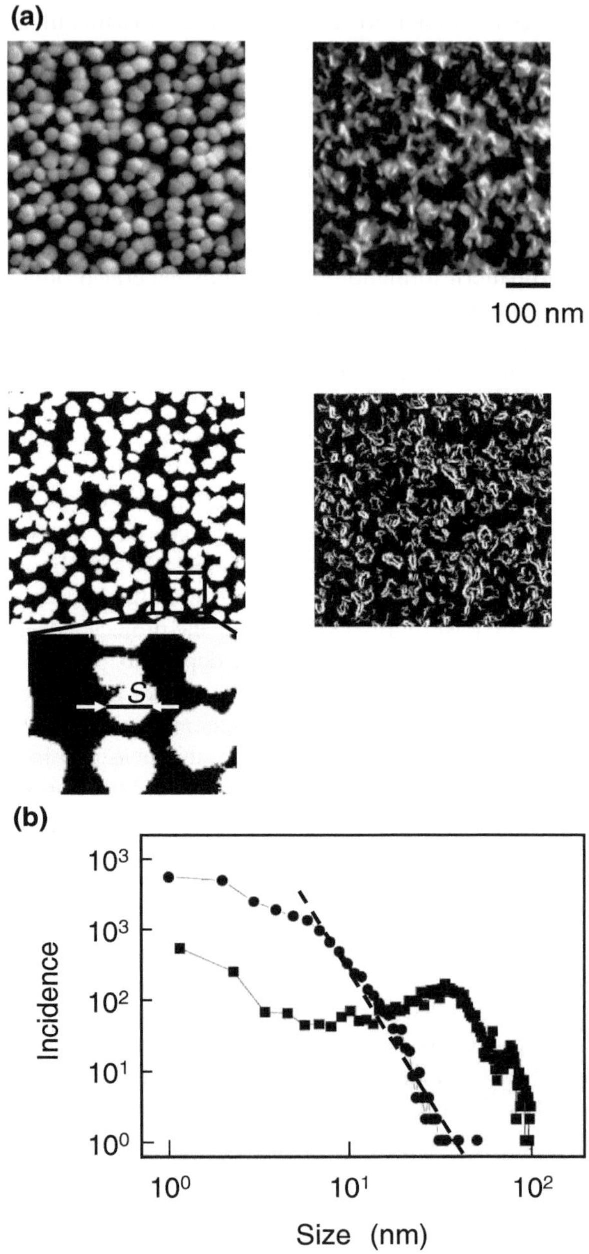

Fig. 8.14 Profiles of Zn nanoparticles deposited on the apexes of the ZnO nanorods. **a** *Upper left* and *right* pictures show scanning electron microscope images of the Zn nanoparticles grown without and with light illumination, respectively. *Lower left* and *right* pictures show images digitized from the upper images. **b** Incidence of the scale calculated from the lower pictures in (**a**). Closed squares and circles indicate the structures formed without and with light illumination, respectively. The broken line represents a power-law distribution fitted to the closed circles

in some cases, and this will be a serious problem in the future. In contrast to these, the method using DPPs relies on autonomous annihilation or generation, which is completely different from the deterministic methods of conventional materials science and technology.

The methods in Sects. 6.3, 7.2, and 7.3 are advantageous because they can be applied to large-area substrates, e.g., a glass surface, a large electrode of a photo-voltaic device, and a large pn-junction area in an LED. However, since the autonomous annihilation and generation of DPPs take place in a nanometric area on the surface of or inside the material, deterministic theoretical models of these processes inevitably involve an extremely long computing time. In order to decrease the computing time, novel statistical or mathematical scientific models are required. The following subsections review some of these models.

8.4.1 Formation of Nanomaterials

This subsection reviews stochastic models for describing, as an example, unique autonomous size- and position-controlled nanoparticle formation by using DPPs.

(a) Size control of a zinc oxide nanoparticle

A sol-gel method has been popularly used to grow ZnO nanoparticles in a solution. In order to decrease fluctuations in their sizes, they are illuminated by light while growing [24]. If the photon energy of the incident light is tuned to be lower than the bandgap energy E_g of smaller ZnO nanoparticles, they do not absorb this light, and thus, their growth is not impeded by the light. However, because this photon energy is higher than E_g of larger ZnO nanoparticles, they absorb the light, causing a certain amount of ZnO to be desorbed from the surface of the larger ZnO nanoparticles. As a result of this size-dependent resonant absorption of light, the fabricated ZnO nanoparticles are fixed to a specific size, which depends on the photon energy of the incident light, resulting in reduced size fluctuations of the ZnO nanoparticles. Experiments have demonstrated that the size fluctuation was decreased to 18 % by light illumination, whereas it is as large as 23 % in the case of the conventional sol-gel method. Figure 8.15a shows the measured size distribution of the ZnO nanoparticles grown without light illumination, which has a symmetric profile. On the other hand, Fig. 8.15b shows the distribution with illumination, which is asymmetric and the peak size is smaller than that of Fig. 8.15a. These features can be analyzed by using a stochastic model, as is reviewed in the following.

In the absence of light illumination, the formation process is represented as a stochastic pile-up model [25], as is schematically illustrated in Fig. 8.16a. An elemental material that constitutes a nanoparticle is represented by a square-shaped block. Such blocks are grown, or stacked one after another, with a piling probability p; accordingly, the piling fails with a probability $1 - p$. In other words, if the height of the pile at step t is expressed as $s(t)$, the piling probability is given by

$$P \langle s(t+1) = s(t) + 1 \mid s(t) \rangle = p, \tag{8.1a}$$

$$P \langle s(t+1) = s(t) \mid s(t) \rangle = 1 - p. \tag{8.1b}$$

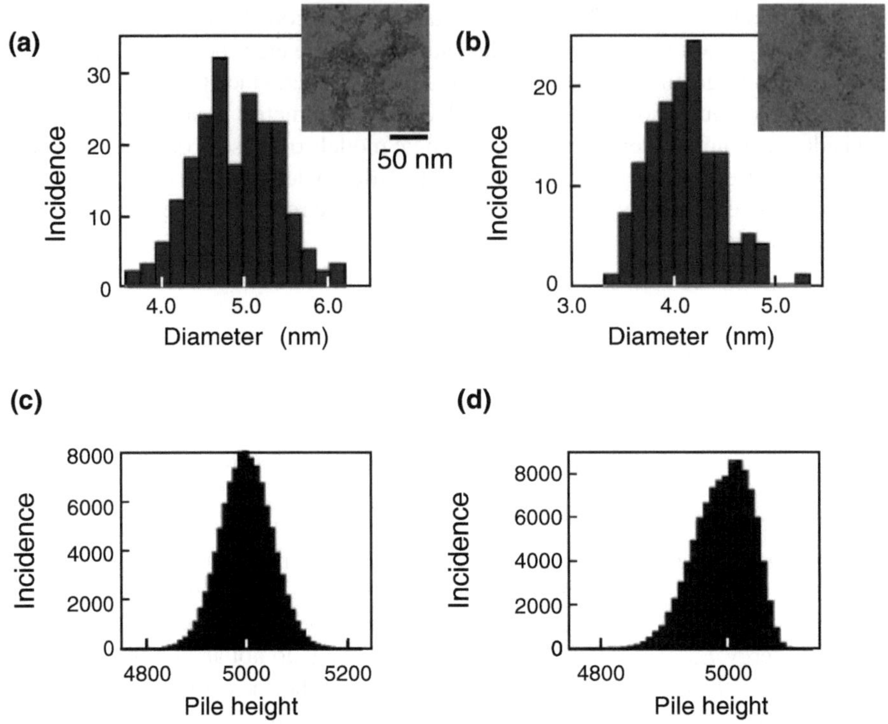

Fig. 8.15 Size distributions of the ZnO nanoparticles. **a** and **b** represent the measured results of the ZnO nanoparticles grown without and with light illumination, respectively. Their scanning electron microscope images are also shown. **c** and **d** represent the calculated results for the ZnO nanoparticles grown without and with light illumination, respectively

Since this is equivalent to a random walk with drift, the resultant heights of the piles exhibit a normal distribution after repeating this process with an initial condition $s(0) = 0$. The results shown in Fig. 8.15c were obtained by repeating 10,000 steps for 100,000 different trials. This result agrees with the experimental result of Fig. 8.15a.

On the other hand, the effect of light illumination on the formation process in the stochastic model is expressed as follows: Since desorption is induced at a particular diameter of nanoparticles, the piling probability p is assumed to be a function of the diameter, namely the height of the pile. For simplicity, p is assumed to decrease linearly beyond a certain total pile height R, as schematically illustrated in Fig. 8.16b. In other words, the desorption is more likely to be induced beyond a certain pile size due to the size-dependent resonant absorption mentioned above. That is, the probability p in Eq. (8.2) is replaced with the following size-dependent probability:

$$p[s(t)] = \begin{cases} c & s(t) \leq R \\ c - \alpha s(t) & s(t) \geq R \end{cases} \tag{8.2}$$

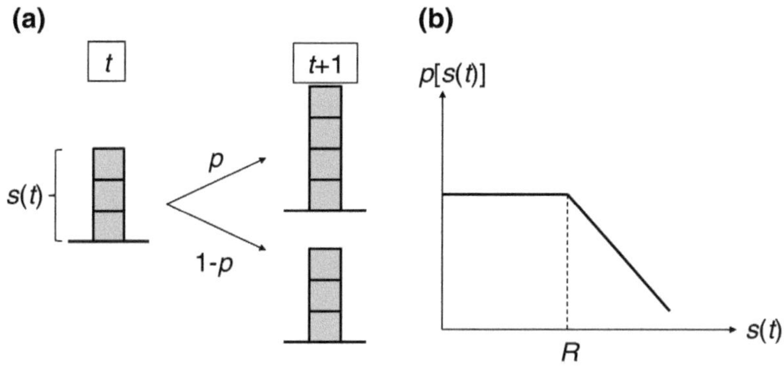

Fig. 8.16 Stochastic pile-up model. **a** Pile up with probability p. **b** Dependence of probability p on the pile height $s(t)$

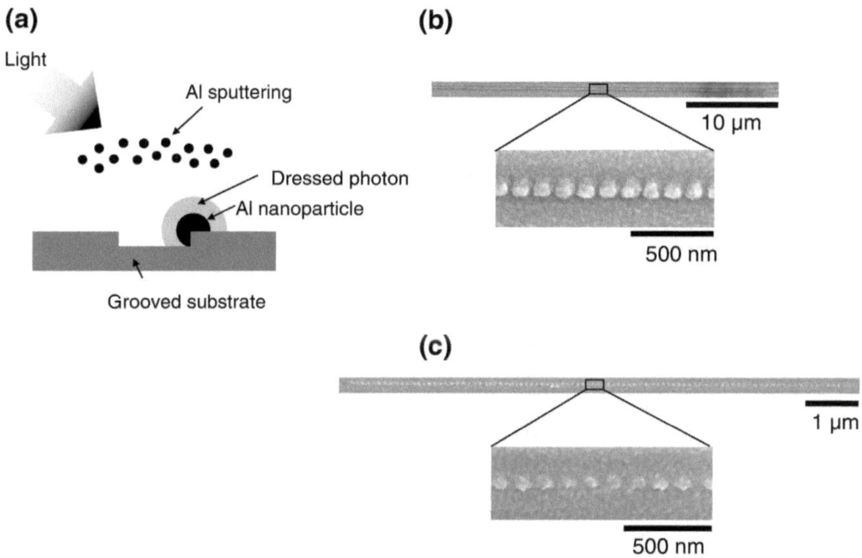

Fig. 8.17 Deposition of Al nanoparticles on a grooved substrate. **a** Cross-sectional profile of the substrate. **b** and **c** show scanning electron microscope images of Al nanoparticles, where the photon energies of the light irradiated in the process of deposition are 2.33 and 2.62 eV, respectively

where c and α are constants. With such a stochastic model, the resultant incidence distribution of the piles is skewed towards larger sizes. In the calculated results shown in Fig. 8.15d, the values of and are assumed to be 1/2 and 1/250, respectively. The distribution is asymmetric and is consistent with the experimental results of Fig. 8.15b.

(b) Autonomous formation of an ultra-long array of metallic nanoparticles

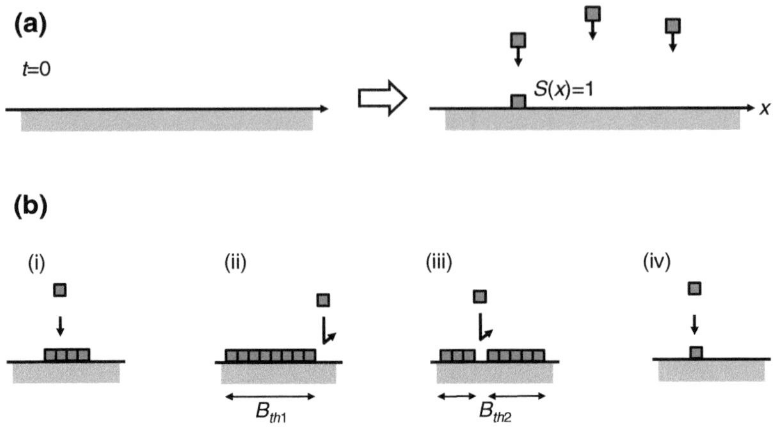

Fig. 8.18 Stochastic model of array formation. **a** Temporal evolution of deposition. **b** Four cases of deposition at position x on the substrate

Al nanoparticles can be deposited on a grooved SiO_2 substrate by RF sputtering, as schematically illustrated in Fig. 8.17a. By irradiating light with a photon energy of 2.33 eV (wavelength, 532 nm; light power, 50 mW) onto the ridgeline of the groove for generating DPs, an array of Al nanoparticles of uniform diameter (average diameter, 100 nm) and uniform separation (average separation, 28 nm) is formed at the ridgeline, as is demonstrated by the SEM image in Fig. 8.17b [26]. Since this formation occurs in the region where the ridgeline is illuminated by the light, the length of the array can be as long as 100 μm; i.e., the number of deposited Al nanoparticles can be as large as 780. For comparison, the SEM image in Fig. 8.17c shows an ultra-long array of Al nanoparticles formed by a similar method involving irradiating light with a photon energy of 2.62 eV (wavelength, 473 nm; light power, 100 mW). In this case, the average values of the diameter and separation are 84 and 49 nm, respectively. The experimental results of Figs. 8.17 suggest that autonomous material formation and arrangement are possible by utilizing the interaction between the DPs and Al nanoparticles. This can be applied to the fabrication of, e.g., the energy transmitter reviewed in Sect. 5.1.1.

To model this array formation [25], a one-dimensional (1-D) horizontal system that mimics the ridgeline of the groove is assumed. More specifically, it consists of an array of N pixels identified by an index i $(1 \leq i \leq N)$. An elemental material to be deposited onto the system is schematically represented by a square-shaped block. As depicted in Fig. 8.18a, the initial condition is a flat surface without any blocks.

At every iteration cycle, the position x at which a block arrives is randomly chosen. The success of the deposition at x is determined by the following rules. The occupation by a block at position x is denoted by $S(x)$: $S(x)=1$ when a block occupies a position x, and $S(x)=0$ when there is no block at position x. Also, the term "cluster" is used to mean multiple blocks consecutively located along the ridgeline. A single isolated block in the system is also called a "cluster".

(1) When the randomly chosen position x belongs to one of the clusters, namely, $S(x)$=1, the value of $S(x)$ is maintained at 1. (Fig. 8.18b–i)
(2) Even if $S(x)$=0, when the chosen position x belongs to a "neighbor" of a cluster with a size larger than a particular value B_{th1}, the deposition is inhibited. That is, $S(x)$ is maintained at 0. (Fig. 8.18b–ii)
(3) Even if $S(x)$=0, when the chosen position x has blocks on both sides and the total number of connected blocks is larger than B_{th2}, deposition is inhibited. That is, $S(x)$ is maintained at 0. (Fig. 8.18b–iii)
(4) In other cases, the deposition at position x succeeds; namely, $S(x)$=1. (Fig. 8.18b–iv)

These rules correspond to the physical effect of the size-dependent resonance between the material and the light that facilitates desorption of the particles. Since the size of the DP field in the vicinity of a nanostructure depends on the material size, the DP promotes material desorption and inhibits material deposition, beyond a certain size of nanoparticles, which is characterized as rule (2) given above. Also, even when a single cluster size is small, a desorption effect should be induced overall if several such clusters are located in close proximity. Such an effect is represented as rule (3). One remark here is that no more than one block is piled at a single position x; that is to say, $S(x)$ takes binary values only, since the main purpose of this modeling is to know how the clusters are formed in the 1-D system.

Numerical simulations can evaluate the incidence of the cluster size and the cluster-to-cluster interval between two neighboring clusters, which correspond to the diameter and the separation of the Al nanoparticle in Figs. 8.17b and c, respectively. Figures 8.19a and b summarize the evolution of these two values at t = 100, 1,000, and 100,000, where N was fixed to 1,000. For the threshold values in rules (2) and (3), B_{th1} and B_{th2} were assumed to be 8 and 12, respectively. These figures clearly show that the size and the interval converged to representative values, which are consistent with the experimental observations shown in Figs. 8.17b and c.

By comparing Figs. 8.17b and c, it is found that the particle diameter (Fig. 8.17c) formed by a higher photon energy is smaller than that (Fig. 8.17b) formed by a lower photon energy. This difference is attributed to the fact that the higher photon energy leads to desorption at smaller diameters [26, 27]. On the other hand, the separation is larger when they are formed by the higher photon energy, which has not yet been obviously explained. In the following, the reason for the larger separation will be analyzed.

It is assumed that a stronger light-matter size-dependent resonant interaction is induced at a higher photon energy, which more strongly induces desorption, or inhibits deposition, in the neighboring clusters. This effect can be taken into account by modifying the stochastic model described above: Instead of blocking the deposition at the neighboring position by rule (2), it is assumed that the distant neighbors are also inhibited, i.e.: (2′) Even if $S(x)$=0, when x sees a cluster with a size larger than a particular value B_{th1} within an area (a) between $x-3$ and $x-1$ or (b) between $x+1$ and $x+3$, the deposition is inhibited. That is, $S(x)$ is maintained at 0.

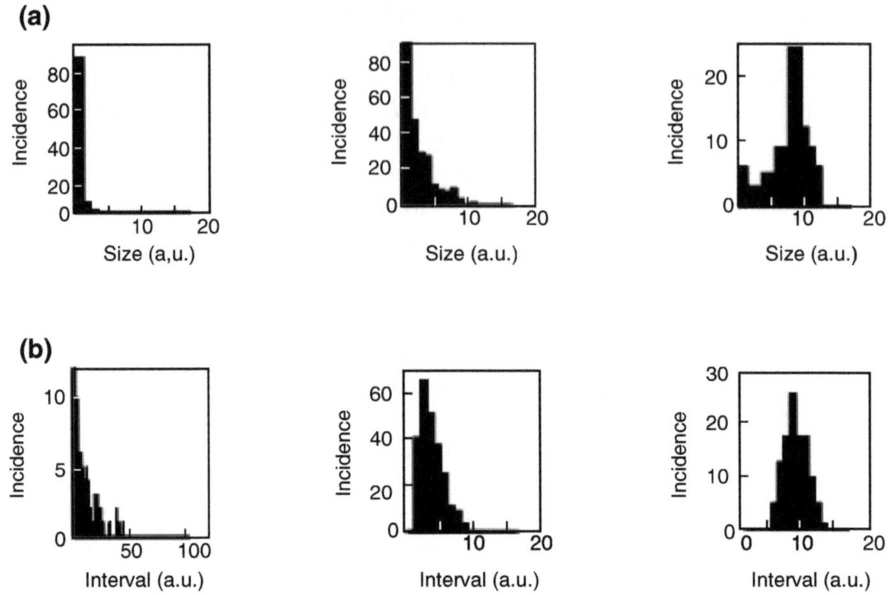

Fig. 8.19 Calculated results of the array formed by light with low photon energy. **a** and **b** show the incidences of size and interval of the cluster, respectively. Number of depositions are t = 100 (*left*), 1,000 (*center*), and 100,000 (*right*)

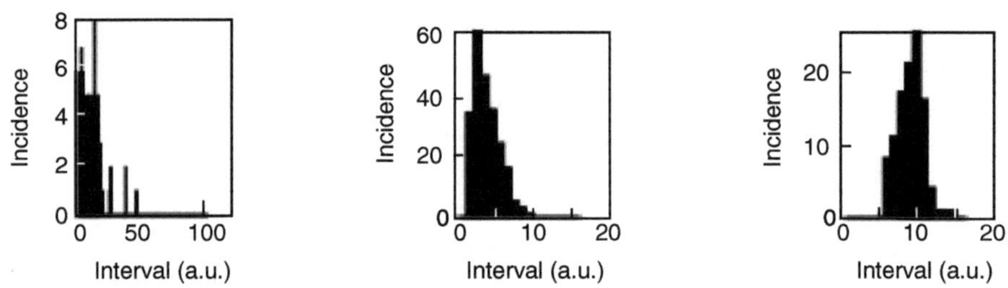

Fig. 8.20 Calculated results of the interval of the clusters in the array formed by light with high photon energy. Number of depositions are t = 100 (*left*), 1,000 (*center*), and 100,000 (*right*)

While preserving the B_{th1} and B_{th2} values with the previous example, the cluster interval statistics evolve as shown in Fig. 8.20. This figure shows that the cluster interval converges to a maximum of 10, which is larger than the previous case, which converged to 8, as shown in Fig. 8.19b. This is consistent with the experimental observations, and thus, it is confirmed that the larger separation is obtained with higher photon energy.

8.4.2 Statistical Modeling of Morphology

In Sect. 7.2.1, the DPPs and the reverse-bias voltage determined the amount of Ag particles flowing into or out from a specific position on the Ag film surface, and autonomous fabrication terminated when the spatial distribution of the DPP field reached a stationary state. The morphology of the Ag film surface varies from moment to moment until the system reaches the stationary state. Comparing devices B and C, the amount of Ag particle flow is larger in device C because the irradiated light power for fabrication is higher for device C. Thus, it is assumed that the system of device C reaches the stationary state within a shorter time.

In order to confirm the difference between devices B and C, Fig. 8.21 shows the incidence patterns as a function of the size of the Ag clusters, which were obtained by analyzing the SEM images of Fig. 7.18 [28]. Here, the term "size" means the area of the two-dimensional image of the Ag cluster, whereas the lower parts of Fig. 7.18b and c show histograms as a function of diameter. Figure 8.21a shows the measured results for device A. The incidence pattern can be approximated by a Poisson distribution and takes a maximum at the cluster area of $0.5 \times 10^4\,\text{nm}^2$. The incidence patterns for devices B and C (Figs. 8.21b and c, respectively) are very different from that of Fig. 8.21a, exhibiting exponential decreases with increasing cluster area. Furthermore, as indicated by the downward arrows in Figs. 8.21b and c, the incidence pattern takes a local maximum at a characteristic value of the area. This area is smaller for the higher light power irradiated in the fabrication process. It should be noted that the average diameters of Ag clusters in Figs. 7.18a and b are 90 and 86 nm, respectively, which is smaller for the higher irradiated light power. In summary, the morphology features depend on the irradiated light power.

In order to explain the morphology features described above, a simple stochastic model is constructed here. First, a two-dimensional (2-D) $M \times M$ square grid cell structure is considered, where a cell, also called a pixel, is specified by $P = (p_x, p_y)$. Each cell is assigned a variable $h(P)$, and $h(P) = 1$ means that the Ag film surface has bumps as a result of the inflow of Ag particles. The areas where the surface is flat are represented by $h(P) = 0$. In Fig. 8.22a, the pixels with $h(P) = 1$ are indicated by black cells, while those with $h(P) = 0$ are indicated by white ones.

The deposition process is simulated as follows: Initially, a completely flat surface is assumed, namely, $h(P) = 0$ for all P. First, a cell P is randomly chosen, and a particle arrives at P. Second, it is determined whether the particle successfully lands on a cell or is repulsed by taking account of the positively charged Ag particle to be sputtered and the positive holes on the Ag film surface. A pseudo footprint denoted by Q_P is calculated, to be defined below, in order to evaluate this effect in the stochastic modeling.

If the calculated value of Q_P is smaller than or equal to a threshold Z, and if the surface is flat ($h(P) = 0$), an arriving Ag particle is able to land on the cell P; that is, the value of $h(P)$ changes from 0 to 1. In contrast if Q_P is larger than Z, the arriving Ag particle is deflected outside the system, representing repulsion between the positively charged Ag particle and the positive holes on the Ag film.

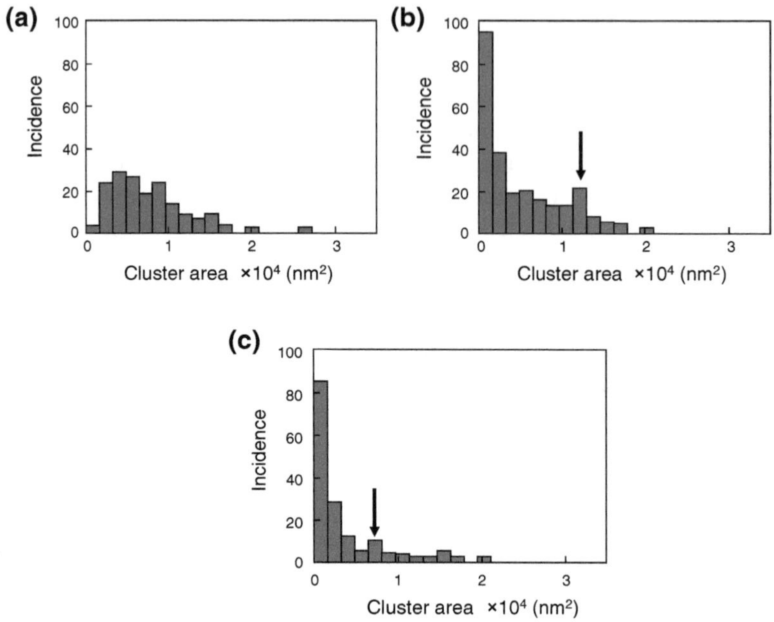

Fig. 8.21 Measured incidences of the Ag cluster size. **a**, **b**, and **c** represent the values for devices A, B, and C, respectively. Downward arrows represent the positions of the local maxima

If Q_P is smaller than or equal to a threshold Z but the point P is already occupied ($h(P) = 1$), the arriving particle will land in a free, randomly chosen neighbor, representing a drift process.

The pseudo footprint metric in each square grid cell corresponds to the sum of the areas of its eight neighbors. More precisely, the pseudo footprint at P is defined by

$$Q_P = \sum_{i=\{-1,0,1\},j=\{-1,0,1\}} S_P^{(i,j)}, \tag{8.3}$$

where $S_P^{(i,j)}$ represents the total number of occupied cells, or area, connected to the cell $(p_x + i, p_y + j)$ $(i, j = -1, 0, +1, i = j \neq 0)$, either in horizontal ($x$) or vertical ($y$) neighbors, as schematically explained in Fig. 8.22a. For example, in the case of Fig. 8.22b, $S_P^{(-1.+1)} = 3$ for the cell at the top-left corner because it is occupied and is connected to two occupied cells. On the other hand, $S_P^{(-1,-1)} = 1$ for the occupied cell at the bottom-left corner because it is not connected to any occupied cells. Therefore, from Eq. (8.3), the pseudo footprint is given by $Q_P = S_P^{(-1.+1)} + S_P^{(-1,-1)} = 4$. Similarly in the case of Fig. 8.22c, Q_P is 7 because $S_P^{(-1,+1)} = 3$, $S_P^{(0,+1)} = 3$, and $S_P^{(-1,-1)} = 1$. Also for Fig. 8.22d, Q_P is 21.

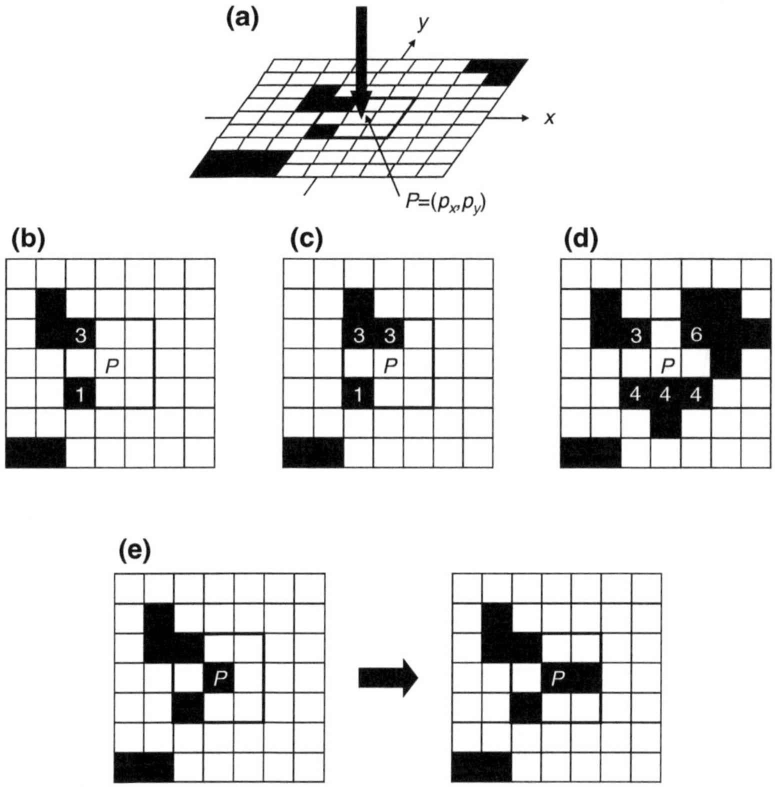

Fig. 8.22 Stochastic model. **a** Definition of the pseudo footprint. **b**, **c**, and **d** represent examples of calculating the pseudo footprint. **e** Comparison of the spatial patterns before and after the Ag nanoparticle flows in

When an arriving Ag particle is not repulsed from the system, but the point P is occupied, the particle lands in a randomly chosen neighboring cell. The left part of Fig. 8.22e represents one such example where $Q_P = 4$. Here, suppose that this Q_P is smaller than the threshold Z. Since the point P is occupied, a free neighboring cell is randomly chosen. For example, the system is updated as shown in the right part of Fig. 8.22e, where a newly arriving Ag particle lands to the right of the point P, in other words, $h\left(p_x + 1, p_y\right)$ changes from 0 to 1. Such a rule represents the drift process occurring on the Ag film surface.

By iteratively applying the stochastic process described above in T cycles from a flat initial state, a variety of resultant spatial patterns are generated. They also depend on the threshold Z. By setting $M = 16$, $Z = 10$, and $T = 300$, Fig. 8.23a shows three examples of the spatial patterns generated in different trials. To examine the statistical properties, the incidence pattern of the mean number of each cluster in N different samples was evaluated, as shown in Fig. 8.23b, where the number of repetition cycles is $T = 1,000$. Closed squares, triangles, and circles respectively

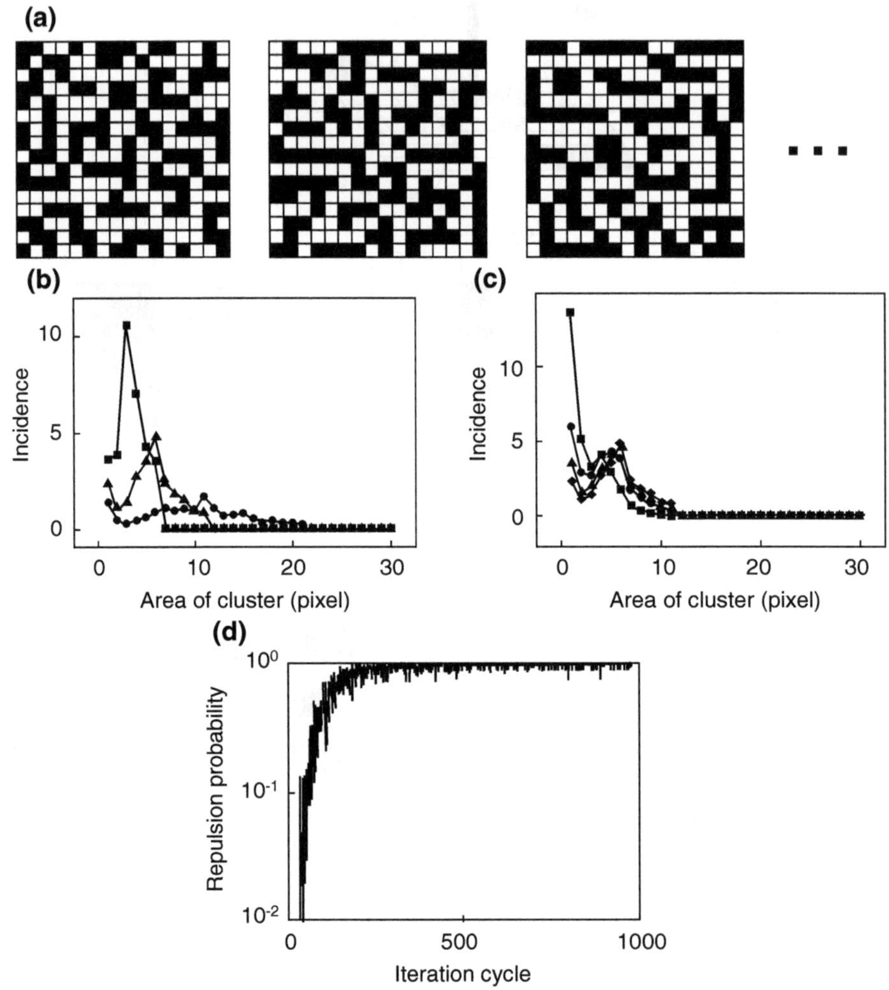

Fig. 8.23 Spatial pattern generation by the stochastic model. **a** Example of generated pattern. **b** Incidence of the cluster size. Closed squares, triangles, and circles represent the cases of $Z = 5$, 10, and 20, respectively. **c** Incidence of the cluster size. Closed squares, circles, triangles, and diamonds represent the cases of $T = 100$, 200, 300, and 1000, respectively. **d** Relation between the number of iteration cycles and repulsion probability

represent the incidence of the clusters in the system with different thresholds Z=5,10, and 20.

The incidence pattern exhibited different characters depending on the threshold value Z; with smaller Z, the cluster area yielding the local maximum incidence shifted towards a smaller value in Fig. 8.23b, which agrees with the experimentally observed behavior shown in Fig. 8.21 where higher irradiation light power produced smaller clusters. In other words, higher power light irradiation more likely induces repulsion, leading to a local maximum at a smaller cluster area. This supports the physical interpretation that the pseudo footprint appropriately represents the repulsion due to the DPPs. That is to say, the pseudo footprint reflects the concentration of the positive holes, the associated DPPs around a cluster, and spatial inhomogeneity.

Figure 8.23c shows the evolution of the incidence patterns of the cluster sizes as the number of repetitions T increases. Closed squares, circles, triangles, and diamonds in this figure respectively show the incidences when T was $\times 100^0$, 200, 300, and 1,000. It can be clearly observed that the peak-like incidence clusters emerge as T increases. Figure 8.23d characterizes the occurrence of repulsion at cycle T among $N = 100$ trials. In other words, it shows the time evolution of the probability of repulsion. The probability increases as the number of iteration cycles increases; a repulsion probability of 0.8 or higher is observed after the number of iteration cycles reaches around 300.

Since the present stochastic model includes a threshold value Z, strictly speaking, it is not so-called self-organized criticality [29]. However, since Fig. 8.23 shows that a flat surface autonomously converges to various kinds of pattern in a self-organized manner while exhibiting common statistical properties, this convergence can be regarded as a kind of self-organized critical phenomenon due to DPPs.

The statistical model can be effectively employed not only in the topic of the present section but also in the autonomous generation of DPPs reviewed in Sect. 7.1 and 7.3, especially to analyze the optimum surface morphology to realize the highest efficiency of the energy up-conversion. Furthermore, it can also be applied to study the autonomous annihilation of DPPs reviewed in Sect. 6.3.

References

1. J. Lim, T. Yatsui, M. Ohtsu, IEICE Trans. Electron. **E88-C**, 1832 (2005)
2. M. Naya, S. Mononobe, R. Uma Maheswari, T. Saiki, M. Ohtsu, Opt. Commun. **124**, 9 (1996)
3. M. Naruse, T. Yatsui, W. Nomura, N. Hirose, M. Ohtsu, Opt. Express **13**, 9265 (2005)
4. N. Tate, W. Nomura, T. Yatsui, M. Naruse, M. Ohtsu, Appl. Phys. B **96**, 1 (2009)
5. M. Naruse, T. Yatsui, T. Kawazoe, Y. Akao, M. Ohtsu, IEEE Trans. Nanotechnol. **7**, 14 (2008)
6. M. Naruse, T. Yatsui, J.H. Kim, M. Ohtsu, Appl. Phys. Express **1**, 062004 (2008)
7. M. Naruse, T. Inoue, H. Hori, Jpn. J. Appl. Phys. **46**, 6095 (2007)
8. N. Tate, W. Nomura, T. Yatsui, M. Naruse, M. Ohtsu, Opt. Express **16**, 607 (2008)
9. N. Tate, M. Naruse, T. Yatsui, T. Kawazoe, M. Hoga, Y. Ohyagi, T. Fukuyama, M. Kitamura, M. Ohtsu, Opt. Express **18**, 7497 (2010)
10. T. Matsumoto, J. Appl. Phys. Jpn. **80**, 30 (2011)
11. M. Naruse, T. Yatsui, H. Hori, M. Yasui, M. Ohtsu, J. Appl. Phys. **103**, 113525 (2008)
12. N. Tate, H. Sugiyama, M. Naruse, W. Nomura, T. Yatsui, T. Kawazoe, M. Ohtsu, Opt. Express **17**, 11113 (2009)
13. N. Tate, H. Tokoro, K. Takeda, W. Nomura, T. Yatsui, T. Kawazoe, M. Naruse, S. Ohkoshi, M. Ohtsu, Appl. Phys. B **98**, 685 (2009)
14. O. Sato, S. Hayashi, Y. Einaga, Z.Z. Gu, Bull. Chem. Soc. Jpn. **76**, 443 (2003)
15. H. Tokoro, T. Matsuda, T. Nuida, Y. Morimoto, K. Ohyama, E.D.L.D. Dangui, K. Boukheddaden, S. Ohkoshi, Chem. Mater. **20**, 423 (2008)
16. S. Ohkoshi, H. Tokoro, M. Utsunomiya, M. Mizuno, M. Abe, K. Hashimoto, J. Phys. Chem. B **106**, 2423 (2002)

17. H. Tokoro, S. Ohkoshi, T. Matsuda, K. Hashimoto, Inorg. Chem. **43**, 5231 (2004)
18. H. Tokoro, T. Matsuda, K. Hashimoto, S. Ohkoshi, J. Appl. Phys. **97**, 10M508 (2005)
19. J. Tanida, Y. Ichioka, Appl. Opt. **27**, 2926 (1988)
20. M. Ohtsu, T. Kawazoe, T. Yatsui, M. Naruse, IEEE J. Select. Top. Quantum Electron. **14**, 1404 (2008)
21. B. Lee, J. Kang, K.-Y. Kim, Proc. SPIE **4803**, 220 (2002)
22. N. Tate, M. Naruse, W. Nomura, T. Kawazoe, T. Yatsui, M. Hoga, Y. Ohyagi, Y. Sekine, H. Fujita, M. Ohtsu, Opt. Express **19**, 18260 (2011)
23. M. Naruse, T. Yatsui, H. Hori, K. Kitamura, M. Ohtsu, Opt. Express **15**, 11790 (2007)
24. Y. Liu, T. Morishima, T. Yatsui, T. Kawazoe, M. Ohtsu, Nanotechnol **22**, 215605 (2011)
25. M. Naruse, Y. Liu, W. Nomura, T. Yatsui, M. Aida, L.B. Kish, M. Ohtsu, Appl. Phys. Lett. **100**, 193106 (2012)
26. T. Yatsui, W. Nomura, M. Ohtsu, Nano Lett. **5**, 2548 (2005)
27. T. Yatsui, S. Takubo, J. Lim, W. Nomura, M. Kourogi, M. Ohtsu, Appl. Phys. Lett. **83**, 1716 (2003)
28. M. Naruse, T. Kawazoe, T. Yatsui, N. Tate, M. Ohtsu, Appl. Phys. B **105**, 185 (2011)
29. P. Bak, C. Tang, K. Wiezenfeld, Phys. Rev. A **38**, 364 (1988)

Chapter 9
Summary and Future Outlook

Disce libens.
Decimus Magnus Ausonius, *Epistulae*

This chapter summarizes the discussions in each chapter of this book and presents the future outlook for dressed-photon (DP) technology. Section 9.1 summarizes the discussions in Chaps. 2–8, and highlights the differences between technologies that exploit phenomena involving DPs and those involving propagating light. Differences in their theoretical treatments are also summarized. The future outlook for DP technology will be given in Sect. 9.2.

9.1 Summary

This book deals with light–matter interactions in a nanometric space whose size is much smaller than the wavelength of light. When attempting to use conventional theories for formulating this interaction, a fatal problem was that a virtual cavity could not be defined. Furthermore, due to the uncertainty relation $\Delta k \cdot \Delta x \geq 1$ between the uncertainty Δk of the wave-number k of the light and that Δx of its position x, it was found that the wave-number had a large uncertainty ($\Delta k \gg k$) due to the sub-wavelength size ($\Delta x \ll \lambda$) of the nanometric space. This large uncertainty placed the topic discussed in this book outside the scope of conventional classical and quantum optics, and, furthermore, discussions based on the dispersion relation of materials were not valid either. In order to overcome these difficulties, Chap. 2 analyzed the interactions between photons with an infinite number of modes and electron–hole pairs with an infinite number of energy levels. As a result, a physical picture of the DP was derived, which was the virtual photon dressing the energy of the electron–hole pair. This picture showed that the DP field is expressed by temporally and spatially modulated photons in nanometric space. The temporal modulation feature was represented by the modulation sidebands, i.e., an infinite series of photon

M. Ohtsu, *Dressed Photons*, Nano-Optics and Nanophotonics,
DOI: 10.1007/978-3-642-39569-7_9,

eigen-energies. Furthermore, a dual relation was found; namely, the electron–hole pair also dressed the photon energy, and its eigen-energy was modulated. This dual relation was a consequence of the inherent features of the light–matter interaction in nanometric space, namely, that a virtual cavity cannot be defined and that the wave-number and momentum are uncertain (i.e., non-conserved).

Furthermore, since an actual nanometric system (composed of nanomaterials and DPs) is always surrounded by a macroscopic system (composed of macroscopic materials and electromagnetic fields), energy transfer between the nanometric and macroscopic systems has to be considered when analyzing the interaction between the nanomaterials. For this consideration, Chap. 2 used the projection operator method and derived the effective energy of the DP-mediated interaction between the nanomaterials. As a result, it was found that this energy is represented by the Yukawa function. This function also represents the spatial modulation feature of the DPs.

From the analysis mentioned above, size-dependent resonance was found; that is to say, the efficiency of the energy transfer between nanomaterials depends on the size of the nanomaterials under interaction. It should be noted that this resonance is unrelated to diffraction which governs the conventional wave-optical phenomena. Furthermore, since the DP is localized in nanometric space, the long-wavelength approximation, which is valid for conventional light–matter interactions, is not valid for the DP-mediated interactions. As a result of this invalidity, an electric dipole-forbidden transition turned out to be allowed in the case of the DP-mediated interactions, as was described in Chap. 3.

Here, one should carefully consider the "size" mentioned above. In the case of a spherical material, for example, its size is represented by the diameter. However, even though it looks like a sphere when it is viewed from the far field, one can find surface roughness when one views it in the near field. That is, the shape and size depend on the separation between the material and the observer. The hierarchy described in Chap. 2 originated from this separation-dependency of the shape and size. If this spherical material is divided into small spheres in order to represent the roughness by the superposition of these spheres, these smaller spheres also exhibit the separation-dependency. Based on this repeated separation-dependency at smaller and smaller scales, the concept of hierarchy was established by assuming that the spatial features of the divided parts are equivalent to those of the original material. However, this division cannot be repeated infinitely. That is to say, after the size of the material decreases to a specific value, its features can become very different from those of the original material, and also different from those of a single atom. The range in which this specific size appears has been called mesoscopic, and has been the subject of intense study recently.

The above-mentioned hierarchy, which is a common concept in modern science, requires employing an appropriate classical or quantum theoretical model, depending on the size of the material under study. The origin of this concept can be traced back as far as the proposal by Democritus, an ancient Greek philosopher. He called the indivisible minimum unit of a material that finally appears after the matter is divided repeatedly "atomos" in the ancient Greek language. This is the origin of the word "atom" used in modern science. Since then, modern science has already found that even the atom can be divided into a nucleus and electrons. However, since the features of these constituent elements are completely different from those of the atom, one does not have to pay any attention to nuclei and electrons when one studies the features of a material with a size larger than an atom, as was dealt with in this book.

Chapters 6 and 7 dealt with macroscopic materials for fabrication and energy conversion using DPs, respectively. Since the position of the generated DP and its energy localization in these materials are random, their analyses required stochastic or mathematical science methods, as described in Chap. 8. In particular, because the surface profile and internal structure of the macroscopic material vary from moment to moment in the process of fabrication using DPs, temporal and spatial features of these variations have to be analyzed dynamically. Furthermore, energy transfer via DPs takes place between the equal-sized materials due to size-dependent resonance, enabling tamper-resistance and bio-mimetic operation of DP devices, as described in Chap. 5.

When analyzing the conventional light scattering phenomenon in a macroscopic material, it has been sufficient to study one phonon. In contrast, Chap. 4 discussed the coherent phonon, which is composed of an infinite number of phonons. This coherent phonon assists in exciting the electron in the adjacent nanomaterial instead of merely increasing the material temperature, which enables phonon-assisted excitation of the electrons. Therefore, for analyzing the optical excitation of the nanomaterial, it is essential to consider the intrinsic quantum state of the nanomaterial, which is expressed by the direct product of the quantum states of the electron and the coherent phonon. As a result, this intrinsic quantum state means that an infinite number of energy levels are distributed in the energy bandgap between the valence and conduction energy bands of an electron in a semiconductor. This quasi-continuous energy distribution originates from the modulation of the eigen-energy of the electron–hole pair as a result of the coupling between the DP and the coherent phonon. By exploiting phonon-assisted excitation and de-excitation, Chap. 7 showed that a semiconductor can absorb light even though its photon energy is lower than the bandgap energy of the material and, furthermore, that an indirect transition-type semiconductor can emit light efficiently.

Conventional optical device technology has relied on materials science and technology over a period of many years since its beginning. That is, novel materials with appropriate bandgap energy have been explored and developed for device operation, and a method of contacting heterogeneous materials had to be developed by solving the difficult problem of lattice mismatching. Furthermore, rare and toxic materials had to be used, which have caused problems in terms of resource conservation and environmental protection. DP-based technology has solved these problems and has succeeded in constructing novel optical devices using abundant and non-toxic materials. This is because the DP is a virtual photon that dresses the material energy; stated another way, the DP represents a quasi-particle in which a photon and matter are fused. As a result, this gave birth to a novel technology exploiting DPs, called "light–matter fusion technology".

9.2 Future Outlook

It should be pointed out that there still remain problems to be solved for gaining a deeper understanding of the DP and finding novel applications in order to achieve qualitative innovations. These include:

1. Improving the accuracy of the physical picture of the quasi-particle representing the coupled state of a photon, an electron, and a phonon in a nanometric space.

2. Elucidating the details of the energy transfer and dissipation between nanomaterials, mediated by DPs or DPPs.

3. Elucidating the details of phonon-assisted light–matter interactions in nanometric space.

4. Elucidating the physical origins of autonomy and hierarchy.

5. Developing technical methods of controlling the generation and annihilation of DPs or DPPs more efficiently.

6. Exploiting the advanced methods of statistical mechanics, mathematical science, and numerical simulation to analyze and solve Problems 1–5 above.

Some remarks should be made here on Problem 6: Since the shapes, sizes, and composition of the materials fabricated by DPs may look random due to local interactions between the DPs and nanometric parts of the material, one does not have to analyze these local interactions individually. Instead, analysis based on a mathematical scientific model can be more efficient for practical applications, as was demonstrated in Sect. 8.4. Although such methods have been rarely employed for conventional materials technology, the need for this kind of model will grow in order to control light–matter interactions mediated by DPs and to establish suitable criteria for designing the devices and systems reviewed in Chaps. 5–8, so as to advance light–matter fusion technology.

Generic technology can be established if it is based on an original concept and drives an existing technology out from the market. In the early stages of DP science and technology, many successes were obtained by using sharpened fiber probes. However, the use of a probe limits the positions where the DPs are generated, which as a result limits the expansion of its applications. It should be pointed out that DPs can be generated in nanometric parts on the surface of and inside the material if they are illuminated by light, as has been reviewed in this book. Such versatile generation of DPs suggests the possibility of many more applications as compared with using a probe. Thus, generic technology is expected to be established by developing more advanced control methods for DP generation, as well as annihilation. Problems 1–6 listed above will need to be solved in order to establish such technology.

Light–matter interactions in nanometric space, energy transfer between nanomaterials, and energy dissipation were studied based on concepts from optical science, quantum field theory, and condensed matter physics. As a result, a physical picture of the DP was established, which allowed us to describe energy transfer even from and to electric dipole-forbidden energy levels. Furthermore, coupling between DPs and multi-mode coherent phonons was found, which led to the development of novel technologies for optical devices, fabrication, energy conversion, and information security. Generic technology is expected to be established in the near future via further studies of DPs, e.g., by investigating the possibility of coupling DPs not only with phonons but also with other types of primary excitations. Moreover, it is expected that a novel optical technology, called light–matter fusion technology, could be established.

To end this chapter, several references are listed for further study [1–8].

References

1. M. Ohtsu (ed.), *Near-Field Nano/Atom Optics and Technology* (Springer-Verlag, Berlin, 1998)
2. M. Ohtsu, H. Hori, *Near-Field Nano-Optics* (Kluwer Academic, New York, 1999)
3. M. Ohtsu, K. Kobayashi, *Optical Near Fields* (Springer-Verlag, Berlin, 2004)
4. M. Ohtsu, K. Kobayashi, T. Kawazoe, T. Yatsui, M. Naruse, *Principles of Nanophotonics* (CRC Press, Boca Raton, 2008)
5. M. Ohtsu (ed.), *Nanophotonics and Nanofabrication* (Wiley-VCH, Weinheim, 2009)
6. M. Ohtsu (ed.), *Progress in Nano-Electro-Optics I–VII* (Springer-Verlag, Berlin, 2003–2010)
7. M. Ohtsu (ed.), *Progress in Nanophotonics I, II* (Springer-Verlag, Berlin, 2011–2013)
8. M. Ohtsu (ed.), *Handbook of Nano-Optics and Nanophotonics* (Springer-Verlag, Berlin, 2013)

Appendix A
Multipolar Hamiltonian

There are two ways to describe the interaction between an electromagnetic field and a charged particle. One is to use the minimal coupling Hamiltonian, and the other is to employ the multipolar Hamiltonian. These two Hamiltonians are related to each other by a unitary transformation, and there are, in principle, no problems regardless of which is adopted [1–3]. The multipolar Hamiltonian has a simple form without the static Coulomb interaction and can exactly describe the retardation effects by exchanging transverse photons, which are photons possessing only a polarization component perpendicular to the wave-vector $\boldsymbol{k}$. Since these features are advantageous for the discussion in Chap. 2, this appendix reviews the multipolar Hamiltonian.

Let us consider a charged particle system confined in a nanomaterial. In the following, we choose two nanomaterials as an example and look for an appropriate Hamiltonian. When the wavelength of the electromagnetic wave is much longer than the size of the nanomaterial, the vector potential $\boldsymbol{A}(\boldsymbol{R})$ at the center position $\boldsymbol{R}$ of the nanomaterial is the same as $\boldsymbol{A}(\boldsymbol{q})$, independent of the position $\boldsymbol{q}$ of an electric charge in the nanomaterial:

$$\boldsymbol{A}(\boldsymbol{q}) = \boldsymbol{A}(\boldsymbol{R}). \tag{A.1}$$

From Eq. (A.1), it follows that the magnetic flux density is zero ($\boldsymbol{B} = \nabla \times \boldsymbol{A} = \boldsymbol{0}$), and thus, the interaction between the charged particle and magnetic field can be neglected. Moreover, one can take only the electric dipole interaction into account because the magnetic dipole and higher multipoles can be neglected. In addition, by assuming that the electron exchange interaction is also negligible, the Lagrangian, L, for the system can be written as

$$L = L_{mol} + L_{rad} + L_{\text{int}}, \tag{A.2a}$$

where

$$L_{mol} = \sum_{\varsigma} \left\{ \sum_{\alpha} \frac{m_\alpha \dot{\boldsymbol{q}}_\alpha^2(\varsigma)}{2} - V(\varsigma) \right\}, \tag{A.2b}$$

M. Ohtsu, *Dressed Photons*, Nano-Optics and Nanophotonics,
DOI: 10.1007/978-3-642-39569-7,

$$L_{rad} = \frac{\varepsilon_0}{2} \int \left\{ \dot{A}^2 - c^2 (\nabla \times A)^2 \right\} d^3 r, \tag{A.2c}$$

and

$$L_{\mathrm{int}} = \sum_{\varsigma} \sum_{\alpha} e \dot{q}_{\alpha} (\varsigma) \cdot A (R_{\varsigma}) - V_{\mathrm{int}}. \tag{A.2d}$$

Here, · represents the time derivative. The index ς is used for distinguishing the nanomaterials 1 and 2, and α is used to specify a charged particle in a nanomaterial. ε_0, c, and e represent the dielectric constant in vacuum, the speed of light in vacuum, and the electric charge, respectively. L_{mol} denotes the energy of charged particles with mass m_α and velocity $\dot{q}_\alpha$ in the Coulomb potential $V(\varsigma)$, while L_{rad} denotes the energy of the electromagnetic field in free space. The third term L_{int} denotes the interaction between the charged particle and the electromagnetic field. The Coulomb interaction V_{int} between the nanomaterials 1 and 2 is given by

$$V_{\mathrm{int}} = \frac{1}{4\pi\varepsilon_0 R^3} \left\{ p(1) \cdot p(2) - 3 (p(1) \cdot e_R)(p(2) \cdot e_R) \right\}. \tag{A.3}$$

Here $R = |R| = |R_1 - R_2|$ denotes the center-to-center separation between the nanomaterials 1 and 2, and e_R is R/R, the unit vector along the direction of R. The electric dipole moments of the nanomaterials 1 and 2 are designated by $p(1)$ and $p(2)$, respectively.

In order to simplify the interaction Hamiltonian, the Power–Zienau–Woolley transformation [1, 3][1] is performed on the original Lagrangian L. The result is expressed as

$$L_{mult} = L - \frac{d}{dt} \int P^{\perp}(r) \cdot A(r) \, d^3 r, \tag{A.4}$$

where $P^{\perp}(r)$ is the transverse component of the polarization density $P(r)$.

The polarization density $P(r)$ is

$$P(r) = \sum_{\varsigma,\alpha} e (q_\alpha - R_\varsigma) \left[1 - \frac{1}{2!} \left\{ (q_\alpha - R_\varsigma) \cdot \nabla \right\} + \frac{1}{3!} \left\{ (q_\alpha - R_\varsigma) \cdot \nabla \right\}^2 - \cdots \right] \delta (r - R_\varsigma) \tag{A.5}$$

and only the electric dipole term is retained:

$$P(r) = \sum_{\varsigma,\alpha} e (q_\alpha - R_\varsigma) \delta (r - R_\varsigma) = p(1) \, \delta (r - R_1) + p(2) \, \delta (r - R_2). \tag{A.6}$$

[1] The Power–Zienau–Woolley transformation is a method for deriving the interaction Hamiltonian by using the electric displacement vector and magnetic flux density of the electromagnetic field instead of using the vector potential. Furthermore, measurable macroscopic quantities, such as the electric polarization and magnetization, are used.

Note that the current density is

$$\boldsymbol{j}(\boldsymbol{r}) = \sum_{\varsigma,\alpha} e\dot{\boldsymbol{q}}_\alpha \delta(\boldsymbol{r} - \boldsymbol{R}_\varsigma), \tag{A.7}$$

and the transverse component of the current density is related to the transverse component of the polarization density as follows:

$$\boldsymbol{j}^\perp(\boldsymbol{r}) = \frac{d\boldsymbol{P}^\perp(\boldsymbol{r})}{dt}. \tag{A.8}$$

Using Eqs. (A.7) and (A.8), the interaction Lagrangian L_{int} of Eq. (A.2d) can be rewritten as

$$L_{\text{int}} = \int \boldsymbol{j}^\perp(\boldsymbol{r}) \cdot \boldsymbol{A}(\boldsymbol{r})\, d^3r - V_{\text{int}} = \int \frac{d\boldsymbol{P}^\perp(\boldsymbol{r})}{dt} \cdot \boldsymbol{A}(\boldsymbol{r})\, d^3r - V_{\text{int}}, \tag{A.9}$$

and thus L_{mult} given by Eq. (A.4) becomes

$$\begin{aligned} L_{mult} &= L - \int \frac{d\boldsymbol{P}^\perp(\boldsymbol{r})}{dt} \cdot \boldsymbol{A}(\boldsymbol{r})\, d^3r - \int \boldsymbol{P}^\perp(\boldsymbol{r}) \cdot \dot{\boldsymbol{A}}(\boldsymbol{r})\, d^3r \\ &= L_{mol} + L_{rad} - \int \boldsymbol{P}^\perp(\boldsymbol{r}) \cdot \dot{\boldsymbol{A}}(\boldsymbol{r})\, d^3r - V_{\text{int}}. \end{aligned} \tag{A.10}$$

Here, the momentum $\boldsymbol{p}_\alpha$ conjugate to the coordinate $\boldsymbol{q}_\alpha$ and the vector potential $\boldsymbol{A}(\boldsymbol{r})$ conjugate to the momentum $\boldsymbol{\Pi}(\boldsymbol{r})$ are defined by

$$\boldsymbol{p}_\alpha = \frac{\partial L_{mult}}{\partial \dot{\boldsymbol{q}}_\alpha} = \frac{\partial L_{mol}}{\partial \dot{\boldsymbol{q}}_\alpha} = m_\alpha \dot{\boldsymbol{q}}_\alpha, \tag{A.11a}$$

$$\begin{aligned} \boldsymbol{\Pi}(\boldsymbol{r}) &= \frac{\partial L_{mult}}{\partial \dot{\boldsymbol{A}}(\boldsymbol{r})} = \frac{\partial L_{rad}}{\partial \dot{\boldsymbol{A}}(\boldsymbol{r})} - \frac{\partial}{\partial \dot{\boldsymbol{A}}(\boldsymbol{r})} \int \boldsymbol{P}^\perp(\boldsymbol{r}) \cdot \dot{\boldsymbol{A}}(\boldsymbol{r})\, d^3r \\ &= \varepsilon_0 \dot{\boldsymbol{A}}(\boldsymbol{r}) - \boldsymbol{P}^\perp(\boldsymbol{r}) = -\varepsilon_0 \boldsymbol{E}^\perp(\boldsymbol{r}) - \boldsymbol{P}^\perp(\boldsymbol{r}). \end{aligned} \tag{A.11b}$$

From the relation between the electric field $\boldsymbol{E}(\boldsymbol{r})$ and electric displacement vector $\boldsymbol{D}(\boldsymbol{r})$, their transverse components also satisfy

$$\boldsymbol{D}^\perp(\boldsymbol{r}) = \varepsilon_0 \boldsymbol{E}^\perp(\boldsymbol{r}) + \boldsymbol{P}^\perp(\boldsymbol{r}) \tag{A.12}$$

and thus the momentum $\boldsymbol{\Pi}(\boldsymbol{r})$ can be rewritten as

$$\boldsymbol{\Pi}(\boldsymbol{r}) = -\boldsymbol{D}^\perp(\boldsymbol{r}). \tag{A.13}$$

By putting them together, canonical transformation of the Lagrangian L_{mult} gives a new Hamiltonian H_{mult}:

$$\begin{aligned} H_{mult} &= \sum_{\zeta,\alpha} \boldsymbol{p}_\alpha(\zeta)\cdot\dot{\boldsymbol{q}}_\alpha(\varsigma) + \int \boldsymbol{\Pi}(\boldsymbol{r})\cdot\dot{\boldsymbol{A}}(\boldsymbol{r})\,d^3r - L_{mult} \\ &= \sum_\zeta \left\{ \sum_\alpha \frac{p_\alpha^2(\zeta)}{2m_\alpha} + V(\zeta) \right\} + \left\{ \frac{1}{2}\int \left[\frac{\boldsymbol{\Pi}^2(\boldsymbol{r})}{\varepsilon_0} + \varepsilon_0 c^2 (\nabla\times\boldsymbol{A}(\boldsymbol{r}))^2 \right] d^3r \right\} \\ &\quad + \frac{1}{\varepsilon_0}\int \boldsymbol{P}^\perp(\boldsymbol{r})\cdot\boldsymbol{\Pi}(\boldsymbol{r})\,d^3r + \frac{1}{2\varepsilon_0}\int \left|\boldsymbol{P}^\perp(\boldsymbol{r})\right|^2 d^3r + V_{\mathrm{int}}. \end{aligned} \tag{A.14}$$

It is possible to simplify Eq. (A.14) by separating $(1/2\varepsilon_0)\int \left|\boldsymbol{P}^\perp(\boldsymbol{r})\right|^2 d^3r$ into two parts: an inter-nanomaterial part and an intra-nanomaterial part. In order to consider the inter-nanomaterial part

$$\frac{1}{2\varepsilon_0}\int \boldsymbol{P}_1{}^\perp(\boldsymbol{r})\cdot\boldsymbol{P}_2{}^\perp(\boldsymbol{r})\,d^3r, \tag{A.15}$$

by noting

$$\boldsymbol{P}_2(\boldsymbol{r}) = \boldsymbol{P}_2{}^\parallel(\boldsymbol{r}) + \boldsymbol{P}_2{}^\perp(\boldsymbol{r}),\ \boldsymbol{P}_1{}^\perp(\boldsymbol{r})\cdot\boldsymbol{P}_2{}^\parallel(\boldsymbol{r}) = 0 \tag{A.16}$$

(the symbol $\parallel$ represents the longitudinal component of the polarization, i.e., the component parallel to the wave-vector $\boldsymbol{k}$) and

$$\boldsymbol{P}_1{}^\perp(\boldsymbol{r})\cdot\boldsymbol{P}_2{}^\perp(\boldsymbol{r}) = \boldsymbol{P}_1{}^\perp(\boldsymbol{r})\cdot\left\{\boldsymbol{P}_2{}^\parallel(\boldsymbol{r}) + \boldsymbol{P}_2{}^\perp(\boldsymbol{r})\right\} = \boldsymbol{P}_1{}^\perp(\boldsymbol{r})\cdot\boldsymbol{P}_2(\boldsymbol{r}), \tag{A.17}$$

Equation (A.15) is rewritten as follows:

$$\begin{aligned} \frac{1}{\varepsilon_0}\int \boldsymbol{P}_1{}^\perp(\boldsymbol{r})\cdot\boldsymbol{P}_2{}^\perp(\boldsymbol{r})\,d^3r &= \frac{1}{\varepsilon_0}\int \boldsymbol{P}_1{}^\perp(\boldsymbol{r})\cdot\boldsymbol{P}_2(\boldsymbol{r})\,d^3r \\ &= \frac{1}{\varepsilon_0} p_i(1)\,p_j(2)\int \delta_{ij}^\perp(\boldsymbol{r}-\boldsymbol{R}_1)\,\delta(\boldsymbol{r}-\boldsymbol{R}_2)\,d^3r \\ &= \frac{1}{\varepsilon_0} p_i(1)\,p_j(2)\,\delta_{ij}^\perp(\boldsymbol{r}-\boldsymbol{R}_1-\boldsymbol{R}_2) \\ &= -\frac{p_i(1)\,p_j(2)}{4\pi\varepsilon_0 R^3}\left(\delta_{ij} - 3\hat{e}_{Ri}\hat{e}_{Rj}\right) \\ &= -\frac{1}{4\pi\varepsilon_0 R^3}\left\{\boldsymbol{p}(1)\cdot\boldsymbol{p}(2) - 3\,(\boldsymbol{p}(1)\cdot\boldsymbol{e}_R)\,(\boldsymbol{p}(2)\cdot\boldsymbol{e}_R)\right\}, \end{aligned} \tag{A.18}$$

where $\hat{e}_{Ri}$ and $\hat{e}_{Rj}$ are the i-th and j-th Cartesian components of the unit vector $\boldsymbol{e}_R\,(\equiv \boldsymbol{R}/R)$. Equation (A.16) was used in the first row. Also, the following identities for the Dirac δ function and the δ-dyadics were used in the third line:

$$\begin{aligned}\delta_{ij}\delta(\boldsymbol{r}) &= \delta_{ij}^{\parallel}(\boldsymbol{r}) + \delta_{ij}^{\perp}(\boldsymbol{r}), \\ \delta_{ij}^{\perp}(\boldsymbol{r}) &= -\delta_{ij}^{\parallel}(\boldsymbol{r}) = -\frac{1}{(2\pi)^3}\int \hat{e}_{ki}\hat{e}_{kj}\exp(i\boldsymbol{k}\cdot\boldsymbol{r})\,d^3r \\ &= \nabla_i\nabla_j\left(\frac{1}{4\pi r}\right) = -\frac{1}{4\pi r^3}\left(\delta_{ij} - 3\hat{e}_{ri}\hat{e}_{rj}\right).\end{aligned} \tag{A.19}$$

Here, $\hat{e}_{ki}$ and $\hat{e}_{kj}$ are the i-th and j-th Cartesian components of the unit vector $\boldsymbol{e}_k$ ($\equiv \boldsymbol{k}/k$). Similarly, $\hat{e}_{ri}$ and $\hat{e}_{rj}$ are the i-th and j-th Cartesian components of the unit vector $\boldsymbol{e}_r$ ($\equiv \boldsymbol{r}/r$). Because exchanging the subscripts 1 and 2 gives the same result as Eq. (A.18), one can derive

$$\frac{1}{2\varepsilon_0}\int \boldsymbol{P}_1{}^{\perp}(\boldsymbol{r})\cdot\boldsymbol{P}_2{}^{\perp}d^3r + V_{\text{int}} = 0 \tag{A.20}$$

by noting Eq. (A.3). This equation suggests that the inter-nanomaterial part given by Eq. (A.15) and V_{int} cancel each other out. Therefore, it is sufficient to consider the intra-nanomaterial part $(1/2\varepsilon_0)\int\left|\boldsymbol{P}_\zeta^{\perp}(\boldsymbol{r})\right|^2 d^3r$ ($\zeta = 1, 2$) only, and Eq. (A.14) can be simplified to

$$\begin{aligned}H_{mult} = &\sum_\zeta\left\{\sum_\alpha\frac{\boldsymbol{p}_\alpha^2(\zeta)}{2m_\alpha} + V(\zeta) + \frac{1}{2\varepsilon_0}\int\left|\boldsymbol{P}_\zeta^{\perp}(\boldsymbol{r})\right|^2 d^3r\right\} \\ &+\left\{\frac{1}{2}\int\left[\frac{\boldsymbol{\Pi}^2(\boldsymbol{r})}{\varepsilon_0} + \varepsilon_0c^2(\nabla\times\boldsymbol{A}(\boldsymbol{r}))^2\right]d^3r\right\} \\ &+\frac{1}{\varepsilon_0}\int\boldsymbol{P}^{\perp}(\boldsymbol{r})\cdot\boldsymbol{\Pi}(\boldsymbol{r})\,d^3r,\end{aligned} \tag{A.21}$$

where each row represents the charged particle motion in each nanomaterial, the free electromagnetic field, and the interaction, respectively. Because the polarization density $\boldsymbol{P}^{\perp}(\boldsymbol{r})$ can be expanded in terms of 2^l multipoles ($l = 1, 2, 3, \cdots$), as shown in Eq. (A.5), H_{mult} of Eq. (A.21) is called the multipolar Hamiltonian.

With the help of Eqs. (A.6) and (A.13), the interaction part in the third row of Eq. (A.21) can be more explicitly written as

$$\begin{aligned}\frac{1}{\varepsilon_0}\int\boldsymbol{P}^{\perp}(\boldsymbol{r})\cdot\boldsymbol{\Pi}(\boldsymbol{r})\,d^3r &= -\frac{1}{\varepsilon_0}\int\boldsymbol{P}^{\perp}(\boldsymbol{r})\cdot\boldsymbol{D}^{\perp}(\boldsymbol{r})\,d^3r \\ &= -\frac{1}{\varepsilon_0}\int\boldsymbol{P}(\boldsymbol{r})\cdot\boldsymbol{D}^{\perp}(\boldsymbol{r})\,d^3r \\ &= -\frac{1}{\varepsilon_0}\left\{\boldsymbol{p}(1)\cdot\boldsymbol{D}^{\perp}(\boldsymbol{R}_1) + \boldsymbol{p}(2)\cdot\boldsymbol{D}^{\perp}(\boldsymbol{R}_2)\right\}\end{aligned} \tag{A.22}$$

by using the electric dipole moment $\boldsymbol{p}$ and the electric displacement vector $\boldsymbol{D}^{\perp}$. The advantage of this expression is that electromagnetic phenomena occurring inside and outside the nanomaterial can be treated equivalently, i.e., the anti-electric field inside the nanomaterial and Coulomb interaction between electrons do not have to be additionally considered, and thus, the static Coulomb interactions between the nanomaterials can be excluded. Further advantages are that the delay effect can be expressed because the interaction is described by the exchange of the transverse photons, and the origin of the electric-dipole–forbidden transition can be clearly interpreted.

When the system under study is quantized, quantities such as $\boldsymbol{p}$ and $\boldsymbol{D}^{\perp}$ should be replaced by the corresponding operators,

$$-\frac{1}{\varepsilon_0}\left\{\hat{\boldsymbol{p}}(1)\cdot\hat{\boldsymbol{D}}^{\perp}(\boldsymbol{R}_1)+\hat{\boldsymbol{p}}(2)\cdot\hat{\boldsymbol{D}}^{\perp}(\boldsymbol{R}_2)\right\}, \tag{A.23}$$

yielding the quantized multipolar Hamiltonian. The third term of Eq. (2.1), and therefore Eq. (2.25), is derived by replacing $\hat{\boldsymbol{D}}^{\perp}(\boldsymbol{R}_1)$ and $\hat{\boldsymbol{D}}^{\perp}(\boldsymbol{R}_2)$ by the electric displacement operators $\hat{\boldsymbol{D}}^{\perp}(\boldsymbol{r}_s)$ and $\hat{\boldsymbol{D}}^{\perp}(\boldsymbol{r}_p)$ at the positions of the electric charges $\boldsymbol{r}_s$ and $\boldsymbol{r}_p$, respectively, and furthermore, by replacing $\hat{\boldsymbol{p}}(1)$ and $\hat{\boldsymbol{p}}(2)$ by $\hat{\boldsymbol{p}}_s$ and $\hat{\boldsymbol{p}}_p$, respectively.

References

1. C. Cohen-Tannoudji, J. Dupont-Roc, G. Grynberg, *Photons and Atoms* (Wiley, New York, 1989)
2. C. Cohen-Tannoudji, J. Dupont-Roc, and G. Grynberg, *Atom-Photon Interactions* (John Wiley & Sons, New York, 1992)
3. D.P. Craig, T. Thirunamachandran, *Molecular Quantum Electrodynamics* (Dover, New York, 1998)

Appendix B
Elementary Excitation and Exciton-polariton

The concept of elementary excitations or quasi-particles has been discussed for a long time [1]. Excited states of a many-body system are considered as a collection of certain fundamental excited states that has been called an elementary excitation. The relation between the momentum $\boldsymbol{p}$ and energy E of the elementary excitation, i.e., $E = E(\boldsymbol{p})$, is referred to as the dispersion relation.

A well-known example of an elementary excitation in a solid is a phonon, which is a quantum of normal modes of crystal vibration. Its motion is collective, which means that the total number of phonons is independent of the number of crystal lattices. The momentum of the phonon is $\boldsymbol{p} = \hbar\boldsymbol{k}$ in terms of the wave-vector $\boldsymbol{k}$ of normal vibration, not the mechanical momentum of an individual crystal lattice. The energy is also given in terms of the angular frequency ω of the normal vibration by $E = \hbar\omega$. Other examples of elementary excitations are plasmons, which correspond to the collective motion of electron density in interacting electron gas; polarons, which originate from the coupling between conduction electrons and optical phonons; and magnons, which correspond to collective modes of spin density waves.

Excitons, which describe the elementary excitation related to an electron–hole pair in a solid, are also well-known. As an extreme case, when the distance between the electron and hole in an exciton (Bohr radius of the exciton) is smaller than the interatomic distance in the crystal, it is called a Frenkel exciton; Wannier Excitons correspond to the opposite extreme case, in which the Bohr radius of the exciton is larger than the interatomic distance in the crystal.

In the following, the light–matter interaction in a macroscopic material is discussed on the basis of the exciton concept. The photon, incident on the macroscopic material is absorbed, creating the exciton. Afterward, this exciton is annihilated, creating a photon. This means that the successive creation and annihilation of the photon and exciton propagate through the macroscopic material. In the other words, the photon and exciton repeat their creation and annihilation in an out-of-phase manner, temporally and spatially. This process indicates that a new steady state with a new dispersion relation and energy is formed in the whole macroscopic material due to the photon–exciton interaction. Normal modes, or elementary excitation modes, representing this coupled oscillation are called polaritons. In particular, they are called

M. Ohtsu, *Dressed Photons*, Nano-Optics and Nanophotonics,

DOI: 10.1007/978-3-642-39569-7,

exciton-polaritons because they originate from the interaction between photons and excitons. The exciton-polariton is a coupled wave generated by the electromagnetic field and the polarization field of the exciton. The situation is analogous to the case in which two coupled oscillations with angular frequencies ω_o and ω_e produce new normal oscillations with angular frequencies Ω_1 and Ω_2.

The following Hamiltonian is used to describe exciton-polaritons:

$$\hat{H} = \hbar\omega_o\hat{a}^\dagger\hat{a} + \hbar\omega_e\hat{b}^\dagger\hat{b} + \hbar D\left(\hat{a} + \hat{a}^\dagger\right)\left(\hat{b} + \hat{b}^\dagger\right). \tag{B.1}$$

Since a macroscopic material is dealt with here, a virtual cavity can be defined for quantization. Therefore, the first term, i.e, the non-perturbed Hamiltonian for photons, is equivalent to the incident photon energy $\hbar\omega_o$, which is resonant with the virtual cavity. The second term corresponds to the non-perturbed Hamiltonian for excitons, having the eigenenergy $\hbar\omega_e$. The third term describes the photon–exciton interaction, whose interaction energy is $\hbar D$. The explicit expression for $\hbar D$ is given by Eq. (2.27). $\hat{a}$ and $\hat{a}^\dagger$ are annihilation and creation operators for photons, respectively. $\hat{b}$ and $\hat{b}^\dagger$ are annihilation and creation operators for excitons, which are given by

$$\begin{cases} \hat{b} = \frac{1}{\sqrt{N}}\sum\limits_{l} e^{-i\boldsymbol{k}\cdot\boldsymbol{l}}\hat{b}_l \\ \hat{b}^\dagger = \frac{1}{\sqrt{N}}\sum\limits_{l} e^{i\boldsymbol{k}\cdot\boldsymbol{l}}\hat{b}_l^\dagger \end{cases}. \tag{B.2}$$

Here, N and $\boldsymbol{k}$ are the total number of lattice sites and the wave-vector, respectively. Furthermore, we have

$$\hat{b}_l = \hat{e}_{l,c}\hat{h}_{l,v} \quad \text{and} \quad \hat{b}_l^\dagger = \hat{e}_{l,c}^\dagger\hat{h}_{l,v}^\dagger, \tag{B.3}$$

where $\hat{e}_{l,c}$ and $\hat{e}_{l,c}^\dagger$ are the annihilation and creation operators, respectively, for the electron in the conduction band at the lattice site $\boldsymbol{l}$, whereas $\hat{h}_{l,v}$ and $\hat{h}_{l,v}^\dagger$ are those for the hole in the valence band at the lattice site $\boldsymbol{l}$.

From the Hamiltonian for the exciton-polariton given by Eq. (B.1), one can obtain the eigenstates and eigenenergies of the exciton-polaritons, or the dispersion relation. For simplicity, the rotating wave approximation is adopted to neglect terms $\hat{a}^\dagger\hat{b}^\dagger$ and $\hat{a}\hat{b}$, which represent simultaneous creation or annihilation of a photon and an exciton, resulting in the following Hamiltonian:

$$\hat{H} = \hbar\left(\omega_o\hat{a}^\dagger\hat{a} + \omega_e\hat{b}^\dagger\hat{b}\right) + \hbar D\left(\hat{b}^\dagger\hat{a} + \hat{a}^\dagger\hat{b}\right). \tag{B.4}$$

Next, exciton-polariton creation operators $\hat{\xi}_1^\dagger$ and $\hat{\xi}_2^\dagger$ and annihilation operators $\hat{\xi}_1$ and $\hat{\xi}_2$, corresponding to new eigenfrequencies Ω_1 and Ω_2, respectively, are derived. For this derivation, the Hamiltonian $\hat{H}$ is assumed to be diagonalized in terms of these operators, and Eq. (B.4) is rewritten as

$$\hat{H} = \hbar\left(\Omega_1\hat{\xi}_1^\dagger\hat{\xi}_1 + \Omega_2\hat{\xi}_2^\dagger\hat{\xi}_2\right) = \hbar\left(\hat{b}^\dagger, \hat{a}^\dagger\right) A \begin{pmatrix}\hat{b}\\ \hat{a}\end{pmatrix}$$
$$= \hbar\left(a_{11}\hat{b}^\dagger\hat{b} + a_{12}\hat{b}^\dagger\hat{a} + a_{21}\hat{a}^\dagger\hat{b} + a_{22}\hat{a}^\dagger\hat{a}\right), \tag{B.5}$$

where A is a 2×2 matrix whose elements are given by

$$A = \begin{pmatrix} a_{11} & a_{12} \\ a_{21} & a_{22} \end{pmatrix} = \begin{pmatrix} \omega_e & D \\ D & \omega_o \end{pmatrix}. \tag{B.6}$$

Applying unitary transformation U

$$\begin{pmatrix}\hat{b}\\ \hat{a}\end{pmatrix} = U \begin{pmatrix}\hat{\xi}_1\\ \hat{\xi}_2\end{pmatrix} \text{with } U = \begin{pmatrix} u_{11} & u_{12} \\ u_{21} & u_{22} \end{pmatrix} \tag{B.7}$$

to Eq. (B.5) yields

$$\hbar\left(\hat{b}^\dagger, \hat{a}^\dagger\right) A \begin{pmatrix}\hat{b}\\ \hat{a}\end{pmatrix} = \hbar\left(\hat{\xi}_1^\dagger, \hat{\xi}_2^\dagger\right) U^\dagger A U \begin{pmatrix}\hat{\xi}_1\\ \hat{\xi}_2\end{pmatrix}. \tag{B.8}$$

Since $U^\dagger AU = U^{-1}AU$ is diagonalized, one can write

$$U^{-1}AU = \begin{pmatrix} \Omega_1 & 0 \\ 0 & \Omega_2 \end{pmatrix} \equiv \Lambda, \tag{B.9}$$

and obtain $AU = U\Lambda$, which reduces to

$$\begin{pmatrix} \omega_e - \Omega_j & D \\ D & \omega_o - \Omega_j \end{pmatrix} \begin{pmatrix} u_{1j} \\ u_{2j} \end{pmatrix} = 0. \tag{B.10}$$

This immediately gives the eigenvalue equation

$$\left(\Omega_j - \omega_e\right)\left(\Omega_j - \omega_o\right) - D^2 = 0, \tag{B.11}$$

and the eigenenergies of the exciton-polariton are

$$\hbar\Omega_j = \hbar\left[\frac{\omega_e + \omega_o}{2} \pm \frac{\sqrt{(\omega_e - \omega_o)^2 + 4D^2}}{2}\right]. \tag{B.12}$$

Equation (B.12) provides a new dispersion relation. Using the dispersion relation of photons $\omega_o = ck$ with $k = |\boldsymbol{k}|$, the eigenenergies of the exciton-polariton can be plotted as a function of k, as shown in Fig. B.1. Here, for simplicity, the exciton dispersion is assumed to be $\hbar\omega_e = \hbar\Omega$, which is independent of k. From Eq. (B.10) and the unitarity of U, the components of the eigenvectors are given by

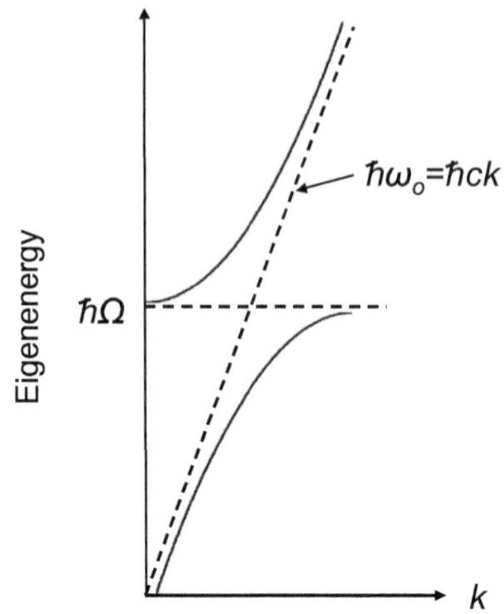

Fig. B.1 Relation between the wavenumber k and the eigenenergy

$$\begin{cases} u_{2j} = -\frac{\omega_e - \Omega_j}{D} u_{1j} \\ u_{1j}^2 + u_{2j}^2 = 1 \end{cases} (j = 1, 2), \tag{B.13}$$

which thus reads

$$\left\{1 + \left(\frac{\omega_e - \Omega_j}{D}\right)^2\right\} u_{1j}^2 = 1. \tag{B.14}$$

Finally, the eigenvectors of the exciton-polariton are given by

$$\begin{cases} u_{1j} = \left\{1 + \left(\frac{\omega_e - \Omega_j}{D}\right)^2\right\}^{-1/2}, \\ u_{2j} = -\left(\frac{\omega_e - \Omega_j}{D}\right)\left\{1 + \left(\frac{\omega_e - \Omega_j}{D}\right)^2\right\}^{-1/2}. \end{cases} \tag{B.15}$$

New steady states for the exciton-polariton can be described by Eqs. (B.12) and (B.15).

Equation (B.12) means that the sum of $\hbar\Omega_1$ and $\hbar\Omega_2$ is equal to the sum of the exciton and photon energies, $\hbar(\omega_e + \omega_o)$, because annihilation and creation of the photon occur in an out-of-phase manner to those of the exciton, as was pointed out at the beginning of this appendix. Furthermore, since the classical model of the exciton-polariton is a coupled wave of the electromagnetic wave of the light and the polarization wave of the exciton, the dependence of its amplitude on time t and position $\boldsymbol{x}$ is simply expressed by a complex sinusoidal function $\exp\left[i\left(\Omega_j t - \boldsymbol{k}\cdot\boldsymbol{x}\right)\right]$, where Ω_j is the angular frequency in Eq. (B.12). Therefore, if the numbers n of photons and excitons are greater than one, the angular frequency of the coupled wave remains $\hbar\Omega_j$, even though the square of the amplitude of the coupled wave becomes n times greater than that at $n = 1$.

Reference

1. D. Pines, *Elementary Excitation in Solids* (Perseus Books, Reading, Massachusetts, 1999)

Appendix C
Projection Operator and Effective Interaction Operator

C.1 Projection Operator

The total Hamiltonian $\hat{H}$ for the light–matter interaction system is given by

$$\hat{H} = \hat{H}_0 + \hat{V}, \tag{C.1}$$

where $\hat{H}_0$ and $\hat{V}$ describe the unperturbed and interaction Hamiltonians, respectively. The state $|\phi_{Pj}\rangle$ in Eq. (2.32) is the eigenstate of $\hat{H}_0$. If the eigenstate and eigenenergy of $\hat{H}$ are written as $|\psi_j\rangle$ and E_j, respectively, the following Schrödinger equation holds:

$$\hat{H}\,|\psi_j\rangle = E_j\,|\psi_j\rangle\,, \tag{C.2}$$

where the subscript j is used as a quantum number to distinguish each eigenstate. In a similar way, by denoting the eigenstate of $\hat{H}_0$ by $|\phi_j\rangle$, a projection operator $\hat{P}$ is defined as

$$\hat{P} = \sum_j |\phi_j\rangle\langle\phi_j|\,. \tag{C.3}$$

Applying the projection operator to an arbitrary state $|\psi\rangle$ yields

$$\hat{P}\,|\psi\rangle = \sum_j |\phi_j\rangle\langle\phi_j \mid \psi\rangle\,. \tag{C.4}$$

The right-hand side of this equation is represented by the linear superposition of the eigenstates $|\phi_j\rangle$ because the inner product $\langle\phi_j \mid \psi\rangle$ is a constant. This suggests that the projection operator transforms the arbitrary state $|\psi\rangle$ into the P space spanned by the eigenstate $|\phi_j\rangle$. In Eq. (C.3), the projection operator was defined based on steady states of the Schrödinger equation. The time-dependent approach of the projection operator method is reviewed in Ref. [1].

M. Ohtsu, *Dressed Photons*, Nano-Optics and Nanophotonics,
DOI: 10.1007/978-3-642-39569-7, © Springer-Verlag Berlin Heidelberg 2014

Because the eigenstate $|\phi_j\rangle$ is orthonormalized, the projection operator $\hat{P}$ satisfies the following relation:

$$\hat{P} = \hat{P}^{\dagger}, \tag{C.5a}$$

$$\hat{P}^2 = \hat{P}, \tag{C.5b}$$

where $\hat{P}^{\dagger}$ is the Hermitian conjugate operator of $\hat{P}$. Equation (C.5a) means that $\hat{P}$ is a Hermitian operator.

The complimentary operator $\hat{Q}$ given by

$$\hat{Q} = 1 - \hat{P} \tag{C.6a}$$

reads

$$\hat{Q} = \hat{Q}^{\dagger}, \tag{C.6b}$$

$$\hat{Q}^2 = \hat{Q}. \tag{C.6c}$$

Any state in the P space is orthogonal to any state in the Q space, and thus one has

$$\hat{P}\hat{Q} = \hat{Q}\hat{P} = 0. \tag{C.7}$$

Noting that $|\phi_j\rangle$ is an eigenstate of $\hat{H}_0$, the commutations between the projection operators and $\hat{H}_0$ are

$$\left[\hat{P}, \hat{H}_0\right] = \hat{P}\hat{H}_0 - \hat{H}_0\hat{P} = 0, \tag{C.8a}$$

$$\left[\hat{Q}, \hat{H}_0\right] = \hat{Q}\hat{H}_0 - \hat{H}_0\hat{Q} = 0. \tag{C.8b}$$

C.2 Effective Interaction Operator

The expectation value of an arbitrary physical quantity is expressed as $\langle\psi|\,\hat{O}\,|\psi\rangle$, where $\hat{O}$ is the corresponding operator, and $|\psi\rangle$ is the state of the system under discussion. In order to derive this value using only the states in the P space, i.e., to express the expectation value as $\langle\phi_i|\,\hat{O}_{eff}\,|\phi_j\rangle$, a new operator $\hat{O}_{eff}$, called an effective operator, should be derived.

The following discussion considers the eigenstates $\left|\psi_j\right\rangle$ of $\hat{H}$ instead of the arbitrary state $|\psi\rangle$ because $|\psi\rangle$ is given by the linear superposition of $|\psi_j\rangle (j = 1, 2, 3, \ldots)$. The eigenstates $|\psi_j\rangle$ are divided into two groups, the state $\left|\psi_j^{(P)}\right\rangle$ in the P space and the state $\left|\psi_j^{(Q)}\right\rangle$ in the Q space, which are defined as follows:

$$\left|\psi_j^{(P)}\right\rangle = \hat{P}\,|\psi_j\rangle, \quad \left|\psi_j^{(Q)}\right\rangle = \hat{Q}\,|\psi_j\rangle. \tag{C.9}$$

By noting that $\hat{P} + \hat{Q} = 1$, we have

$$|\psi_j\rangle = \left(\hat{P} + \hat{Q}\right)\left|\psi_j\right\rangle = \hat{P}\left|\psi_j\right\rangle + \hat{Q}\left|\psi_j\right\rangle = \left|\psi_j^{(P)}\right\rangle + \left|\psi_j^{(Q)}\right\rangle, \tag{C.10}$$

and from Eqs. (C.5b) and (C.6c), we have

$$\hat{P}\left|\psi_j^{(P)}\right\rangle = \hat{P}\hat{P}\left|\psi_j\right\rangle = \hat{P}\left|\psi_j\right\rangle = \left|\psi_j^{(P)}\right\rangle, \tag{C.11a}$$

$$\hat{Q}\left|\psi_j^{(Q)}\right\rangle = \hat{Q}\hat{Q}\left|\psi_j\right\rangle = \hat{Q}\left|\psi_j\right\rangle = \left|\psi_j^{(Q)}\right\rangle. \tag{C.11b}$$

Inserting Eqs. (C.11a) and (C.11b) into Eq. (C.10) yields

$$\left|\psi_j\right\rangle = \hat{P}\left|\psi_j^{(P)}\right\rangle + \hat{Q}\left|\psi_j^{(Q)}\right\rangle. \tag{C.12}$$

Since Eqs. (C.1) and (C.2) give

$$\left(E_j - \hat{H}_0\right)|\psi_j\rangle = \hat{V}|\psi_j\rangle, \tag{C.13a}$$

inserting Eq. (C.12) into Eq. (C.13a) yields

$$\left(E_j - \hat{H}_0\right)\hat{P}\left|\psi_j^{(P)}\right\rangle + \left(E_j - \hat{H}_0\right)\hat{Q}\left|\psi_j^{(Q)}\right\rangle = \hat{V}\hat{P}\left|\psi_j^{(P)}\right\rangle + \hat{V}\hat{Q}\left|\psi_j^{(Q)}\right\rangle. \tag{C.13b}$$

Applying $\hat{P}$ to Eq. (C.13b) from the left and using Eqs. (C.5b) and (C.7) gives

$$\left(E_j - \hat{H}_0\right)\hat{P}\left|\psi_j^{(P)}\right\rangle = \hat{P}\hat{V}\hat{P}\left|\psi_j^{(P)}\right\rangle + \hat{P}\hat{V}\hat{Q}\left|\psi_j^{(Q)}\right\rangle. \tag{C.14}$$

Similarly applying $\hat{Q}$ to Eq. (C.13b) from the left and using Eqs. (C.6b) and (C.7), Eq. (C.13b) can be rewritten as

$$\left(E_j - \hat{H}_0\right)\hat{Q}\left|\psi_j^{(Q)}\right\rangle = \hat{Q}\hat{V}\hat{P}\left|\psi_j^{(P)}\right\rangle + \hat{Q}\hat{V}\hat{Q}\left|\psi_j^{(Q)}\right\rangle. \tag{C.15}$$

By moving the second term on the right-hand side of Eq. (C.15) to the left-hand side, it is possible to formally express $\hat{Q}\left|\psi_j^{(Q)}\right\rangle$ as

$$\begin{aligned}
\hat{Q}\left|\psi_j^{(Q)}\right\rangle &= \left(E_j - \hat{H}_0 - \hat{Q}\hat{V}\right)^{-1}\hat{Q}\hat{V}\hat{P}\left|\psi_j^{(P)}\right\rangle \\
&= \left\{\left(E_j - \hat{H}_0\right)\left[1 - \left(E_j - \hat{H}_0\right)^{-1}\hat{Q}\hat{V}\right]\right\}^{-1}\hat{Q}\hat{V}\hat{P}\left|\psi_j^{(P)}\right\rangle \\
&= \hat{J}\left(E_j - \hat{H}_0\right)^{-1}\hat{Q}\hat{V}\hat{P}\left|\psi_j^{(P)}\right\rangle,
\end{aligned} \tag{C.16}$$

where the operator $\hat{J}$ is defined by

$$\hat{J} = \left[1 - \left(E_j - \hat{H}_0\right)^{-1} \hat{Q}\hat{V}\right]^{-1}. \tag{C.17}$$

By inserting Eq. (C.16) into Eq. (C.14), one obtains the following equation for $\hat{P}\left|\psi_j^{(P)}\right\rangle$:

$$\begin{aligned}\left(E_j - \hat{H}_0\right)\hat{P}\left|\psi_j^{(P)}\right\rangle &= \hat{P}\hat{V}\hat{P}\left|\psi_j^{(P)}\right\rangle + \hat{P}\hat{V}\hat{J}\left(E_j - \hat{H}_0\right)^{-1}\hat{Q}\hat{V}\hat{P}\left|\psi_j^{(P)}\right\rangle \\ &= \hat{P}\hat{V}\hat{J}\left\{\hat{J}^{-1} + \left(E_j - \hat{H}_0\right)^{-1}\hat{Q}\hat{V}\right\}\hat{P}\left|\psi_j^{(P)}\right\rangle.\end{aligned} \tag{C.18}$$

Since

$$\hat{J}^{-1} = 1 - \left(E_j - \hat{H}_0\right)^{-1}\hat{Q}\hat{V} \tag{C.19}$$

is derived from Eq. (C.17), inserting this equation into {} on the second row of Eq. (C.18) yields

$$\left(E_j - \hat{H}_0\right)\hat{P}\left|\psi_j^{(P)}\right\rangle = \hat{P}\hat{V}\hat{J}\hat{P}\left|\psi_j^{(P)}\right\rangle, \tag{C.20}$$

which is the equation that $\left|\psi_j^{(P)}\right\rangle$ must satisfy. On the other hand, inserting Eq. (C.16) into the second term on the right-hand side of Eq. (C.12) yields the following equation for $|\psi_j\rangle$:

$$\begin{aligned}|\psi_j\rangle &= \hat{P}\left|\psi_j^{(P)}\right\rangle + \hat{J}\left(E_j - \hat{H}_0\right)^{-1}\hat{Q}\hat{V}\hat{P}\left|\psi_j^{(P)}\right\rangle \\ &= \hat{J}\left\{\hat{J}^{-1} + \left(E_j - \hat{H}_0\right)^{-1}\hat{Q}\hat{V}\right\}\hat{P}\left|\psi_j^{(P)}\right\rangle \\ &= \hat{J}\hat{P}\left|\psi_j^{(P)}\right\rangle,\end{aligned} \tag{C.21}$$

where Eq. (C.19) was used to derive the third row.

Noting the normalization condition for $|\psi_j\rangle$, inserting Eq. (C.21) into $\langle\psi_j \mid \psi_j\rangle = 1$ gives

$$\left\langle\psi_j^{(P)}\right|\hat{P}\hat{J}^{\dagger}\hat{J}\hat{P}\left|\psi_j^{(P)}\right\rangle = 1. \tag{C.22a}$$

This can be rewritten as

$$\left\langle\psi_j^{(P)}\right|\left(\hat{P}\hat{J}^{\dagger}\hat{J}\hat{P}\right)^{1/2}\left(\hat{P}\hat{J}^{\dagger}\hat{J}\hat{P}\right)^{1/2}\left|\psi_j^{(P)}\right\rangle = 1, \tag{C.22b}$$

which suggests that $\left(\hat{P}\hat{J}^{\dagger}\hat{J}\hat{P}\right)^{-1/2}\left|\psi_j^{(P)}\right\rangle$ should be considered as $\left|\psi_j^{(P)}\right\rangle$ in order to normalize $\left|\psi_j^{(P)}\right\rangle$. Following this suggestion, Eq. (C.21) can be rewritten as

$$|\psi_j\rangle = \hat{J}\hat{P}\left(\hat{P}\hat{J}^{\dagger}\hat{J}\hat{P}\right)^{-1/2}\left|\psi_j^{(P)}\right\rangle, \tag{C.22c}$$

where $\left|\psi_j^{(P)}\right\rangle$ has already been normalized, as was described above.

Using Eq. (C.22c), one can derive the effective operator $\hat{O}_{eff}$ for an arbitrary operator $\hat{O}$ by the following relation [2-4]:

$$\langle\psi_i|\,\hat{O}\,|\psi_j\rangle = \left\langle\psi_i^{(P)}\right|\hat{O}_{eff}\left|\psi_j^{(P)}\right\rangle. \tag{C.23}$$

Inserting Eq. (C.22c) into the left-hand side of Eq.(C.23) and comparing it with the right-hand side leads to

$$\hat{O}_{eff} = \left(\hat{P}\hat{J}^{\dagger}\hat{J}\hat{P}\right)^{-1/2}\left(\hat{P}\hat{J}^{\dagger}\hat{O}\hat{J}\hat{P}\right)\left(\hat{P}\hat{J}^{\dagger}\hat{J}\hat{P}\right)^{-1/2}. \tag{C.24}$$

By replacing $\hat{O}$ with the bare interaction operator $\hat{V}$ in Eq. (C.1), the effective interaction operator $\hat{V}_{eff}$ is written as

$$\hat{V}_{eff} = \left(\hat{P}\hat{J}^{\dagger}\hat{J}\hat{P}\right)^{-1/2}\left(\hat{P}\hat{J}^{\dagger}\hat{V}\hat{J}\hat{P}\right)\left(\hat{P}\hat{J}^{\dagger}\hat{J}\hat{P}\right)^{-1/2}. \tag{C.25}$$

This is what we are searching for. Once the bare interaction operator $\hat{V}$ is given, it only remains to obtain the unknown operator $\hat{J}$ for deriving $\hat{V}_{eff}$.

In order to obtain an explicit form of the operator $\hat{J}$, the states $|\phi_{Pj}\rangle$ in the P space and their eigenvalues have to be used. For this purpose, the operator $\left[\hat{J},\hat{H}_0\right]\hat{P}$ is considered and is applied to $|\psi_j\rangle$. This yields

$$\begin{aligned}\left[\hat{J},\hat{H}_0\right]\hat{P}\,|\psi_j\rangle &= \left(\hat{J}\hat{H}_0 - \hat{H}_0\hat{J}\right)\hat{P}\,|\psi_j\rangle \qquad \text{(C.26a)}\\ &= \left\{\left(E_j - \hat{H}_0\right)\hat{J} - \hat{J}\left(E_j - \hat{H}_0\right)\right\}\hat{P}\,|\psi_j\rangle.\end{aligned}$$

Using Eqs. (C.1), (C.2), (C.11a), and (C.21), the first term $\left(E_j - \hat{H}_0\right)$ on the second row is replaced by $\hat{V}$, which gives

$$\left[\hat{J},\hat{H}_0\right]\hat{P}\,|\psi_j\rangle = \hat{V}\hat{J}\hat{P}\,|\psi_j\rangle - \hat{J}\left(E_j - \hat{H}_0\right)\hat{P}\,|\psi_j\rangle. \tag{C.26b}$$

Also, by using Eqs. (C.1), (C.2), (C.11a), and (C.12), the second term of Eq. (C.26b), except the operator $\hat{J}$, can be rewritten as

$$\begin{aligned}\left(E_j - \hat{H}_0\right)\hat{P}\left|\psi_j\right\rangle &= \left(E_j - \hat{H}_0\right)\hat{P}\left|\psi_j^{(P)}\right\rangle \\ &= \hat{P}\hat{V}\hat{P}\left|\psi_j^{(P)}\right\rangle + \hat{P}\hat{V}\hat{Q}\left|\psi_j^{(Q)}\right\rangle.\end{aligned} \tag{C.27a}$$

Inserting Eq. (C.16) into $\hat{Q}\left|\psi_j^{(Q)}\right\rangle$ in the second term on the second row gives

$$\begin{aligned}\left(E_j - \hat{H}_0\right)\hat{P}\left|\psi_j\right\rangle &= \hat{P}\hat{V}\hat{P}\left|\psi_j^{(P)}\right\rangle + \hat{P}\hat{V}\hat{J}\left(E_j - \hat{H}_0\right)^{-1}\hat{Q}\hat{V}\hat{P}\left|\psi_j^{(P)}\right\rangle \\ &= \hat{P}\hat{V}\hat{J}\left\{\hat{J}^{-1} + \left(E_j - \hat{H}_0\right)^{-1}\hat{Q}\hat{V}\right\}\hat{P}\left|\psi_j^{(P)}\right\rangle.\end{aligned} \tag{C.27b}$$

Furthermore, inserting Eq. (C.19) into the second row, one can rewrite it as

$$\left(E_j - \hat{H}_0\right)\hat{P}\left|\psi_j\right\rangle = \hat{P}\hat{V}\hat{J}\hat{P}\left|\psi_j^{(P)}\right\rangle. \tag{C.27c}$$

Since

$$\left(E_j - \hat{H}_0\right)\hat{P}\left|\psi_j\right\rangle = \hat{P}\hat{V}\hat{J}\hat{P}\left|\psi_j\right\rangle \tag{C.27d}$$

is obtained by inserting Eq. (C.11a) into the right-hand side, inserting Eq. (C.27d) into the second term on the right-hand side of Eq. (C.26b) gives

$$\left[\hat{J}, \hat{H}_0\right]\hat{P}\left|\psi_j\right\rangle = \hat{V}\hat{J}\hat{P}\left|\psi_j\right\rangle - \hat{J}\hat{P}\hat{V}\hat{J}\hat{P}\left|\psi_j\right\rangle. \tag{C.28a}$$

Therefore, for the operator $\hat{J}$, we have

$$\left[\hat{J}, \hat{H}_0\right]\hat{P} = \hat{V}\hat{J}\hat{P} - \hat{J}\hat{P}\hat{V}\hat{J}\hat{P}, \tag{C.28b}$$

where all operators involved are known, except $\hat{J}$.

C.3 Approximate Expression

In order to solve Eq. (C.28b) perturbatively, a polynomial form

$$\hat{J} = \sum_{n=0}^{\infty} g^n \hat{J}^{(n)} \tag{C.29}$$

is assumed, where $\hat{J}^{(n)}$ contains n operators $\hat{V}$. First, by noting that the first term of Eq. (C.17) is unity, and that the relation $1 = \hat{P}+\hat{Q}$ holds due to Eq. (C.6a), one finds that

$$\hat{J}^{(0)} = \hat{P}. \tag{C.30}$$

Next, by inserting Eqs. (C.29) and (C.30) into Eq. (C.28b) and equating terms of order g^n on both sides, one successively obtains $\hat{J}^{(1)}, \hat{J}^{(2)}, \ldots, \hat{J}^{(n)}$. For example, by multiplying both sides of Eq. (C.28b) by $\hat{Q}$ from their left, the relation

$$\hat{Q}\left[\hat{J}^{(1)}, \hat{H}_0\right]\hat{P} = \hat{Q}\hat{V}\hat{J}^{(0)}\hat{P} - \hat{Q}\hat{J}^{(0)}\hat{P}\hat{V}\hat{J}^{(0)}\hat{P} \tag{C.31a}$$

is obtained. By inserting Eq. (C.30) into this equation, it is written as

$$\hat{Q}\left[\hat{J}^{(1)}, \hat{H}_0\right]\hat{P} = \hat{Q}\hat{V}\hat{P}^2 - \hat{Q}\hat{P}^2\hat{V}\hat{P}^2 = \hat{Q}\hat{V}\hat{P}, \tag{C.31b}$$

where Eqs. (C.5b) and (C.7) were used for deriving the right-hand side of this equation. The matrix element of Eq. (C.31b) with $\langle\psi_i|$ and $|\psi_j\rangle$ can be written as

$$\langle\psi_i|\,\hat{Q}\left[\hat{J}^{(1)}, \hat{H}_0\right]\hat{P}\,|\psi_j\rangle = \langle\psi_i|\,\hat{Q}\hat{V}\hat{P}\,|\psi_j\rangle. \tag{C.32}$$

By noting

$$\hat{H}_0\hat{P}\,|\psi_j\rangle = \hat{H}_0\hat{P}\left|\psi_j^{(P)}\right\rangle = \hat{P}\hat{H}_0\left|\psi_j^{(P)}\right\rangle = \hat{P}E_P^0\left|\psi_j^{(P)}\right\rangle = E_P^0\hat{P}\,|\psi_j\rangle, \tag{C.33a}$$

$$\hat{H}_0\hat{Q}\,|\psi_j\rangle = \hat{H}_0\hat{Q}\left|\psi_j^{(Q)}\right\rangle = \hat{Q}\hat{H}_0\left|\psi_j^{(Q)}\right\rangle = \hat{Q}E_Q^0\left|\psi_j^{(Q)}\right\rangle = E_Q^0\hat{Q}\,|\psi_j\rangle, \tag{C.33b}$$

the left-hand side of Eq. (C.32) is rewritten as

$$\begin{aligned}\langle\psi_i|\,\hat{Q}\left(\hat{J}^{(1)}\hat{H}_0 - \hat{H}_0\hat{J}^{(1)}\right)\hat{P}\,|\psi_j\rangle &= \langle\psi_i|\left(\hat{Q}\hat{J}^{(1)}E_P^0\hat{P} - \hat{Q}E_Q^0\hat{J}^{(1)}\hat{P}\right)|\psi_j\rangle \\ &= \langle\psi_i|\left\{\hat{Q}\hat{J}^{(1)}\left(E_P^0 - E_Q^0\right)\hat{P}\right\}|\psi_j\rangle.\end{aligned} \tag{C.34}$$

On the other hand, the right-hand side of Eq. (C.32) is rewritten as

$$\langle\psi_i|\,\hat{Q}\hat{V}\hat{P}\,|\psi_j\rangle = \langle\psi_i|\,\hat{Q}^2\hat{V}\hat{P}^2\,|\psi_j\rangle \tag{C.35}$$

by using Eqs. (C.5b) and (C.6c). Inserting them into Eq. (C.32) and comparing both sides yields

$$\hat{Q}\hat{J}^{(1)}\left(E_P^0 - E_Q^0\right)\hat{P} = \hat{Q}^2\hat{V}\hat{P}^2 \tag{C.36}$$

Thus, $\hat{J}^{(1)}$ is

$$\hat{J}^{(1)} = \hat{Q}\hat{V}\left(E_P^0 - E_Q^0\right)^{-1}\hat{P}, \tag{C.37}$$

which contains one $\hat{V}$. Higher orders of $\hat{J}^{(n)}$ are successively given in a similar way.

C.4 Derivation of Eq. (2.30)

Under the first-order approximation $\hat{J} \simeq \hat{J}^{(0)} + \hat{J}^{(1)}$, Eq. (C.25) is expressed as

$$\hat{V}_{eff} \simeq \hat{P}\hat{J}^{\dagger}\hat{V}\hat{J}\hat{P} \simeq \hat{P}\hat{J}^{(0)\dagger}\hat{V}\hat{J}^{(1)}\hat{P} + \hat{P}\hat{J}^{(1)\dagger}\hat{V}\hat{J}^{(0)}\hat{P}. \tag{C.38}$$

Inserting Eqs. (C.30), (C.37), and $\hat{J}^{(1)\dagger} = \hat{P}\left(E_P^0 - E_Q^0\right)^{-1}\hat{V}\hat{Q}$ into this equation gives

$$\begin{aligned}\hat{V}_{eff} &\simeq \hat{P}\hat{V}\hat{Q}\hat{V}\left(\frac{1}{E_P^0 - E_Q^0}\right)\hat{P}\hat{P} + \hat{P}\hat{P}\left(\frac{1}{E_P^0 - E_Q^0}\right)\hat{V}\hat{Q}\hat{V}\hat{P} \\ &= \hat{P}\hat{V}\hat{Q}\hat{V}\left(\frac{1}{E_P^0 - E_Q^0}\right)\hat{P} + \hat{P}\left(\frac{1}{E_P^0 - E_Q^0}\right)\hat{V}\hat{Q}\hat{V}\hat{P},\end{aligned} \tag{C.39}$$

where Eq. (C.5b) was used to derive the second row. Equation (C.39) is rewritten by using Eq. (C.6c) as

$$\hat{V}_{eff} = \hat{P}\hat{V}\hat{Q}\cdot\hat{Q}\hat{V}\left(\frac{1}{E_P^0 - E_Q^0}\right)\hat{P} + \hat{P}\left(\frac{1}{E_P^0 - E_Q^0}\right)\hat{V}\hat{Q}\cdot\hat{Q}\hat{V}\hat{P}, \tag{C.40}$$

where $\hat{Q}\hat{V}\hat{P}$ and $\hat{P}\hat{V}\hat{Q}$ mean that $\hat{V}$ is screened in the P and Q spaces. Using this equation, the effective interaction energy of Eq. (2.29) is given by

$$\begin{aligned}V_{eff} = \langle\phi_{Pf}|&\left\{\hat{P}\hat{V}\hat{Q}\cdot\hat{Q}\hat{V}\left(\frac{1}{E_P^0 - E_Q^0}\right)\hat{P}\right. \\ &\left.+\hat{P}\left(\frac{1}{E_P^0 - E_Q^0}\right)\hat{V}\hat{Q}\cdot\hat{Q}\hat{V}\hat{P}\right\}|\phi_{Pi}\rangle.\end{aligned} \tag{C.41}$$

The operator $\hat{P}\hat{V}\hat{Q}$ to the left of the symbol $\cdot$ in the first term is considered to apply to $\langle\phi_{Pf}|$. On the other hand, the operator $\hat{Q}\hat{V}\left(E_P^0 - E_Q^0\right)^{-1}\hat{P}$ to the right applies to $|\phi_{Pi}\rangle$. Under these considerations, the eigenenergies E_P^0 and E_Q^0 in this term are rewritten as E_{Pi}^0 and E_{Qj}^0, respectively. Finally, the first term is transformed to

$$\langle \phi_{Pf} | \hat{P}\hat{V}\hat{Q} \cdot \hat{Q}\hat{V} \left(\frac{1}{E_P^0 - E_Q^0} \right) \hat{P} | \phi_{Pi} \rangle$$
$$= \langle \phi_{Pf} | \hat{P}\hat{V}\hat{Q} \cdot \hat{Q}\hat{V}\hat{P} | \phi_{Pi} \rangle \left(\frac{1}{E_{Pi}^0 - E_{Qj}^0} \right). \tag{C.42a}$$

Similarly for the second term, the eigenenergies E_P^0 and E_Q^0 are rewritten as E_{Pf}^0 and E_{Qj}^0, respectively, which transforms the second term to

$$\langle \phi_{Pf} | \hat{P} \left(\frac{1}{E_{Pf}^0 - E_{Qj}^0} \right) \hat{V}\hat{Q} \cdot \hat{Q}\hat{V}\hat{P} | \phi_{Pi} \rangle$$
$$= \left(\frac{1}{E_{Pf}^0 - E_{Qj}^0} \right) \langle \phi_{Pf} | \hat{P}\hat{V}\hat{Q} \cdot \hat{Q}\hat{V}\hat{P} | \phi_{Pi} \rangle . \tag{C.42b}$$

This equation means that Eq. (C.41) was successfully expressed by the screened operators $\hat{Q}\hat{V}\hat{P}$ and $\hat{P}\hat{V}\hat{Q}$.

By noting that the unit operator $\hat{1}$ is expressed by using the basis $\{|\phi_{Qj}\rangle\}$ of the Q space as $\hat{1} = \sum_j |\phi_{Qj}\rangle\langle\phi_{Qj}|$,[2] one can derive

$$V_{eff} = \langle \phi_{Pf} | \sum_j \hat{P}\hat{V}\hat{Q} | \phi_{Qj} \rangle \langle \phi_{Qj} | \hat{Q}\hat{V}\hat{P} | \phi_{Pi} \rangle \left(\frac{1}{E_{Pi}^0 - E_{Qj}^0} + \frac{1}{E_{Pf}^0 - E_{Qj}^0} \right), \tag{C.43}$$

and thus

$$\hat{V}_{eff} = \sum_j \hat{P}\hat{V}\hat{Q} | \phi_{Qj} \rangle \langle \phi_{Qj} | \hat{Q}\hat{V}\hat{P} \left(\frac{1}{E_{Pi}^0 - E_{Qj}^0} + \frac{1}{E_{Pf}^0 - E_{Qj}^0} \right). \tag{C.44}$$

These two equations are nothing but Eqs. (2.31) and (2.30).

References

1. C.R. Willis, R.H. Picard, Phys. Rev. A **9**, 1343 (1974)
2. K. Kobayashi, M. Ohtsu, J. Microsc. **194**, 249 (1999)
3. K. Kobayashi, S. Sangu, H. Ito, M. Ohtsu, Phys. Rev. A **63**, 013806 (2001)
4. H. Hyuga, H. Ohtsubo, Nucl. Phys. A **294**, 348 (1978)

[2] The subscript $j = 1$ represents the states of Eqs. (2.37a) and (2.37b), containing one exciton-polariton state $|1_{(M)}\rangle$. Other values of j represent the states of Eq. (2.36a) and (2.36b), which contain $|n_{(M)}\rangle$.

Appendix D
Transformation from Photon Base to Polariton Base

In this appendix, Eq. (2.28) is derived by diagonalizing the photon–exciton interaction Hamiltonian. Here, the Hamiltonian of the system is expressed as

$$\hat{H} \equiv \sum_k \hat{H}_k, \hat{H}_k = \hbar\omega_k \hat{a}_k^\dagger \hat{a}_k + \hbar\Omega \hat{b}_k^\dagger \hat{b}_k - i\hbar C \left(\hat{b}_{-k} + \hat{b}_k^\dagger\right)\left(\hat{a}_k - \hat{a}_{-k}^\dagger\right), \quad \text{(D.1)}$$

where $\hat{a}_k$ and $\hat{a}_k^\dagger$ are annihilation and creation operators of a phonon with energy $\hbar\omega_k$, and $\hat{b}_k$ and $\hat{b}_k^\dagger$ are annihilation and creation operators of an exciton with energy $\hbar\Omega$. The photon–exciton interaction energy is denoted by $\hbar C$. An exciton-polariton operator $\hat{\xi}_k$ is defined as

$$\hat{\xi}_k = W_k \hat{a}_k + Y_k \hat{a}_{-k}^\dagger + X_k \hat{b}_k + Z_k \hat{b}_{-k}^\dagger, \quad \text{(D.2)}$$

where W_k, Y_k, X_k, and Z_k are the expansion coefficients. It is assumed that $\hat{\xi}_k$ and its Hermitian conjugate $\hat{\xi}_k^\dagger$ obey the boson commutation relation. The Hamiltonian should be diagonalized in the form of $\hat{\xi}_k^\dagger \hat{\xi}_k$, and it follows that

$$\hat{H}_k = \hbar\Omega(k)\, \hat{\xi}_k^\dagger \hat{\xi}_k, \quad \text{(D.3)}$$

and Heisenberg's equation of motion

$$-i\frac{d\hat{\xi}_k}{dt} = \frac{1}{\hbar}\left[\hat{H}, \hat{\xi}_k\right] = -\Omega(k)\, \hat{\xi}_k. \quad \text{(D.4)}$$

Substituting Eq. (D.2) into Eq. (D.4), the left-hand side is written as

$$-iW_k \frac{d\hat{a}_k}{dt} - iY_k \frac{d\hat{a}_{-k}^\dagger}{dt} - iX_k \frac{d\hat{b}_k}{dt} - iZ_k \frac{d\hat{b}_{-k}^\dagger}{dt} \quad \text{(D.5)}$$

M. Ohtsu, *Dressed Photons*, Nano-Optics and Nanophotonics,
DOI: 10.1007/978-3-642-39569-7, © Springer-Verlag Berlin Heidelberg 2014

and the right-hand side as

$$-\Omega(k)\left(W_k\hat{a}_k+Y_k\hat{a}^{\dagger}_{-k}+X_k\hat{b}_k+Z_k\hat{b}^{\dagger}_{-k}\right). \tag{D.6}$$

Using the following Heisenberg's equations of motion

$$-i\frac{d\hat{a}_k}{dt}=\frac{1}{\hbar}\left[\hat{H},\hat{a}_k\right]=-\omega_k\hat{a}_k-iC\left(\hat{b}_k+\hat{b}^{\dagger}_{-k}\right), \tag{D.7a}$$

$$-i\frac{d\hat{a}^{\dagger}_{-k}}{dt}=\frac{1}{\hbar}\left[\hat{H},\hat{a}^{\dagger}_{-k}\right]=\omega_k\hat{a}^{\dagger}_{-k}-iC\left(\hat{b}_k+\hat{b}^{\dagger}_{-k}\right), \tag{D.7b}$$

$$-i\frac{d\hat{b}_k}{dt}=\frac{1}{\hbar}\left[\hat{H},\hat{b}_k\right]=-\Omega\hat{b}_k+iC\left(\hat{a}_k-\hat{a}^{\dagger}_{-k}\right), \tag{D.7c}$$

$$-i\frac{d\hat{b}^{\dagger}_{-k}}{dt}=\frac{1}{\hbar}\left[\hat{H},\hat{b}^{\dagger}_{-k}\right]=\Omega\hat{b}^{\dagger}_{-k}-iC\left(\hat{a}_k-\hat{a}^{\dagger}_{-k}\right), \tag{D.7d}$$

Equation (D.5) is rewritten as

$$\begin{aligned}&\left[-\omega_k\hat{a}_k-iC\left(\hat{b}_k+\hat{b}^{\dagger}_{-k}\right)\right]W_k+\left[\omega_k\hat{a}^{\dagger}_{-k}-iC\left(\hat{b}_k+\hat{b}^{\dagger}_{-k}\right)\right]Y_k\\&+\left[-\Omega\hat{b}_k+iC\left(\hat{a}_k-\hat{a}^{\dagger}_{-k}\right)\right]X_k+\left[\Omega\hat{b}^{\dagger}_{-k}-iC\left(\hat{a}_k-\hat{a}^{\dagger}_{-k}\right)\right]Z_k.\end{aligned} \tag{D.8}$$

Because the operators are linearly independent, from Eqs. (D.6) and (D.8) one has

$$M\begin{pmatrix}W_k\\X_k\\Y_k\\Z_k\end{pmatrix}\equiv\begin{pmatrix}\Omega(k)-\omega_k & iC & 0 & -iC\\ -iC & \Omega(k)-\Omega & -iC & 0\\ 0 & -iC & \Omega(k)+\omega_k & iC\\ -iC & 0 & -iC & \Omega(k)+\Omega\end{pmatrix}\begin{pmatrix}W_k\\X_k\\Y_k\\Z_k\end{pmatrix}=0. \tag{D.9}$$

The conditions that the coefficients W_k, Y_k, X_k, and Z_k are not zero leads to

$$\left\{\Omega^2(k)-\omega_k^2\right\}\left\{\Omega^2(k)-\Omega^2\right\}=4C^2\Omega\omega_k \tag{D.10}$$

as an eigenvalue equation. By setting $\hbar\Omega(k)=E(k)$ and $\hbar\Omega=E_m$, Eq. (D.10) can be rewritten as

$$\left\{E^2(k)-(\hbar\omega_k)^2\right\}\left\{E^2(k)-E_m^2\right\}=4\hbar\omega_kC^2E_m. \tag{D.11}$$

In order to determine the four coefficients, X_k, Y_k, and Z_k are expressed in terms of W_k from Eq. (D.9):

$$Y_k = -\frac{E(k) - \hbar\omega_k}{E(k) + \hbar\omega_k} W_k, \tag{D.12a}$$

$$X_k = -\frac{\{E(k) + E_m\}\{E(k) - \hbar\omega_k\}}{2i\hbar C E_m} W_k, \tag{D.12b}$$

$$Z_k = -\frac{\{E(k) - E_m\}\{E(k) - \hbar\omega_k\}}{2i\hbar C E_m} W_k. \tag{D.12c}$$

The boson commutation relation $\left[\hat{\xi}_k, \hat{\xi}_k^\dagger\right] = 1$ gives the following constraint:

$$|W_k|^2 + |X_k|^2 - |Y_k|^2 - |Z_k|^2 = 1. \tag{D.13}$$

Using Eqs. (D.11) – (D.13), the relation

$$\begin{aligned} W_k &= \frac{E(k) + \hbar\omega_k}{2\sqrt{E(k)\hbar\omega_k}} \sqrt{\frac{E^2(k) - E_m^2}{2E^2(k) - E_m^2 - (\hbar\omega_k)^2}} \\ &= \frac{\Omega(k) + \omega_k}{2\sqrt{\Omega(k)\omega_k}} \sqrt{\frac{\Omega^2(k) - \Omega^2}{2\Omega(k)^2 - \Omega^2 - \omega_k^2}} \end{aligned} \tag{D.14}$$

is obtained. Corresponding to two eigenvalues $E^{(\pm)}(k)$ of Eq. (D.11), the exciton-polariton operator and expansion coefficients are classified by a superscript $(\pm)$, such as $\hat{\xi}_k^{(\pm)}$, $\hat{\xi}_k^{(\pm)\dagger}$, and $W_k^{(\pm)}$. Then, Eq. (D.2) is rewritten as follows:

$$\begin{pmatrix} \hat{\xi}_k^{(+)} \\ \hat{\xi}_k^{(-)} \\ \hat{\xi}_{-k}^{(+)\dagger} \\ \hat{\xi}_{-k}^{(-)\dagger} \end{pmatrix} \equiv \begin{pmatrix} W_k^{(+)} & X_k^{(+)} & Y_k^{(+)} & Z_k^{(+)} \\ W_k^{(-)} & X_k^{(-)} & Y_k^{(-)} & Z_k^{(-)} \\ Y_k^{(+)*} & Z_k^{(+)*} & W_k^{(+)*} & X_k^{(+)*} \\ Y_k^{(-)*} & Z_k^{(-)*} & W_k^{(-)*} & X_k^{(-)*} \end{pmatrix} \begin{pmatrix} \hat{a}_k \\ \hat{b}_k \\ \hat{a}_{-k}^\dagger \\ \hat{b}_{-k}^\dagger \end{pmatrix}, \tag{D.15}$$

which can be inversely transformed to

$$\begin{pmatrix} \hat{a}_k \\ \hat{b}_k \\ \hat{a}_{-k}^\dagger \\ \hat{b}_{-k}^\dagger \end{pmatrix} \equiv \begin{pmatrix} W_k^{(+)*} & W_k^{(-)*} & -Y_k^{(+)} & -Y_k^{(-)} \\ X_k^{(+)*} & X_k^{(-)*} & -Z_k^{(+)} & -Z_k^{(-)} \\ -Y_k^{(+)*} & -Y_k^{(-)*} & W_k^{(+)} & W_k^{(-)} \\ -Z_k^{(+)*} & Z_k^{(-)*} & X_k^{(+)} & X_k^{(-)} \end{pmatrix} \begin{pmatrix} \hat{\xi}_k^{(+)} \\ \hat{\xi}_k^{(-)} \\ \hat{\xi}_{-k}^{(+)\dagger} \\ \hat{\xi}_{-k}^{(-)\dagger} \end{pmatrix}. \tag{D.16}$$

Since the coefficients with the superscrpits $(+)$ and $(-)$ appear equivalently in the matrix on the right-hand side of Eq. (D.16), the superscripts $(\pm)$ can be omitted, giving

$$\begin{cases} \hat{a}_k = W_k^* \hat{\xi}_k - Y_k \hat{\xi}_{-k}^\dagger \\ \hat{a}_{-k}^\dagger = -Y_k^* \hat{\xi}_k + W_k \hat{\xi}_{-k}^\dagger, \end{cases} \tag{D.17}$$

which are substituted for the photon operators in $\hat{\boldsymbol{D}}^{\perp}(\boldsymbol{r})$ in Eq. (2.6), and Eq.(2.6) is then inserted into Eq. (2.25). Using Eqs. (D.12a), (D.12b), (D.12c) and (D.14), one finally obtains

$$K_\alpha(\boldsymbol{k}) = \sum_{\lambda=1}^{2} (\boldsymbol{p}_\alpha \cdot \boldsymbol{e}_{\boldsymbol{k}\lambda}(\boldsymbol{k})) f(k)\, e^{i\boldsymbol{k}\cdot\boldsymbol{r}_\alpha} \tag{D.18}$$

with

$$f(k) = \frac{ck}{\sqrt{\Omega(k)}} \sqrt{\frac{\Omega^2(k) - \Omega^2}{2\Omega^2(k) - \Omega^2 - (ck)^2}}, \tag{D.19}$$

which is Eq. (2.28) to be derived.

Appendix E
Derivation of the Equations for Size-Dependent Resonance

The right-hand side of Eq. (2.78a) is written as

$$V_{eff}(r) = -\frac{p_s p_p}{3(2\pi)\varepsilon_0} W_+ \sum_{\alpha=s}^{p} \frac{\exp\left(-r/a'_\alpha\right)}{a'^2_\alpha r}, \tag{E.1}$$

and the first row of Eq. (2.79) as

$$I\left(R_{sp}\right) = \left|\int \nabla_{r_p} P\left(r_p\right) d^3 r_p\right|^2, \tag{E.2a}$$

$$P\left(r_p\right) = \int V_{eff}\left(\left|\boldsymbol{r}_p - \boldsymbol{r}_s\right|\right) d^3 r_s, \tag{E.2b}$$

where r_{sp} on the left-hand side of the first row of Eq. (2.79) is denoted by R_{sp} on the left-hand side of Eq. (E.2a) to avoid confusion in the notations in the following description.

First, in order to perform the integration of Eq. (E.2b), the z_s-Cartesian axis is fixed along the line connecting an arbitrary position on the sphere p and the center of the sphere s. Then, Eqs. (E.1) and (E.2b) give

$$\begin{aligned} P\left(r_p\right) &= \int \sum_{\alpha=s}^{p} \frac{\exp\left(-\left|\boldsymbol{r}_p - \boldsymbol{r}_s\right|/a'_\alpha\right)}{a'^2_\alpha \left|\boldsymbol{r}_p - \boldsymbol{r}_s\right|} d^3 \boldsymbol{r}_s \\ &= \sum_{\alpha=s}^{p} \int_0^{a_s} dr_s \int_0^{\pi} d\theta_s \int_0^{2\pi} d\phi_s \\ &\times \frac{\exp\left(-\sqrt{\left|\boldsymbol{R}_s - \boldsymbol{r}_p\right|^2 + r_s^2 - 2\left|\boldsymbol{R}_s - \boldsymbol{r}_p\right| r_s \cos\theta_s}/a'_\alpha\right)}{a'^2_\alpha \sqrt{\left|\boldsymbol{R}_s - \boldsymbol{r}_p\right|^2 + r_s^2 - 2\left|\boldsymbol{R}_s - \boldsymbol{r}_p\right| r_s \cos\theta_s}} r_s^2 \sin\theta_s \end{aligned} \tag{E.3}$$

DOI: 10.1007/978-3-642-39569-7,

$$= 2\pi \sum_{\alpha=s}^{p} \int_0^{a_s} dr_s \int_0^{\pi} d\theta_s$$

$$\times \frac{\exp\left(-\sqrt{\left|\boldsymbol{R}_s - \boldsymbol{r}_p\right|^2 + r_s^2 - 2\left|\boldsymbol{R}_s - \boldsymbol{r}_p\right| r_s \cos\theta_s / a'_\alpha}\right)}{a'^2_\alpha \sqrt{\left|\boldsymbol{R}_s - \boldsymbol{r}_p\right|^2 + r_s^2 - 2\left|\boldsymbol{R}_s - \boldsymbol{r}_p\right| r_s \cos\theta_s}} r_s^2 \sin\theta_s.$$

By defining

$$s_s \equiv \sqrt{\left|\boldsymbol{R}_s - \boldsymbol{r}_p\right|^2 + r_s^2 - 2\left|\boldsymbol{R}_s - \boldsymbol{r}_p\right| r_s \cos\theta_s}, \tag{E.4}$$

and using

$$\frac{ds_s}{d\theta_s} = \left|\boldsymbol{R}_s - \boldsymbol{r}_p\right| \frac{r_s}{s_s} \sin\theta_s, \tag{E.5}$$

Equation (E.3) is rewritten as

$$\begin{aligned} P\left(r_p\right) &= \frac{2\pi}{a'^2_\alpha \left|\boldsymbol{R}_s - \boldsymbol{r}_p\right|} \sum_{\alpha=s}^{p} \int_0^{a_s} dr_s r_s \int_{\left|\boldsymbol{R}_s - \boldsymbol{r}_p\right| - r_s}^{\left|\boldsymbol{R}_s - \boldsymbol{r}_p\right| + r_s} ds_s e^{-s_s/a'_\alpha} \\ &= \frac{2\pi}{a'^2_\alpha \left|\boldsymbol{R}_s - \boldsymbol{r}_p\right|} \sum_{\alpha=s}^{p} \int_0^{a_s} dr_s r_s a'_\alpha \\ &\quad \times \left\{ -\exp\left(-\frac{\left|\boldsymbol{R}_s - \boldsymbol{r}_p\right| + r_s}{a'_i}\right) + \exp\left(-\frac{\left|\boldsymbol{R}_s - \boldsymbol{r}_p\right| - r_s}{a'_i}\right) \right\} \\ &= \frac{2\pi}{a'^2_\alpha \left|\boldsymbol{R}_s - \boldsymbol{r}_p\right|} \sum_{\alpha=s}^{p} a'_\alpha \exp\left(-\frac{\left|\boldsymbol{R}_s - \boldsymbol{r}_p\right|}{a'_\alpha}\right) \\ &\quad \times \int_0^{a_s} dr_s r_s \left\{ \exp\left(\frac{r_s}{a'_\alpha}\right) - \exp\left(-\frac{r_s}{a'_\alpha}\right) \right\} \\ &= \frac{4\pi}{a'^2_\alpha \left|\boldsymbol{R}_s - \boldsymbol{r}_p\right|} \sum_{\alpha=s}^{p} a'^3_i \exp\left(-\frac{\left|\boldsymbol{R}_s - \boldsymbol{r}_p\right|}{a'_\alpha}\right) \\ &\quad \times \left\{ \frac{a_s}{a'_\alpha} \cosh\left(\frac{a_s}{a'_\alpha}\right) - \sinh\left(\frac{a_s}{a'_\alpha}\right) \right\}. \end{aligned} \tag{E.6}$$

Next, in order to perform the integration of Eq. (E.2a) with respect to $\boldsymbol{r}_p$, the volume integral is converted to a surface integral, which is expressed as

$$\begin{aligned} \int \nabla_{r_p} P\left(r_p\right) d^3 r_p &= \int_{S_p} P\left(r_p\right) \boldsymbol{n}_p dS_p \\ &= 4\pi \sum_{\alpha=s}^{p} a'^3_i \left\{ \frac{a_s}{a'_\alpha} \cosh\left(\frac{a_s}{a'_\alpha}\right) - \sinh\left(\frac{a_s}{a'_\alpha}\right) \right\} \int_{S_p} \frac{\exp\left(-\frac{\left|\boldsymbol{R}_s - \boldsymbol{r}_p\right|}{a'_\alpha}\right)}{a'^2_\alpha \left|\boldsymbol{R}_s - \boldsymbol{r}_p\right|} \boldsymbol{n}_p dS_p. \end{aligned} \tag{E.7}$$

Here, $\boldsymbol{n}_p$ is the unit vector normal to the surface S_p. By fixing the z_p-Cartesian axis along the line connecting the center $\boldsymbol{R}_s$ of the sphere s and the center $\boldsymbol{R}_p$ of the sphere p, the surface integral in Eq. (E.7) is taken, and the result is given by

$$\begin{aligned}
&\int_{S_p} \frac{\exp\left(-\frac{|\boldsymbol{R}_s-\boldsymbol{r}_p|}{a'_\alpha}\right)}{|\boldsymbol{R}_s-\boldsymbol{r}_p|}\boldsymbol{n}_p dS_p \\
&= \int_0^\pi d\theta_p \int_0^{2\pi} d\phi_p \frac{\exp\left(-\frac{\sqrt{R_{sp}^2+a_p^2-2a_pR_{sp}\cos\theta_p}}{a'_\alpha}\right)}{\sqrt{R_{sp}^2+a_p^2-2a_pR_{sp}\cos\theta_p}} a_p^2 \sin\theta_p \boldsymbol{n}_p \\
&= \frac{a_p}{R_{sp}} \int_{R_{sp}-a_p}^{R_{sp}+a_p} ds_p \mathrm{e}^{-\frac{s_p}{a'_i}} \int_0^{2\pi} d\phi_p \left(\sin\theta_p\cos\phi_p, \sin\theta_p\sin\phi_p, \cos\theta_p\right) \\
&= \frac{a_p}{R_{sp}} \int_{R_{sp}-a_p}^{R_{sp}+a_p} ds_p \mathrm{e}^{-\frac{s_p}{a'_i}} \frac{R_{sp}^2+a_p^2-s_p^2}{2a_pR_{sp}} \hat{\boldsymbol{R}}_{sp} \\
&= \frac{1}{2R_{sp}^2} \int_{R_{sp}-a_p}^{R_{sp}+a_p} ds_p \mathrm{e}^{-\frac{s_p}{a'_i}} \left[R_{sp}^2+a_p^2-s_p^2\right] \hat{\boldsymbol{R}}_{sp}.
\end{aligned} \tag{E.8}$$

where $\hat{\boldsymbol{R}}_{sp}$ is a unit vector along $\boldsymbol{R}_s - \boldsymbol{R}_p$, and the following transformation was used:

$$s_p = \sqrt{R_{sp}^2 + a_p^2 - 2a_pR_{sp}\cos\theta_p} \tag{E.9a}$$

with

$$\frac{ds_p}{d\theta_p} = \frac{a_pR_{sp}}{s_p}\sin\theta_p, \tag{E.9b}$$

$$\cos\theta_p = \frac{R_{sp}^2 + a_p^2 - s_p^2}{2a_pR_{sp}}, \tag{E.9c}$$

Inserting

$$\begin{aligned}
\int_{R_{sp}-a_p}^{R_{sp}+a_p} ds_p e^{-\frac{s_p}{a'_\alpha}} &= -a'_i\left\{\exp\left(-\frac{R_{sp}+a_p}{a'_\alpha}\right) - \exp\left(-\frac{R_{sp}-a_p}{a'_\alpha}\right)\right\} \\
&= 2a'_\alpha e^{-\frac{R_{sp}}{a'_\alpha}} \sinh\left(\frac{a_p}{a'_\alpha}\right)
\end{aligned} \tag{E.10a}$$

and

$$\begin{aligned}
&\int_{R_{sp}-a_p}^{R_{sp}+a_p} ds_p e^{-\frac{s_p}{a'_\alpha}} s_p^2 \\
&= \left[-a'_\alpha s_p^2 \exp\left(-s_p/a'_\alpha\right)\right]_{R_{sp}-a_p}^{R_{sp}+a_p} + 2a'_\alpha \int_{R_{sp}-a_p}^{R_{sp}+a_p} ds_p e^{-\frac{s_p}{a'_\alpha}} s_p
\end{aligned} \tag{E.10b}$$

$$= \left[-a'_\alpha s_p^2 \exp\left(-s_p/a'_\alpha\right) - 2a'^2_\alpha s_p \exp\left(-s_p/a'_\alpha\right)\right]_{R_{sp}-a_p}^{R_{sp}+a_p}$$
$$+ 2a'^2_i \int_{R_{sp}-a_p}^{R_{sp}+a_p} ds_p e^{-\frac{s_p}{a'_\alpha}}$$
$$= \left[-a'_\alpha s_p^2 \exp\left(-s_p/a'_\alpha\right) - 2a'^2_\alpha s_p \exp\left(-s_p/a'_\alpha\right) - 2a'^3_\alpha \exp\left(-s_p/a'_\alpha\right)\right]_{R_{sp}-a_p}^{R_{sp}+a_p}$$
$$= 2a'_\alpha e^{-\frac{R_{sp}}{a'_\alpha}} \left\{\left(R_{sp}^2 + 2R_{sp}a'_\alpha + a_p^2 + 2a'^2_\alpha\right) \sinh\left(\frac{a_p}{a'_\alpha}\right)\right.$$
$$\left. -2a_p\left(R_{sp} + a'_\alpha\right) \cosh\left(\frac{a_p}{a'_\alpha}\right)\right\}$$

into Eq. (E.8) yields

$$\int_{S_p} \frac{\exp\left(-\frac{|\boldsymbol{R}_s - \boldsymbol{r}_p|}{a'_i}\right)}{|\boldsymbol{R}_s - \boldsymbol{r}_p|} \boldsymbol{n}_p dS_p = 2a'^2_i \left\{\frac{a_p}{a'_i} \cosh\left(\frac{a_p}{a'_i}\right) - \sinh\left(\frac{a_p}{a'_i}\right)\right\}$$
$$\times \left(\frac{1}{R_{sp}} + \frac{a'_i}{R_{sp}^2}\right) \mathrm{e}^{-\frac{R_{sp}}{a'_i}} \hat{\boldsymbol{R}}_{sp}. \tag{E.11}$$

Furthermore, inserting this equation into Eq. (E.7) leads to

$$\int \nabla_{r_p} P\left(r_p\right) d^3 r_p = 8\pi \sum_{\alpha=s}^{p} \frac{1}{a'^2_\alpha} a'^6_\alpha \left\{\frac{a_s}{a'_\alpha} \cosh\left(\frac{a_s}{a'_\alpha}\right) - \sinh\left(\frac{a_s}{a'_\alpha}\right)\right\}$$
$$\times \left\{\frac{a_p}{a'_\alpha} \cosh\left(\frac{a_p}{a'_\alpha}\right) - \sinh\left(\frac{a_p}{a'_\alpha}\right)\right\} \left(\frac{1}{a'_\alpha R_{sp}} + \frac{1}{R_{sp}^2}\right) e^{-\frac{R_{sp}}{a'_\alpha}} \hat{\boldsymbol{R}}_{sp}. \tag{E.12}$$

Then, Eq. (E.2a) can be rewritten as

$$I\left(R_{sp}\right) = \left(\frac{p_s p_p}{3\left(2\pi\right)\varepsilon_0} W_+\right)^2$$
$$\times \left[8\pi \sum_{\alpha=s}^{p} a'^4_\alpha \left\{\frac{a_s}{a'_\alpha} \cosh\left(\frac{a_s}{a'_\alpha}\right) - \sinh\left(\frac{a_s}{a'_\alpha}\right)\right\}\right.$$
$$\left. \times \left\{\frac{a_p}{a'_\alpha} \cosh\left(\frac{a_p}{a'_\alpha}\right) - \sinh\left(\frac{a_p}{a'_\alpha}\right)\right\} \left(\frac{1}{a'_\alpha R_{sp}} + \frac{1}{R_{sp}^2}\right) e^{-\frac{R_{sp}}{a'_\alpha}}\right]^2. \tag{E.13}$$

Finally, by replacing R_{sp} with r_{sp}, Eq. (E.13) is reduced to Eq. (2.79).

Appendix F
Energy States of a Semiconductor Quantum Dot

Because electrons, holes, and electron–hole pairs in a QD are confined in a three-dimensional nanometric space, they have unique properties that cannot be achieved in bulk semiconductor materials. For example, they have discrete energy eigenvalues originating from the fact that their wave-functions are confined in the material. This is called the quantum confinement effect. The following sections review fundamental aspects of this effect for spherical or cubic QDs [1, 2].

F.1 One-Particle States

Because a QD contains many electrons, holes, and, electron–hole pairs even though its dimensions are of nanometer order, a many-particle problem has to be solved in order to study the quantum confinement effect. For this purpose, it is useful to employ the envelope function and effective mass approximation on the assumption that the energy eigenvalues of electrons in the periodic lattice, that is, the energy bands, are not appreciably modified through the quantum confinement. This approximation allows us to determine a ground state and excited states of a one-particle (electron or hole) system, and also a ground state of a many-particle problem by successively putting particles into the lowest unoccupied energy levels.

The one-particle wave-function in a QD is given by the product of the one-particle wave-function in a bulk material and the envelope function that satisfies the boundary conditions of the QD. Thus, the eigenstate $|\psi_e\rangle$ of an electron in the QD is expressed as

$$|\psi_e\rangle = \int d^3 r \xi_e(\boldsymbol{r}) \hat{\psi}_e^{\dagger}(\boldsymbol{r}) \left|\Phi_g\right\rangle \tag{F.1}$$

where $\xi_e(\boldsymbol{r})$ is the envelope function of the single electron state, $\hat{\psi}_e^{\dagger}(\boldsymbol{r})$ is the field operator for electron creation, and $\left|\Phi_g\right\rangle$ is the crystal ground state (also called the electronic vacuum state $|0\rangle$). Since no electrons in the conduction band exist in the

M. Ohtsu, *Dressed Photons*, Nano-Optics and Nanophotonics,
DOI: 10.1007/978-3-642-39569-7, © Springer-Verlag Berlin Heidelberg 2014

crystal ground state, applying the field operator $\hat{\psi}_e(\boldsymbol{r})$ for electron annihilation to $|\Phi_g\rangle$ gives the following relation:

$$\hat{\psi}_e(\boldsymbol{r})\,|\Phi_g\rangle = 0. \tag{F.2}$$

The field operators for electron creation and annihilation satisfy the anti-commutation relation for a fermion, namely

$$\left\{\hat{\psi}_e(\boldsymbol{r}'),\hat{\psi}_e^\dagger(\boldsymbol{r})\right\} \equiv \hat{\psi}_e(\boldsymbol{r}')\,\hat{\psi}_e^\dagger(\boldsymbol{r}) + \hat{\psi}_e^\dagger(\boldsymbol{r})\,\hat{\psi}_e(\boldsymbol{r}') = \delta(\boldsymbol{r}-\boldsymbol{r}'), \tag{F.3}$$

where $\delta(\boldsymbol{r}-\boldsymbol{r}')$ is the Dirac delta function. The equation for the envelope function $\xi_e(\boldsymbol{r})$ can be obtained from the Schrödinger equation

$$\hat{H}_e\,|\psi_e\rangle = E_e\,|\psi_e\rangle \tag{F.4}$$

with the Hamiltonian $\hat{H}_e$ for a single electron in the QD

$$\hat{H}_e = \int d^3r\hat{\psi}_e^\dagger(\boldsymbol{r})\left[-\frac{\hbar^2}{2m_e}\nabla^2\right]\hat{\psi}_e(\boldsymbol{r}) + E_g\int d^3r\hat{\psi}_e^\dagger(\boldsymbol{r})\,\hat{\psi}_e(\boldsymbol{r}), \tag{F.5}$$

and the energy eigenvalue E_g. Here, m_e and E_g denote the effective mass of an electron and the bandgap energy of the bulk semiconductor, respectively. Substituting this equation into Eq. (F.4), the left-hand side is given by

$$\begin{aligned}
\hat{H}_e\,|\psi_e\rangle &= \int d^3r'\hat{\psi}_e^\dagger(\boldsymbol{r}')\left[-\frac{\hbar^2}{2m_e}\nabla^2\right]\hat{\psi}_e(\boldsymbol{r}')\int d^3r\xi_e(\boldsymbol{r})\,\hat{\psi}_e^\dagger(\boldsymbol{r})\,|\Phi_g\rangle \\
&\quad + E_g\int d^3r'\hat{\psi}_e^\dagger(\boldsymbol{r}')\,\hat{\psi}_e(\boldsymbol{r}')\int d^3r\xi_e(\boldsymbol{r})\,\hat{\psi}_e^\dagger(\boldsymbol{r})\,|\Phi_g\rangle \\
&= -\frac{\hbar^2}{2m_e}\int d^3r'\int d^3r\delta(\boldsymbol{r}-\boldsymbol{r}')\nabla^2\xi_e(\boldsymbol{r})\,\hat{\psi}_e^\dagger(\boldsymbol{r}')\,|\Phi_g\rangle \\
&\quad + E_g\int d^3r'\int d^3r\delta(\boldsymbol{r}-\boldsymbol{r}')\xi_e(r)\,\hat{\psi}_e^\dagger(\boldsymbol{r}')\,|\Phi_g\rangle \\
&= \int d^3r\left[-\frac{\hbar^2}{2m_e}\nabla^{22}\xi_e(\boldsymbol{r})\right]\hat{\psi}_e^\dagger(\boldsymbol{r})\,|\Phi_g\rangle + E_g\int d^3r\xi_e(\boldsymbol{r})\,\hat{\psi}_e^\dagger(\boldsymbol{r})\,|\Phi_g\rangle
\end{aligned} \tag{F.6}$$

while the right-hand side is given by

$$E_e\,|\psi_e\rangle = E_e\int d^3r\xi_e(\boldsymbol{r})\,\psi_e^\dagger(\boldsymbol{r})\,|\Phi_g\rangle, \tag{F.7}$$

where Eqs. (F.2) and (F.3) were used. By equating Eqs. (F.5) and (F.6), the eigenvalue equation for the envelope function of the one-electron system is derived:

$$-\frac{\hbar^2}{2m_e}\nabla^2\xi_e(\boldsymbol{r}) = \left(E_e - E_g\right)\xi_e(\boldsymbol{r}). \tag{F.8}$$

By replacing the subscript of the envelope function e with h, the eigenvalue equation for the one-hole system is derived:

$$-\frac{\hbar^2}{2m_h}\nabla^2\xi_h(\boldsymbol{r}) = E_h\xi_h(\boldsymbol{r}), \tag{F.9}$$

where $E_g = 0$ was used.

In order to solve these eigenvalue equations, spherical and cubic QDs are considered in the following discussions.

(1) Spherical quantum dot

Assuming that an electron or a hole is confined in a spherical QD with radius R, its boundary condition is expressed as $\xi_e(\boldsymbol{r}) = \xi_h(\boldsymbol{r}) = 0$ for $|\boldsymbol{r}| > R$. Noting that the Laplace operator is written in spherical coordinates as

$$\nabla^2 = \frac{1}{r}\frac{\partial^2}{\partial r^2}r - \frac{\boldsymbol{L}^2}{r^2}, \tag{F.10a}$$

$$\boldsymbol{L}^2 = -\left(\frac{1}{\sin\theta}\frac{\partial}{\partial\theta}\sin\theta\frac{\partial}{\partial\theta} + \frac{1}{\sin^2\theta}\frac{\partial^2}{\partial\phi^2}\right), \tag{F.10b}$$

the envelope function $\xi(\boldsymbol{r})$ can be divided into radial and angular parts as $\xi(\boldsymbol{r}) = f_l(r)\,Y_{lm}(\theta,\phi)$. The subscripts e and h have been removed because the envelope function $\xi(\boldsymbol{r})$ depends only on R and not on specific parameters for the electron and hole. Here, $\boldsymbol{L}$ denotes the operator of the orbital angular momentum and obeys the following eigenvalue equation

$$\boldsymbol{L}^2 Y_{lm}(\theta,\phi) = l(l+1)\,Y_{lm}(\theta,\phi) \tag{F.11}$$

with $|m| \le l$, where the functions $Y_{lm}(\theta,\phi)$ are the spherical harmonics with $l = 0, 1, 2, \ldots$ and $m = 0, \pm 1, \pm 2, \ldots$. The radial part $f_l(r)$ should obey

$$\frac{d^2 f_l}{dr^2} + \frac{2}{r}\frac{df_l}{dr} + \left[\alpha^2 - l(l+1)\right]f_l = 0 \tag{F.12}$$

with

$$\alpha^2 \equiv \frac{2m_e}{\hbar^2}\left(E_e - E_g\right) \text{ or } \frac{2m_e}{\hbar^2}E_h, \tag{F.13}$$

and the solution has the form

$$f_{nl}(r) = \sqrt{\frac{2}{R^3}}\frac{j_l\left(\frac{\alpha_{nl}r}{R}\right)}{j_{l+1}(\alpha_{nl})}, \tag{F.14a}$$

where j_l is a spherical Bessel function of l-th order, and α_{nl} is determined from the boundary conditions as

$$j_l\left(\alpha_{nl}\right) = 0\,(n = 1, 2, 3, \cdots), \text{ and } \alpha_{n0} = n\pi, \alpha_{11} = 4.4934, \ldots . \tag{F.14b}$$

The energy eigenvalues of the electron are discrete and given by

$$E_{e,nlm} = E_g + \frac{\hbar^2}{2m_e}\left(\frac{\alpha_{nl}}{R}\right)^2. \tag{F.15}$$

Those of the hole are

$$E_{h,nlm} = \frac{\hbar^2}{2m_h}\left(\frac{\alpha_{nl}}{R}\right)^2. \tag{F.16}$$

(2) Cubic quantum dot

In the case where an electron or a hole is confined in a cubic QD with side-length L, as the first step, a one-dimensional case is analyzed, where the following one-dimensional well potential

$$V(x) = 0 \quad \text{for } |x| \leq \frac{L}{2} \tag{F.17a}$$

$$V(x) = \infty \quad \text{for } |x| > \frac{L}{2} \tag{F.17b}$$

is assumed. The envelope function $\xi(x)$ obeys the Schrödinger equation

$$\left[-\frac{\hbar^2}{2m}\frac{d^2}{dx^2} + V(x)\right]\xi(x) = E_x\xi(x), \tag{F.18}$$

and the boundary conditions

$$\xi\left(\frac{L}{2}\right) = \xi\left(-\frac{L}{2}\right) = 0 \tag{F.19}$$

are satisfied. The solutions of the equation are given by

$$\begin{cases} \xi_{even}(x) = \sqrt{\frac{2}{L}}\cos(k_x x), \\ \xi_{odd}(x) = \sqrt{\frac{2}{L}}\sin(k_x x), \end{cases} \tag{F.20}$$

and it follows from the boundary conditions that k_x has the following discrete values:

$$\begin{cases} k_x^{even} = \frac{\pi}{L}(2n-1), \\ k_x^{odd} = \frac{\pi}{L}(2n). \end{cases} \quad (n = 1, 2, 3, \ldots) \tag{F.21}$$

Therefore, the energy eigenvalues are also discrete:

$$E_x = \frac{\hbar^2 k_x^2}{2m} = \frac{\hbar^2}{2m}\left(\frac{\pi}{L}n_x\right)^2, \tag{F.22}$$

with $n_x = 1, 2, 3, \ldots$, where one sets $n_x = 2n - 1$ for k_x^{even}, and $n_x = 2n$ for k_x^{odd}. As the second step, the envelope functions $\xi(y)$ and $\xi(z)$ can be similarly obtained by replacing x in Eq. (F.20) with its corresponding y and z, respectively. Finally, the envelope functions for an electron and a hole confined in a three-dimensional potential well have the form $\xi(x)\,\xi(y)\,\xi(z)$. The energy eigenvalues specified by a set of quantum numbers $\left(n_x, n_y, n_z\right)$ are given by

$$E_{n_x,n_y,n_z} = \frac{\hbar^2}{2m}\left(\frac{\pi}{L}\right)^2\left(n_x^2 + n_y^2 + n_z^2\right) \text{ for } n_x, n_y, n_z = 1, 2, 3, \ldots. \tag{F.23}$$

By assuming that the energy eigenvalues of an electron in a periodic potential are not modified drastically, the energies of an electron in the conduction band or valence band are given by

$$E_c = E_g + \frac{\hbar^2 k^2}{2m_c} = E_g + \frac{\hbar^2}{2m_c}\left\{\left(\frac{\pi}{L_x}n_x\right)^2 + \left(\frac{\pi}{L_y}n_y\right)^2 + \left(\frac{\pi}{L_z}n_z\right)^2\right\} \tag{F.24a}$$

and

$$E_v = \frac{\hbar^2 k^2}{2m_v} = \frac{\hbar^2}{2m_v}\left\{\left(\frac{\pi}{L_x}n_x\right)^2 + \left(\frac{\pi}{L_y}n_y\right)^2 + \left(\frac{\pi}{L_z}n_z\right)^2\right\}, \tag{F.24b}$$

respectively.

F.2 Electron–Hole Pair States in a Quantum Dot

In order to analyze the electron–hole pair states in a QD, the eigenstate for an electron–hole pair is expressed as

$$|\psi_{eh}\rangle = \int\int d^3r_e d^3r_h \zeta_{eh}(\boldsymbol{r}_e, \boldsymbol{r}_h)\,\hat{\psi}_e^\dagger(\boldsymbol{r}_e)\,\hat{\psi}_h^\dagger(\boldsymbol{r}_h)\,|\Phi_g\rangle, \tag{F.25}$$

where $\zeta_{eh}(\boldsymbol{r}_e, \boldsymbol{r}_h)$ is the envelope function of the electron–hole pair, $\hat{\psi}_e^\dagger(\boldsymbol{r}_e)$ and $\hat{\psi}_h^\dagger(\boldsymbol{r}_h)$ are the field operators of electron creation in the conduction band and hole creation in the valence band, respectively, and $|\Phi_g\rangle$ is the above-mentioned crystal ground state. The envelope function $\zeta_{eh}(\boldsymbol{r}_e, \boldsymbol{r}_h)$ obeys the following equation:

$$\left[-\frac{\hbar^2}{2m_e}\nabla_e^2-\frac{\hbar^2}{2m_h}\nabla_h^2+V_c+V_{conf}\right]\zeta_{eh}(\boldsymbol{r}_e,\boldsymbol{r}_h)=\left(E-E_g\right)\zeta_{eh}(\boldsymbol{r}_e,\boldsymbol{r}_h) \tag{F.26}$$

with the Coulomb interaction potential V_c and the confinement potential V_{conf}. When the confinement region is a sphere with radius R, $V_{conf}(r)=0$ for $|\boldsymbol{r}|=r\le R$, whereas when it is a cube with side length L, $V_{conf}(x,y,z)=0$ for $-L/2\le x,y,z\le L/2$. Here, it might be useful to examine the electron–hole pair states by comparing the confinement size (R or L) with the Bohr radius a_0, which represents the average distance between the electron and hole in the pair. Noting that the confinement potential is proportional to $1/R^2$ (or $1/L^2$) whereas the Coulomb interaction potential is proportional to $1/R$ (or $1/L$), the following three cases are considered.

(1) $R \ll a_0$

In this case, the Coulomb interaction between an electron and a hole is weak, and each electron (hole) in a pair independently moves in the corresponding electron (hole) confinement potential. In particular, when both the Coulomb and confinement potentials are zero in a perfectly confined area, the lowest energy of an electron–hole pair is given, in terms of energy eigenvalues in the one-particle system already described, by

$$E=E_g+\frac{\pi^2\hbar^2}{2m_eR^2}+\frac{\pi^2\hbar^2}{2m_hR^2}=E_g+\frac{\pi^2\hbar^2}{2m_rR^2}, \tag{F.27}$$

where m_r is the reduced mass for an electron–hole pair defined by

$$\frac{1}{m_r}=\frac{1}{m_e}+\frac{1}{m_h}. \tag{F.28}$$

(2) $R \gg a_0$

Because the Coulomb interaction between an electron and a hole becomes strong in this case, an electron–hole pair can be approximated as a single particle, which is called an exciton. Then, the motion of the center of mass of the exciton is confined within the area R (or L). Defining the mass of the exciton, the center of mass coordinates, and the relative coordinates between the electron and hole as

$$M=m_e+m_h \tag{F.29a}$$

$$\boldsymbol{r}_{CM}=\left(m_e\boldsymbol{r}_e+m_h\boldsymbol{r}_h\right)/M \tag{F.29b}$$

$$\boldsymbol{\beta}=\boldsymbol{r}_e-\boldsymbol{r}_h \tag{F.29c}$$

respectively, the envelope function of the exciton is written as

$$\psi(\boldsymbol{r}_e,\boldsymbol{r}_h)=\phi_\mu(\boldsymbol{\beta})F_\nu(\boldsymbol{r}_{CM}), \tag{F.30}$$

where, in particular,

$$F_v(\boldsymbol{r}_{CM}) = \sqrt{\frac{2}{R^3}} \frac{j_l\left(\frac{\alpha_{nl} r_{CM}}{R}\right)}{j_{l+1}(\alpha_{nl})} Y_{lm}(\Omega_{CM}) \tag{F.31}$$

and

$$\phi_{\mu=1s}(\boldsymbol{\beta}) = \frac{1}{\sqrt{\pi a_0^3}} \exp\left(-\frac{\beta}{a_0}\right) \tag{F.32}$$

for the spherical boundary conditions, similar to the one-particle system. Here, the solid angle Ω_{CM} for $\boldsymbol{r}_{CM}$ was used, and $\phi_\mu(\boldsymbol{\beta})$ is assumed to be the function for the lowest energy state (the 1s state in the case of an electron in a hydrogen atom). The energy eigenvalues of the states specified by the quantum numbers (n, l) are

$$E_{nl} = E_g + E_{ex} + \frac{\hbar^2 \alpha_{nl}^2}{2MR^2} \quad (n = 1, 2, 3, \ldots), \tag{F.33}$$

where E_{ex} is the exciton binding energy in the bulk system.

Similarly, for the cubic boundary conditions, the envelope function for the center of mass is expressed as

$$F_v(\boldsymbol{r}_{CM}) = \sqrt{\frac{8}{L^3}} \begin{cases} \cos\left(\frac{\pi}{L}(2n_x - 1)x_{CM}\right)\cos\left(\frac{\pi}{L}(2n_y - 1)y_{CM}\right) \\ \qquad \times \cos\left(\frac{\pi}{L}(2n_z - 1)z_{CM}\right), \\ \sin\left(\frac{2\pi}{L}n_x x_{CM}\right)\sin\left(\frac{2\pi}{L}n_y y_{CM}\right)\sin\left(\frac{2\pi}{L}n_z z_{CM}\right). \end{cases} \tag{F.34}$$

The function for the relative motion is the same as Eq. (F.32). The energy eigenvalues are expressed in a similar way as

$$E_{n_x,n_y,n_z} = E_g + E_{ex} + \frac{\pi^2\hbar^2}{2ML^2}\left(n_x^2 + n_y^2 + n_z^2\right) \tag{F.35}$$
$$(n_x, n_y, n_z = 1, 2, 3, \ldots).$$

Note that the motion of the center of mass is confined to a sphere of radius $R - \eta a_0$ or a cube of side-length $L - \eta a_0$, where the factor η is of the order of unity and depends on the electron–hole mass ratio [3]. This is called dead-layer correction.

(3) $R \simeq a_0$

The situation in this case is more complicated than those of (F.1) and (F.2). Let the Bohr radii of an electron and a hole be a_e and a_h, respectively, and suppose that the confinement size is larger than a_h and smaller than a_e. Then one may assume that a hole can move in an average potential created by a free electron confined within a QD, and approximate the envelope function of the exciton by the product of those of the electron $\xi_{nlm}(\boldsymbol{r}_e)$ and hole $\psi_h(\boldsymbol{r}_h)$ as

$$\psi(\boldsymbol{r}_e, \boldsymbol{r}_h) = \xi_{nlm}(\boldsymbol{r}_e)\,\psi_h(\boldsymbol{r}_h). \tag{F.36}$$

Using the orthonormalization of $\xi_{nlm}(\boldsymbol{r}_e)$, the equation for the envelope function of the hole is written as

$$\begin{aligned}&\left[-\frac{\hbar^2}{2m_h}\nabla_h^2-\int d\boldsymbol{r}_e|\xi_{nlm}(\boldsymbol{r}_e)|^2V_c\right]\psi_h(\boldsymbol{r}_h)\\&=\left(E-E_g-\frac{\hbar^2}{2m_e}\frac{\alpha_{nl}^2}{R^2}\right)\psi_h(\boldsymbol{r}_h),\end{aligned}\tag{F.37}$$

where the spherical confinement is assumed to be $V_{conf}=0$ within the confinement region. When the cubic boundary conditions are used, the envelope function $\xi_{nlm}(\boldsymbol{r}_e)$ of an electron and the discrete energy $\left(\hbar^2/2m_e\right)\alpha_{nl}^2/R^2$ should be replaced with $\xi_{n_xn_yn_z}(\boldsymbol{r}_e)$ and $\left(\hbar^2/2m_e\right)(\pi/L)^2\left(n_x^2+n_y^2+n_z^2\right)$, respectively. In both cases, the second term on the left-hand side of Eq. (F.37) shows the Coulomb potential for the hole averaged by the electron. Numerical calculations are required to solve these equations.

F.3 Electric-Dipole–Forbidden Transition

On the basis of the above discussion, the behavior of an electron–hole pair when it is excited by a DP or by propagating light is now examined. In order to make the difference clear, case (2) in Sect. F.2 is examined. It is then convenient to use a Wannier function basis, which is a complete set of orthogonal functions representing electrons localized at an atomic site $\boldsymbol{R}$. The Wannier function $w_{b\boldsymbol{R}}(\boldsymbol{r})$ is defined by

$$w_{b\boldsymbol{R}}(\boldsymbol{r})\equiv\frac{1}{\sqrt{N}}\sum_{\boldsymbol{k}}\exp(-i\boldsymbol{k}\cdot\boldsymbol{R})\,\psi_{b\boldsymbol{k}}(\boldsymbol{r}),\tag{F.38}$$

with the Bloch function $\psi_{b\boldsymbol{k}}(\boldsymbol{r})$, which is a plane wave modulated by the periodicity of the lattice and is obtained from a linear combination of electron wave-functions in an isolated atom at an arbitrary site. Here, N is the number of constituent atoms. The Wannier functions for different bands (identified by the subscript b) and different sites $\boldsymbol{R}$ are orthogonal, which follows from

$$\begin{aligned}&\int w^*{}_{b\boldsymbol{R}}(\boldsymbol{r})w_{b'\boldsymbol{R}'}(\boldsymbol{r})\,d^3r\\&=\frac{1}{N}\sum_{\boldsymbol{k},\boldsymbol{k}'}\exp\left[i\left(\boldsymbol{k}\cdot\boldsymbol{R}-\boldsymbol{k}'\cdot\boldsymbol{R}'\right)\right]\int\psi^*{}_{b\boldsymbol{k}}(\boldsymbol{r})\psi_{b'\boldsymbol{k}'}(\boldsymbol{r})\,d^3r\\&=\frac{1}{N}\sum_{\boldsymbol{k},\boldsymbol{k}'}\exp\left[i\left(\boldsymbol{k}\cdot\boldsymbol{R}-\boldsymbol{k}'\cdot\boldsymbol{R}'\right)\right]\delta_{bb'}\delta_{\boldsymbol{k}\boldsymbol{k}'}\\&=\frac{1}{N}\sum_{\boldsymbol{k},\boldsymbol{k}'}\exp\left[i\boldsymbol{k}\cdot\left(\boldsymbol{R}-\boldsymbol{R}'\right)\right]\delta_{bb'}=\delta_{bb'}\delta_{\boldsymbol{R}\boldsymbol{R}'},\end{aligned}\tag{F.39}$$

The field operator of electron creation in the conduction band and that of hole creation in the valence band can be expressed in terms of the Wannier bases. In case (2), where an electron–hole pair (an exciton) is confined in a QD, the exciton state $|\Phi_\nu\rangle$ specified by the quantum numbers $\nu = (m, \mu)$ is represented as a superposition of electron states at $\boldsymbol{R}$ and hole states at $\boldsymbol{R}'$:

$$|\Phi_\nu\rangle = \sum_{\boldsymbol{R},\boldsymbol{R}'} F_m(\boldsymbol{R}_{CM})\,\varphi_\mu(\boldsymbol{\beta})\,\hat{e}^\dagger_{c\boldsymbol{R}}\hat{e}_{v\boldsymbol{R}'}\left|\Phi_g\right\rangle. \tag{F.40}$$

Here, $F_m(\boldsymbol{R}_{CM})$ is the motion of the center of mass of the exciton specified by a set of quantum numbers $\boldsymbol{m} = \left(m_x, m_y, m_z\right)$, whereas $\varphi_\mu(\boldsymbol{\beta})$ is the relative motion specified by the quantum number μ, and the product of them represents the envelope function of the exciton. The creation operator of an electron at $\boldsymbol{R}$ in the conduction band is denoted by $\hat{e}^\dagger_{c\boldsymbol{R}}$, and the annihilation operator of an electron at $\boldsymbol{R}'$ in the valence band is represented by $\hat{e}_{v\boldsymbol{R}'}$. is the crystal ground operator.

F.3.1 Excitation by Dressed Photons

In order to derive the effective interaction energy between two QDs based on Eq. (2.31), the matrix element representing the transition from the exciton state $|\Phi_\nu\rangle$ to the crystal ground state $\left|\Phi_g\right\rangle$ is calculated as

$$\begin{aligned}\left\langle\Phi_g\right|\hat{V}\left|\Phi_\nu\right\rangle &= \sum_{\boldsymbol{k},\lambda}\sum_{\boldsymbol{R},\boldsymbol{R}'} F_m(\boldsymbol{R}_{CM})\,\varphi_\mu(\boldsymbol{\beta}) \\ &\quad\times\left(\hat{\xi}(\boldsymbol{k})\,g_{v\boldsymbol{R}'c\boldsymbol{R},\boldsymbol{k}\lambda} - \hat{\xi}^\dagger(\boldsymbol{k})\,g_{v\boldsymbol{R}'c\boldsymbol{R},-\boldsymbol{k}\lambda}\right)\end{aligned} \tag{F.41}$$

with

$$g_{v\boldsymbol{R}'c\boldsymbol{R},\boldsymbol{k}\lambda} = -i\sqrt{\frac{\hbar}{2\varepsilon_0 V}}f(k)\int w^*_{v\boldsymbol{R}'}(\boldsymbol{r})\,\boldsymbol{p}(\boldsymbol{r})\,w_{c\boldsymbol{R}}(\boldsymbol{r})\cdot\boldsymbol{e}_\lambda(\boldsymbol{k})\,e^{i\boldsymbol{k}\cdot\boldsymbol{r}}d^3r, \tag{F.42}$$

which was derived using the fact that the expectation values of $\left\langle\Phi_g\right|\hat{e}^\dagger_{v\boldsymbol{R}_1}\hat{e}_{c\boldsymbol{R}_2}\hat{e}^\dagger_{c\boldsymbol{R}}\hat{e}_{v\boldsymbol{R}'}\left|\Phi_g\right\rangle$ are not zero only if $\boldsymbol{R}' = \boldsymbol{R}_1$ and $\boldsymbol{R} = \boldsymbol{R}_2$ hold. By transforming the spatial integral to the sum of the unit cells and by noting the spatial locality of the Wannier functions, one finds that Eq. (F.42) is proportional to $\delta_{\boldsymbol{R}\boldsymbol{R}'}$. Defining the electric dipole moment for each unit cell as

$$\boldsymbol{p}_{cv} = \int_{UC} w^*_{v\boldsymbol{R}'}(\boldsymbol{r})\,\boldsymbol{p}(\boldsymbol{r})\,w_{c\boldsymbol{R}}(\boldsymbol{r})d^3r \tag{F.43}$$

and noting that it is the same as that of the bulk material, independent of the site $\boldsymbol{R}$, the final form of Eq. (F.41) is

$$\langle \Phi_g | \hat{V} | \Phi_v \rangle = -i\sqrt{\frac{\hbar}{2\varepsilon_0 V}} \sum_{\boldsymbol{k}} \sum_{\lambda=1}^{2} \sum_{\boldsymbol{R}} f(k) \left[\boldsymbol{p}_{cv} \cdot \boldsymbol{e}_\lambda(\boldsymbol{k}) \right] F_m(\boldsymbol{R})\, \varphi_\mu(0) \times \left\{ \hat{\xi}(\boldsymbol{k})\, e^{i\boldsymbol{k}\cdot\boldsymbol{R}} - \hat{\xi}^\dagger(\boldsymbol{k})\, e^{-i\boldsymbol{k}\cdot\boldsymbol{R}} \right\}. \tag{F.44}$$

Here it should be noted that the long-wavelength approximation $e^{\pm i\boldsymbol{k}\cdot\boldsymbol{R}} \simeq 1$ is not applied because the energy of the DP is localized in a nanometric space.

According to the effective interaction energy between the two QDs given by Eq. (2.31), the initial and final states in the P space are set to $|\phi_{Pi}\rangle = \left|\Phi^A_{m\mu}\right\rangle \left|\Phi^B_g\right\rangle |0\rangle$ and $\left|\phi_{Pf}\right\rangle = \left|\Phi^A_g\right\rangle \left|\Phi^B_{m'\mu'}\right\rangle |0\rangle$, respectively. As the intermediate states $\left|\phi_{Qj}\right\rangle$ in the Q space, $\left|\Phi^A_g\right\rangle \left|\Phi^B_g\right\rangle |\boldsymbol{k}\rangle$ and $\left|\Phi^A_{m\mu}\right\rangle \left|\Phi^B_{m'\mu'}\right\rangle |\boldsymbol{k}\rangle$ are employed, where $\boldsymbol{k}$ is the wave-vector of the exciton-polariton. The superscripts A and B are used to label the two QDs. Using Eq. (F.44), one can rewrite Eq. (2.31) as

$$V_{\mathit{eff}} = \varphi^A_\mu(0)\, \varphi^{B*}_{\mu'}(0) \times \int\!\!\int F^A_m(\boldsymbol{R}_A)\, F^{B*}_{m'}(\boldsymbol{R}_B) \left[Y_A(\boldsymbol{R}_A - \boldsymbol{R}_B) + Y_B(\boldsymbol{R}_A - \boldsymbol{R}_B) \right] d^3R_A d^3R_B, \tag{F.45}$$

where $\sum_{\boldsymbol{R}}$ in Eq. (F.44) was transformed to the integral form. The integral Kernels $Y_\alpha(\boldsymbol{R}_{AB})$ with $\boldsymbol{R}_{AB} = \boldsymbol{R}_A - \boldsymbol{R}_B$, which connect the two spatially isolated envelope functions $F^A_m(\boldsymbol{R}_A)$ and $F^B_{m'}(\boldsymbol{R}_B)$, are defined by

$$Y_\alpha(\boldsymbol{R}_{AB}) = -\frac{\hbar^2}{(2\pi)^3 \varepsilon_0} \sum_{\lambda=1}^{2} \int \left[\boldsymbol{p}^A_{cv} \cdot \boldsymbol{e}_\lambda(\boldsymbol{k}) \right] \left[\boldsymbol{p}^B_{cv} \cdot \boldsymbol{e}_\lambda(\boldsymbol{k}) \right] \hbar f^2(k) \times \left\{ \frac{1}{E(k) + E_\alpha} + \frac{1}{E(k) - E_\alpha} \right\} e^{i\boldsymbol{k}\cdot\boldsymbol{R}_{AB}} d\boldsymbol{k}. \tag{F.46}$$

Here, the electric dipole moment $\boldsymbol{p}^\alpha_{cv}$ for QD$_\alpha$ ($\alpha = A, B$) is defined by Eq. (F.43), and E_α denotes the exciton energy in QD$_\alpha$. Then Eq. (F.46) can be rewritten in the same way as described in Chap. 2, for example, corresponding to Eq. (2.75):

$$Y_\alpha(\boldsymbol{R}_{AB}) = -\frac{p^A_{cv} p^B_{cv}}{3(2\pi)\varepsilon_0} \left(W_{\alpha+} \Delta^2_{\alpha+} \frac{e^{-\Delta_{\alpha+} R_{AB}}}{R_{AB}} - W_{\alpha-} \Delta^2_{\alpha-} \frac{e^{-\Delta_{\alpha-} R_{AB}}}{R_{AB}} \right), \tag{F.47}$$

where $R_{AB} = |\boldsymbol{R}_{AB}|$ was used.

F.3.2 Excitation by Propagating Light

Since the electric displacement vector of the propagating light is spatially homogeneous in the QDs, the long-wavelength approximation $e^{\pm i\boldsymbol{k}\cdot\boldsymbol{R}} \simeq 1$ is applied. Then, the transition matrix elements can be written in separated form in terms of $\boldsymbol{R}$ and $(\boldsymbol{k}, \lambda)$ as

$$\begin{aligned}
&\left\langle \Phi_g \right| \hat{V} \left| \Phi_\nu^\alpha \right\rangle \\
&= -i\sqrt{\frac{\hbar}{2\varepsilon_0 V}} \sum_R F_m^\alpha(\boldsymbol{R})\, \varphi_\mu^\alpha(0) \sum_{\boldsymbol{k}} \sum_{\lambda=1}^{2} f(k) \left[\boldsymbol{p}_{cv} \cdot \boldsymbol{e}_\lambda(\boldsymbol{k})\right] \left\{\hat{\xi}(\boldsymbol{k}) - \hat{\xi}^\dagger(\boldsymbol{k})\right\} \\
&= -i\sqrt{\frac{\hbar}{2\varepsilon_0 V}} \left[\int F_m^\alpha(\boldsymbol{R})\, d\boldsymbol{R}\right] \varphi_\mu^\alpha(0) \sum_{\boldsymbol{k}} \sum_{\lambda=1}^{2} f(k) \left[\boldsymbol{p}_{cv} \cdot \boldsymbol{e}_\lambda(\boldsymbol{k})\right] \left\{\hat{\xi}(\boldsymbol{k}) - \hat{\xi}^\dagger(\boldsymbol{k})\right\}.
\end{aligned} \tag{F.48}$$

The integral $\int F_m^\alpha(\boldsymbol{R})\, d\boldsymbol{R}$, as well as $\boldsymbol{p}_{cv}$, provide a criterion for whether the electric dipole transition between the crystal ground state $\left|\Phi_g\right\rangle$ and the exciton state $\left|\Phi_\nu^\alpha\right\rangle$ in the QD$_\alpha$ specified by the quantum number $\nu = (\boldsymbol{m}, \mu)$ is allowed or forbidden. Thus, it follows that the electric dipole transition is forbidden if the spatial integral $\int F_m^\alpha(\boldsymbol{R})\, d\boldsymbol{R}$ is zero, whereas it is allowed if the integral is not zero. Since the spatial integral for a spherical QD, as an example, is given by

$$\begin{aligned}
\int d^3r F_m(\boldsymbol{r}) &= \sqrt{\frac{2}{R^3}} \int_0^R r^2 dr \frac{j_l\left(\frac{\alpha_{nl} r}{R}\right)}{j_{l+1}(\alpha_{nl})} \int\int \sin\theta d\theta d\phi Y_{lm}(\theta, \phi) \\
&= \frac{1}{n}\sqrt{\frac{2R^3}{\pi^2}} \delta_{l0}\delta_{m0},
\end{aligned} \tag{F.49}$$

the transition only to the state specified by $l = m = 0$ is allowed. Similarly when all of the integrands are even functions, the nonzero result

$$\begin{aligned}
&\int d^3r F_m(\boldsymbol{r}) \\
&= \sqrt{\frac{8}{L^3}} \int_{-\frac{L}{2}}^{\frac{L}{2}} dx \cos\left(\frac{(2n_x - 1)\pi x}{L}\right) \\
&\quad \times \int_{-\frac{L}{2}}^{\frac{L}{2}} dy \cos\left(\frac{(2n_y - 1)\pi y}{L}\right) \int_{-\frac{L}{2}}^{\frac{L}{2}} dz \cos\left(\frac{(2n_z - 1)\pi z}{L}\right) \\
&= \sqrt{\frac{512L^3}{\pi^6}} \frac{1}{(2n_x - 1)} \frac{1}{(2n_y - 1)} \frac{1}{(2n_z - 1)} \\
&\quad \times \sin\left(\frac{(2n_x - 1)\pi}{2}\right) \sin\left(\frac{(2n_y - 1)\pi}{2}\right) \sin\left(\frac{(2n_z - 1)\pi}{2}\right) \\
&= \sqrt{\frac{512L^3}{\pi^6}} \frac{1}{(2n_x - 1)} \frac{1}{(2n_y - 1)} \frac{1}{(2n_z - 1)}
\end{aligned} \tag{F.50}$$

is obtained for a cubic QD, and thus the transitions are allowed if all of (n_x, n_y, n_z) are odd, whereas they are forbidden if any one of (n_x, n_y, n_z) is even.

References

1. H. Haug, S.W. Koch, *Quantum Theory of the Optical and Electronic Properties of Semiconductors*, 4th edn. (World Scientific Publishing, Singapore, 2004)
2. U. Waggon, *Optical Properties of Semiconductor QDs* (Springer, Berlin, 1997)
3. Y. Matsumoto, T. Takagahara, *Semiconductor QDs* (Springer, Berlin, 2002)

Appendix G
Solutions of the Quantum Master Equations for the Density Matrix Operators

G.1 The Case of the Two Quantum Dots

In order to derive Eqs. (3.18a)–(3.18d) in Chap. 3 for the case of $n = 0$, $\rho_{12}(t) - \rho_{21}(t)$ is denoted by $\Delta\rho_{12}(t)$. Then, Eq. (3.18b) can be rewritten as

$$\frac{d\Delta\rho_{12}(t)}{dt} = 2iU(r)[\rho_{11}(t) - \rho_{22}(t)] - \gamma\Delta\rho_{12}(t). \tag{G.1}$$

Equations (3.18a) and (3.18c) are also rewritten as

$$\frac{d\rho_{11}(t)}{dt} = iU(r)\Delta\rho_{12}(t), \tag{G.2}$$

$$\frac{d\rho_{22}(t)}{dt} = -iU(r)\Delta\rho_{12}(t) - 2\gamma\rho_{22}(t) + 2\gamma\rho_{33}(t). \tag{G.3}$$

Laplace transformations of these equations give

$$s\rho_{11}(s) - \rho_{11}(0) = iU(r)\Delta\rho_{12}(s), \tag{G.4}$$

$$\Delta\rho_{12}(s) - \Delta\rho_{12}(0) = 2iU(r)[\rho_{11}(t) - \rho_{22}(t)] - \gamma\Delta\rho_{12}(t) \tag{G.5}$$

$$s\rho_{22}(s) - \rho_{22}(0) = -iU(r)\Delta\rho_{12}(s) - 2\gamma\rho_{22}(s). \tag{G.6}$$

Under the initial conditions of $\rho_{11}(0) = 1$, $\Delta\rho_{12}(0) = 0$, and $\rho_{22}(0) = 0$, these equations can be solved simultaneously, and the solutions are

$$\rho_{11}(s) = \frac{s^2 + 3\gamma s + 2\left(U^2 + \gamma^2\right)}{(s+\gamma)\left(s^2 + 2\gamma s + 4U^2\right)}, \tag{G.7}$$

M. Ohtsu, *Dressed Photons*, Nano-Optics and Nanophotonics,
DOI: 10.1007/978-3-642-39569-7, © Springer-Verlag Berlin Heidelberg 2014

$$\rho_{22}(s) = \frac{2U^2}{(s+\gamma)\left(s^2+2\gamma s+4U^2\right)}, \tag{G.8}$$

$$\Delta\rho_{12}(s) = \frac{2iU(s+2\gamma)}{(s+\gamma)\left(s^2+2\gamma s+4U^2\right)}. \tag{G.9}$$

By defining a parameter

$$Z \equiv \sqrt{(\gamma/2)^2 - U^2}, \tag{G.10}$$

the term $s^2 + 2\gamma s + 4U^2$ in the denominators on the right-hand sides of Eqs. (G.7)–(G.9) is transformed to $(s+\gamma+2Z)(s+\gamma-2Z)$. Then, the inverse-Laplace transformations of Eqs. (G.7)–(G.9) give

$$\rho_{11}(t) = \frac{1}{8Z^2}e^{-(2Z+\gamma)t}\left[-2U^2\left(1+e^{2Zt}\right)+\gamma\left\{2Z\left(-1+e^{4Zt}\right)+\gamma\left(1+e^{4Zt}\right)\right\}\right], \tag{G.11}$$

$$\rho_{22}(t) = -\frac{U^2}{4Z^2}e^{-(2Z+\gamma)t}\left(-1+e^{2Zt}\right)^2, \tag{G.12}$$

$$\Delta\rho_{12}(t) = \frac{iU}{4Z^2}e^{-(2Z+\gamma)t}\left(-1+e^{2Zt}\right)\left\{2Z\left(1+e^{2Zt}\right)+\gamma\left(-1+e^{2Zt}\right)\right\}. \tag{G.13}$$

By rewriting the exponential functions in these equations as hyperbolic sinusoidal and cosinusoidal functions, Eqs. (3.19a)–(3.19c) in Chap. 3 are derived.

G.2 XOR Logic Gate Composed of Three Quantum Dots

Laplace transformations of Eqs. (3.23a)–(3.23e) in Chap. 3 give

$$s\rho_{S_1,S_1}(s) - \rho_{S_1,S_1}(0) = i\sqrt{2}U'\left\{\rho_{S_1,P_1'}(s) - \rho_{P_1',S_1}(s)\right\}, \tag{G.14a}$$

$$s\rho_{S_1,P_1'}(s) - \rho_{S_1,P_1'}(0) = \left\{i(\Delta\Omega - U) - \frac{\gamma}{2}\right\}\rho_{S_1,P_1'}(s) + i\sqrt{2}U'\left\{\rho_{S_1,S_1}(s) - \rho_{P_1',P_1'}(s)\right\}, \tag{G.14b}$$

$$s\rho_{P_1',S_1}(s) - \rho_{P_1',S_1}(0) = -\left\{i(\Delta\Omega - U) + \frac{\gamma}{2}\right\}\rho_{P_1',S_1}(s) - i\sqrt{2}U'\left\{\rho_{S_1,S_1}(s) - \rho_{P_1',P_1'}(s)\right\}, \tag{G.14c}$$

$$s\rho_{P_1',P_1'}(s) - \rho_{P_1',P_1'}(0) = -\gamma\rho_{P_1',P_1'}(s) - i\sqrt{2}U'\left\{\rho_{S_1,P_1'}(s) - \rho_{P_1',S_1}(s)\right\}, \tag{G.14d}$$

$$s\rho_{P_1,P_1}(s) = \gamma\rho_{P_1',P_1'}(s). \tag{G.14e}$$

These are solved simultaneously under the initial conditions

$$\rho_{S_1,S_1}(0) = 1/2,\ \rho_{S_1,P_1'}(0) = \rho_{P_1',S_1}(0) = \rho_{P_1',P_1'}(0) = \rho_{P_1,P_1}(0) = 0. \tag{G.15}$$

By defining

$$\omega_\pm \equiv -\frac{1}{\sqrt{2}}\sqrt{(\Delta\Omega - U)^2 + W_+W_- \pm \sqrt{\{(\Delta\Omega - U)^2 + W_-^2\}\{(\Delta\Omega - U)^2 + W_+^2\}}}, \quad (G.16)$$

$$W_\pm \equiv 2\sqrt{2}U' \pm \frac{\gamma}{2} \quad (G.17)$$

the solution of $\rho_{P_1,P_1}(t)$ on the left-hand side of Eq. (G.14e) is expressed as

$$\begin{aligned}\rho_{P_1,P_1}(s) = &-\frac{4iU'^2\gamma}{(2s+\gamma+2i\omega_-)(-i\gamma+2\omega_-)\left(\omega_-^2-\omega_+^2\right)} \quad (G.18)\\ &+\frac{4iU'^2\gamma}{(2s+\gamma-2i\omega_-)(i\gamma+2\omega_-)\left(\omega_-^2-\omega_+^2\right)}\\ &+\frac{4iU'^2\gamma}{(2s+\gamma+2i\omega_+)(-i\gamma+2\omega_+)\left(\omega_-^2-\omega_+^2\right)}\\ &-\frac{4iU'^2\gamma}{(2s+\gamma-2i\omega_+)(i\gamma+2\omega_+)\left(\omega_-^2-\omega_+^2\right)}+\frac{16U'^2\gamma^2}{s\left(\gamma^2+4\omega_-^2\right)\left(\gamma^2+4\omega_+^2\right)}.\end{aligned}$$

Since the inverse-Laplace transformations of the first and second terms in Eq. (G.18) are

$$\frac{2e^{-(\gamma+2i\omega_-)\frac{t}{2}}U'^2\gamma\left\{\left(1+e^{2i\omega_-t}\right)\gamma+2i\left(-1+e^{2i\omega_-t}\right)\omega_-\right\}}{\left(\gamma^2+4\omega_-^2\right)\left(\omega_-^2-\omega_+^2\right)}, \quad (G.19)$$

replacement of the exponential functions in this equation with sinusoidal and cosinusoidal functions gives

$$\begin{aligned}&-\frac{4U'^2}{\omega_+^2-\omega_-^2}e^{-\frac{\gamma t}{2}}\frac{\gamma}{\sqrt{\gamma^2+4\omega_-^2}}\left(\frac{\gamma}{\sqrt{\gamma^2+4\omega_-^2}}\cos(\omega_-t)-\frac{2\omega_-}{\sqrt{\gamma^2+4\omega_-^2}}\sin(\omega_-t)\right) \quad (G.20)\\ &=-\frac{4U'^2}{\omega_+^2-\omega_-^2}e^{-\frac{\gamma t}{2}}\cos(\phi_-)\cos(\omega_-t+\phi_-),\end{aligned}$$

where

$$\phi_- = \tan^{-1}\left(\frac{2\omega_-}{\gamma}\right) \quad (G.21)$$

was used. Similarly, since the inverse-Laplace transformations of the third and fourth terms are

$$\frac{2e^{-\frac{t}{2}(\gamma+2i\omega_+)}U'^2\gamma\left\{\left(1+e^{2i\omega_+t}\right)\gamma+2i\left(-1+e^{2i\omega_+t}\right)\omega_+\right\}}{\left(\gamma^2+4\omega_+^2\right)\left(-\omega_-^2+\omega_+^2\right)}, \tag{G.22}$$

replacement of the exponential functions in this equation with sinusoidal and cosinusoidal functions gives

$$\begin{aligned}&\frac{4U'^2}{\omega_+^2-\omega_-^2}e^{-\frac{\gamma t}{2}}\frac{\gamma}{\sqrt{\gamma^2+4\omega_+^2}}\\&\times\left(\frac{\gamma}{\sqrt{\gamma^2+4\omega_+^2}}\cos\left(\omega_+t\right)-\frac{2\omega_+}{\sqrt{\gamma^2+4\omega_+^2}}\sin\left(\omega_+t\right)\right)\\&=\frac{4U'^2}{\omega_+^2-\omega_-^2}e^{-\frac{\gamma t}{2}}\cos\left(\phi_+\right)\cos\left(\omega_+t+\phi_+\right),\end{aligned} \tag{G.23}$$

where

$$\phi_+=\tan^{-1}\left(\frac{2\omega_+}{\gamma}\right) \tag{G.24}$$

was used. The inverse-Laplace transformation of the fifth term is

$$\frac{16U'^2\gamma^2}{\left(\gamma^2+4\omega_-^2\right)\left(\gamma^2+4\omega_+^2\right)}, \tag{G.25}$$

which is found to be equal to 1/2 by noting the definition of given by Eq. (G.16). Finally, summation of the five terms given above derives Eq. (3.25) in Chap. 3.

G.3 AND Logic Gate Composed of Three Quantum Dots

Laplace transformations of Eqs. (3.33a)–(3.33d) in Chap. 3 give

$$s\rho_{S_2',S_2'}(s)-\rho_{S_2',S_2'}(0)=i\sqrt{2}U'\left\{\rho_{S_2',P_2'}(s)-\rho_{P_2',S_2'}(s)\right\}-\gamma\rho_{S_2',S_2'}(s), \tag{G.26a}$$

$$\begin{aligned}s\rho_{S_2',P_2'}(s)-\rho_{S_2',P_2'}(0)=&\left\{i\left(\Delta\Omega+U\right)+\tfrac{\gamma}{2}\right\}\rho_{S_2',P_2'}(s)\\&+i\sqrt{2}U'\left\{\rho_{S_2',S_2'}(s)-\rho_{P_2',P_2'}(s)\right\},\end{aligned} \tag{G.26b}$$

$$\begin{aligned}s\rho_{P_2',S_2'}(s)-\rho_{P_2',S_2'}(0)=&\left\{i\left(\Delta\Omega+U\right)-\tfrac{\gamma}{2}\right\}\rho_{P_2',S_2'}(s)\\&-i\sqrt{2}U'\left\{\rho_{S_2',S_2'}(s)-\rho_{P_2',P_2'}(s)\right\},\end{aligned} \tag{G.26c}$$

$$s\rho_{P'_2,P'_2}(s) - \rho_{P'_2,P'_2}(0) = -i\sqrt{2}U' \left\{\rho_{S'_2,P'_2}(s) - \rho_{P'_2,S'_2}(s)\right\}. \tag{G.26d}$$

The Laplace transformation of the first row of Eq. (3.34) is expressed as

$$\rho_{S_2,S_2}(s) + \rho_{P_2,P_2}(s) = \frac{\gamma}{s}\rho_{S'_2,S'_2}(s). \tag{G.26e}$$

Equation (G.26e) is derived by solving the simultaneous equations of Eqs. (G.26a)–(G.26d) under the initial conditions

$$\rho_{S'_2,S'_2}(0) = 0, \quad \rho_{S'_2,P'_2}(0) = \rho_{P'_2,S'_2}(0) = 0, \quad \rho_{P'_2,P'_2} = 1. \tag{G.27}$$

Inserting the solution of into Eq. (G.26e) yields

$$\begin{aligned}\rho_{S_2,S_2}(s) + \rho_{P_2,P_2}(s) = {} & 8U'^2\gamma(2s+\gamma) \\ & /s\left\{4s^4 + 8s^3\gamma + 8U'^2\gamma^2 + 4(\Delta\Omega)^2 s(s+\gamma) + 8(\Delta\Omega)sU(s+\gamma)\right. \\ & \left.+ s\gamma\left(4U^2 + 32U'^2 + \gamma^2\right) + s^2\left(4U^2 + 32U'^2 + 5\gamma^2\right)\right\}.\end{aligned} \tag{G.28}$$

By defining

$$\omega'_\pm \equiv \frac{1}{\sqrt{2}}\sqrt{(\Delta\Omega + U)^2 + W_+W_- \pm \sqrt{\left\{(\Delta\Omega + U)^2 + W_-^2\right\}\left\{(\Delta\Omega + U)^2 + W_+^2\right\}}}, \tag{G.29}$$

$$W_\pm = 2\sqrt{2}U' \pm \frac{\gamma}{2}, \tag{G.30}$$

Equation (G.28) is rewritten as

$$\begin{aligned}\rho_{S_2,S_2}(s) + \rho_{P_2,P_2}(s) = & -\frac{8iU'^2\gamma}{(2s+\gamma+2i\omega'_-)(-i\gamma+2\omega'_-)(\omega'^2_- - \omega'^2_+)} \\ & + \frac{8iU'^2\gamma}{(2s+\gamma-2i\omega'_-)(i\gamma+2\omega'_-)(\omega'^2_- - \omega'^2_+)} \\ & + \frac{8iU'^2\gamma}{(2s+\gamma+2i\omega'_+)(-i\gamma+2\omega'_+)(\omega'^2_- - \omega'^2_+)} \\ & - \frac{8iU'^2\gamma}{(2s+\gamma-2i\omega'_+)(i\gamma+2\omega'_+)(\omega'^2_- - \omega'^2_+)} \\ & + \frac{32U'^2\gamma^2}{s(\gamma^2+4\omega'^2_-)(\gamma^2+4\omega'^2_+)}.\end{aligned} \tag{G.31}$$

Since the inverse-Laplace transformations of the first and second terms on the right-hand side are

$$\frac{4e^{-(\gamma+2i\omega'_-)\frac{t}{2}}U'^2\gamma\left\{\left(1+e^{2i\omega'_-t}\right)\gamma+2i\left(-1+e^{2i\omega'_-t}\right)\omega'_-\right\}}{\left(\gamma^2+4\omega'^2_-\right)\left(\omega'^2_- - \omega'^2_+\right)}, \tag{G.32}$$

replacement of the exponential functions in this equations with sinusoidal and cosinusoidal functions gives

$$-\frac{8U'^2}{\omega'^2_+ - \omega'^2_-}e^{-\left(\frac{\gamma}{2}\right)t}\cos\left(\phi'_-\right)\cos\left(\omega'_-t+\phi'_-\right), \tag{G.33}$$

where

$$\phi'_- = \tan^{-1}\left(\frac{2\omega'_-}{\gamma}\right) \tag{G.34}$$

was used. Similarly, since the inverse-Laplace transformations of the third and fourth terms are

$$\frac{4e^{-\frac{t}{2}(\gamma+2i\omega'_+)}U'^2\gamma\left\{\left(1+e^{2i\omega'_+t}\right)\gamma+2i\left(-1+e^{2i\omega'_+t}\right)\omega'_+\right\}}{\left(\gamma^2+4\omega'^2_+\right)\left(-\omega'^2_- + \omega'^2_+\right)}, \tag{G.35}$$

replacement of the exponential functions in this equation with sinusoidal and cosinusoidal functions give

$$\frac{8U'^2}{\omega'^2_+ - \omega'^2_-}e^{-\frac{\gamma t}{2}}\cos\left(\phi'_+\right)\cos\left(\omega'_+t+\phi'_+\right), \tag{G.36}$$

where

$$\phi'_+ = \tan^{-1}\left(\frac{2\omega'_+}{\gamma}\right) \tag{G.37}$$

was used. The inverse-Laplace transformation of the fifth term is

$$\frac{32U'^2\gamma^2}{\left(\gamma^2+4\omega'^2_-\right)\left(\gamma^2+4\omega'^2_+\right)}, \tag{G.38}$$

which is found to be equal to 1 by noting the definition of $\omega'_\pm$ given by Eq. (G.29). Finally, summation of the five terms given above derives Eq. (3.34) in Chap. 3.

Appendix H
Derivation of Equations in Chap. 4

H.1 Unitary Transformation

This section reviews how to diagonalize the Hamiltonian

$$\hat{H} = \hat{H}_0 + \hat{V} \tag{H.1}$$

where $\hat{H}_0$ and $\hat{V}$ are unperturbed and interaction Hamiltonians, respectively [1]. In order to transform $\hat{H}$ to the diagonalized Hamiltonian

$$\tilde{H} = \hat{U}\hat{H}\hat{U}^{\dagger} \tag{H.2}$$

by a unitary transformation, the unitary operators $\hat{U}$ and $\hat{U}^{\dagger}$ are defined by

$$\hat{U} \equiv \mathrm{e}^{\hat{S}}, \tag{H.3a}$$

$$\hat{U}^{\dagger} = \hat{U}^{-1}, \tag{H.3b}$$

where $\hat{S}$ is an anti-Hermitian operator satisfying

$$\hat{S}^{\dagger} = -\hat{S}. \tag{H.4}$$

Inserting Eqs. (H.3a) and (H.3b) into Eq. (H.2) and performing polynomial expansion gives

$$\tilde{H} = \hat{U}\hat{H}\hat{U}^{\dagger} = e^{\hat{S}}\hat{H}e^{-\hat{S}} = \left(1 + \hat{S} + \frac{1}{2!}\hat{S}^2 + \cdots\right)\hat{H}\left(1 - \hat{S} + \frac{1}{2!}\hat{S}^2 + \cdots\right) \tag{H.5}$$

$$= \hat{H} + \hat{S}\hat{H} - \hat{H}\hat{S} + \frac{1}{2!}\left(\hat{S}^2\hat{H} - 2\hat{S}\hat{H}\hat{S} + \hat{H}\hat{S}^2\right) + \cdots$$

M. Ohtsu, *Dressed Photons*, Nano-Optics and Nanophotonics,
DOI: 10.1007/978-3-642-39569-7, © Springer-Verlag Berlin Heidelberg 2014

$$= \hat{H} + \left[\hat{S}, \hat{H}\right] + \frac{1}{2!}\left[\hat{S}, \left[\hat{S}, \hat{H}\right]\right] + \cdots$$
$$= \hat{H}_0 + \hat{V} + \left[\hat{S}, \hat{H}_0\right] + \left[\hat{S}, \hat{V}\right] + \frac{1}{2!}\left[\hat{S}, \left[\hat{S}, \hat{H}_0\right]\right] + \cdots.$$

If the anti-Hermitian operator $\hat{S}$ satisfies the relation

$$\hat{V} = -\left[\hat{S}, \hat{H}_0\right], \tag{H.6}$$

Equation (H.5) is reduced to

$$\tilde{H} = \hat{H}_0 - \frac{1}{2}\left[\hat{S}, \left[\hat{S}, \hat{H}_0\right]\right] + \cdots. \tag{H.7}$$

In the case where the value of the interaction Hamiltonian $\hat{V}$ is sufficiently small, Eq. (H.7) can be treated perturbatively. That is, by neglecting the terms higher than the second order of $\hat{S}$ in this equation, the Hamiltonian $\hat{H}$ can be diagonalized as

$$\tilde{H} = \hat{U}\hat{H}\hat{U}^{\dagger} = e^{\hat{S}}\hat{H}e^{-\hat{S}} \simeq \hat{H}_0. \tag{H.8}$$

Based on the formulation given above, the Hamiltonian for the DP–phonon interaction

$$\hat{H}' = \sum_{i=1}^{N} \hbar\omega\tilde{a}_i^{\dagger}\tilde{a}_i + \sum_{p-1}^{N} \hbar\Omega_p\hat{c}_p^{\dagger}\hat{c}_p + \sum_{i=1}^{N}\sum_{p=1}^{N} \hbar\chi_{ip}\tilde{a}_i^{\dagger}\tilde{a}_i\left(\hat{c}_p^{\dagger} + \hat{c}_p\right) \tag{H.9}$$

is diagonalized. Here, the fourth term of Eq. (4.44) in Chap. 4, representing DP-hopping, has been excluded. Operators in this equation satisfy the boson commutation relation as follows:

$$\left[\tilde{a}_i, \tilde{a}_j^{\dagger}\right] = \delta_{ij}, \tag{H.10}$$

$$\left[\hat{c}_p, \hat{c}_q^{\dagger}\right] = \delta_{pq}, \tag{H.11}$$

$$\begin{aligned} \left[\tilde{a}_i, \hat{c}_p\right] &= \left[\tilde{a}_i, \hat{c}_p^{\dagger}\right] = \left[\tilde{a}_i^{\dagger}, \hat{c}_p\right] = \left[\tilde{a}_i^{\dagger}, \hat{c}_q^{\dagger}\right] = 0, \\ \left[\tilde{a}_i, \tilde{a}_j\right] &= \left[\tilde{a}_i^{\dagger}, \tilde{a}_j^{\dagger}\right] = \left[\hat{c}_p, \hat{c}_q\right] = \left[\hat{c}_p^{\dagger}, \hat{c}_q^{\dagger}\right] = 0. \end{aligned} \tag{H.12}$$

The anti-Hermitian operator $\hat{S}$ to be used for diagonalization is given by

$$\hat{S} = \sum_{i=1}^{N}\sum_{p=1}^{N} f_{ip}\tilde{a}_i^{\dagger}\tilde{a}_i\left(\hat{c}_p^{\dagger} - \hat{c}_p\right). \tag{H.13}$$

Denoting the first and second terms of Eq. (H.9) by $\hat{H}_0$, and the third term by $\hat{V}$, it is found from Eq. (H.6) that the coefficient f_{ip} in Eq. (H.13) is equal to χ_{ip}/Ω_p. As a result, one can derive

$$\hat{S} = \sum_{i=1}^{N}\sum_{p=1}^{N} \frac{\chi_{ip}}{\Omega_p} \tilde{a}_i^{\dagger} \tilde{a}_i \left(\hat{c}_p^{\dagger} - \hat{c}_p \right), \tag{H.14}$$

which corresponds to Eq. (4.26) in Chap. 4.

Since the commutation relation between $\hat{S}$ and the Hamiltonian of Eq. (H.9) can be calculated analytically, the Hamiltonian for the DP–phonon interaction can be diagonalized without using any perturbative methods. In order to diagonalize it, the operator function

$$\hat{F}_i(t) = e^{t\hat{S}} \tilde{a}_i^{\dagger} e^{-t\hat{S}} \tag{H.15}$$

is defined. First, differentiating it with respect to t yields

$$\begin{aligned} \frac{d}{dt}\hat{F}_i(t) &= e^{t\hat{S}} \left(\hat{S}\tilde{a}_i^{\dagger} - \tilde{a}_i^{\dagger}\hat{S} \right) e^{-t\hat{S}} \\ &= e^{t\hat{S}} \sum_{p=1}^{N} \frac{\chi_{ip}}{\Omega_p} \tilde{a}_i^{\dagger} \left(\hat{c}_p^{\dagger} - \hat{c}_p \right) e^{-t\hat{S}} = \hat{F}_i(t) \sum_{p=1}^{N} \frac{\chi_{ip}}{\Omega_p} \left(\hat{c}_p^{\dagger} - \hat{c}_p \right). \end{aligned} \tag{H.16}$$

Second, solving this differential equation by using the initial condition $\hat{F}_i(0) = \tilde{a}_i^{\dagger}$ yields

$$\hat{F}_i(t) = \tilde{a}_i^{\dagger} \exp\left\{ t \sum_{p=1}^{N} \frac{\chi_{ip}}{\Omega_p} \left(\hat{c}_p^{\dagger} - \hat{c}_p \right) \right\} \tag{H.17}$$

and therefore

$$\hat{\alpha}_i^{\dagger} = \hat{U}^{\dagger} \tilde{a}_i^{\dagger} \hat{U} = \hat{F}_i(-1) = \tilde{a}_i^{\dagger} \exp\left\{ -\sum_{p=1}^{N} \frac{\chi_{ip}}{\Omega_p} \left(\hat{c}_p^{\dagger} - \hat{c}_p \right) \right\}. \tag{H.18}$$

Similarly, for the operator function

$$\hat{G}_p(t) = e^{t\hat{S}} \hat{c}_p^{\dagger} e^{-t\hat{S}}, \tag{H.19a}$$

one derives

$$\frac{d}{dt}\hat{G}_p(t) = -\sum_{i=1}^{N} \frac{\chi_{ip}}{\Omega_p} \tilde{a}_i^{\dagger} \tilde{a}_i, \tag{H.19b}$$

and therefore

$$\hat{G}_p(t) = \hat{c}_p^\dagger - t\sum_{i=1}^{N}\frac{\chi_{ip}}{\Omega_p}\tilde{a}_i^\dagger\tilde{a}_i, \tag{H.20}$$

$$\hat{\beta}_p^\dagger = \hat{G}_p(-1) = \hat{c}_p^\dagger + \sum_{i=1}^{N}\frac{\chi_{ip}}{\Omega_p}\tilde{a}_i^\dagger\tilde{a}_i. \tag{H.21}$$

These derivations correspond to the proof for Eqs. (4.30) and (4.31) in Chap. 4.

In order to derive the diagonalized Hamiltonian $\tilde{H}'$ by using the equations derived above, the operators $\tilde{a}_i^\dagger$ and $\hat{c}_p^\dagger$ are replaced with $\hat{F}_i(1)$ and $\hat{G}_p(1)$, respectively. Using them, one derives

$$\begin{aligned}\tilde{H}' &= \hat{U}\hat{H}'\hat{U}^\dagger \\ &= \sum_{i=1}^{N}\hbar\omega\tilde{a}_i^\dagger\tilde{a}_i + \sum_{p=1}^{N}\hbar\Omega_p\left(\hat{c}_p^\dagger - \sum_{i=1}^{N}\frac{\chi_{ip}}{\Omega_p}\tilde{a}_i^\dagger\tilde{a}_i\right)\left(\hat{c}_p - \sum_{j=1}^{N}\frac{\chi_{jp}}{\Omega_p}\tilde{a}_j^\dagger\tilde{a}_j\right) \\ &+ \sum_{i=1}^{N}\sum_{p=1}^{N}\hbar\chi_{ip}\tilde{a}_i^\dagger\tilde{a}_i\left(\hat{c}_p + \hat{c}_p^\dagger - 2\sum_{j=1}^{N}\frac{\chi_{jp}}{\Omega_p}\tilde{a}_j^\dagger\tilde{a}_j\right) \\ &= \sum_{i=1}^{N}\hbar\omega\tilde{a}_i^\dagger\tilde{a}_i + \sum_{p=1}^{N}\hbar\Omega_p\hat{c}_p^\dagger\hat{c}_p - \sum_{i=1}^{N}\sum_{j=1}^{N}\sum_{p=1}^{N}\hbar\frac{\chi_{ip}\chi_{jp}}{\Omega_p}\tilde{a}_i^\dagger\tilde{a}_i\tilde{a}_j^\dagger\tilde{a}_j.\end{aligned} \tag{H.22}$$

The original Hamiltonian $\hat{H}'$ can be derived by the inverse transformation, which is expressed by using the transformed operators $\hat{\alpha}_i^\dagger$, $\hat{\alpha}_i$, $\hat{\beta}_p^\dagger$, and $\hat{\beta}_p$ as

$$\begin{aligned}\hat{H}' &= \hat{U}^\dagger\tilde{H}'\hat{U} \\ &= \sum_{i=1}^{N}\hbar\omega\hat{\alpha}_i^\dagger\hat{\alpha}_i + \sum_{p=1}^{N}\hbar\Omega_p\hat{\beta}_p^\dagger\hat{\beta}_p - \sum_{i=1}^{N}\sum_{j=1}^{N}\sum_{p=1}^{N}\hbar\frac{\chi_{ip}\chi_{jp}}{\Omega_p}\hat{\alpha}_i^\dagger\hat{\alpha}_i\hat{\alpha}_j^\dagger\hat{\alpha}_j.\end{aligned} \tag{H.23}$$

Since the term for DP-hopping can also be transformed by expressing $\tilde{a}_i^\dagger$ in Eq. (H.18) by $\hat{\alpha}_i^\dagger$ and by replacing $\hat{c}_p^\dagger - \hat{c}_p$ with $\hat{\beta}_p^\dagger - \hat{\beta}_p$, the Hamiltonian and the hopping operator can be derived, as shown by Eqs. (4.35) and (4.36) in Chap. 4, respectively.

H.2 Coherent State

In the coherent state $|\gamma\rangle$, an infinite number of quasi-particles cohere with each other, which can be expressed by using annihilation ($\hat{c}$) and creation ($\hat{c}^\dagger$) operators as

$$|\gamma\rangle = e^{\gamma(\hat{c}^\dagger - \hat{c})}|0\rangle. \tag{H.24}$$

Here, for simplicity, quasi-particles of the single mode are considered and the coefficient γ is assumed to take real numbers. Differentiating the operator function

$$f(\gamma) = e^{\gamma\left(\hat{c}^{\dagger}-\hat{c}\right)} \tag{H.25}$$

with respect to γ yields[3]

$$\frac{df}{d\gamma} = \hat{c}^{\dagger} f - f\hat{c} = \left(\hat{c}^{\dagger} - \hat{c} + \gamma\right) f. \tag{H.26}$$

Equating the solution of this differential equation,

$$f = e^{\gamma\left(\hat{c}^{\dagger}-\hat{c}\right)} e^{\frac{1}{2}\gamma^2}, \tag{H.27}$$

with Eq. (H.25), gives the relation

$$e^{\gamma\left(\hat{c}^{\dagger}-\hat{c}\right)} = e^{-\frac{1}{2}\gamma^2} e^{\gamma\hat{c}^{\dagger}} e^{-\gamma\hat{c}}. \tag{H.28}$$

Therefore, the coherent state can be expressed also as

$$|\gamma\rangle = e^{-\frac{1}{2}\gamma} e^{\gamma\hat{c}^{\dagger}} |0\rangle . \tag{H.29}$$

Polynomial expansion of the exponential function gives

$$|\gamma\rangle = e^{-\frac{1}{2}\gamma^2} e^{\gamma\hat{c}^{\dagger}} |0\rangle = e^{-\frac{1}{2}\gamma^2} \sum_{n=0}^{\infty} \frac{\left(\gamma\hat{c}^{\dagger}\right)^n}{n!} |0\rangle = e^{-\frac{1}{2}\gamma^2} \sum_{n=0}^{\infty} \frac{\gamma^n}{\sqrt{n!}} |n\rangle . \tag{H.30}$$

---

[3] If the operators $\hat{A}$ and $\hat{B}$ satisfy the commutation relation

$$\left[\hat{A}, \hat{B}\right] = 1, \tag{a}$$

the commutation relation between $\hat{A}$ and $\hat{B}^n$ is derived by mathematical induction as

$$\left[\hat{A}, \hat{B}^n\right] = n\hat{B}^{n-1} = \frac{d}{d\hat{B}}\hat{B}^n. \tag{b}$$

Therefore, after expanding an operator function $f\left(\hat{B}\right)$ into a power series of $\hat{B}$, use of Eq. (b) gives

$$\left[\hat{A}, f\left(\hat{B}\right)\right] = \frac{d}{d\hat{B}} f\left(\hat{B}\right). \tag{c}$$

In the case of $\hat{A} = \hat{c}$ and $\hat{B} = \hat{c}^{\dagger} - \hat{c}$, Eq. (c) can be employed because Eq. (a) holds, and one obtains

$$\hat{c}f - f\hat{c} = \gamma f. \tag{d}$$

By inserting $f\hat{c} = \left(-\hat{c} + \gamma\right) f$, derived from this equation, into the second term of the middle of Eq. (H.26), the right-hand side is derived.

From the right-hand side of this equation, it is confirmed that an infinite number of quasi-particles cohere in the coherent state.

The coherent state has several features, which are expressed as

$$\hat{c}\,|\gamma\rangle = \gamma\,|\gamma\rangle\,, \tag{H.31}$$

$$\langle\gamma \mid \gamma\rangle = 1, \tag{H.32}$$

$$\langle N\rangle = \langle\gamma|\,\hat{c}^{\dagger}\hat{c}\,|\gamma\rangle = \gamma^2, \tag{H.33}$$

$$\Delta N = \sqrt{\langle N^2\rangle - \langle N\rangle^2} = \sqrt{\langle N\rangle} = |\gamma|\,. \tag{H.34}$$

Equation (H.31) means that the coherent state is an eigenstate of the annihilation operator, which can be confirmed by applying the operators on the both sides of Eq. (d) in the footnote [(1)] to the vacuum state $|0\rangle$. Furthermore, since the coherent state is not the eigenstate of the number operator $\hat{N}\left(=\hat{c}^{\dagger}\hat{c}\right)$, the standard deviation of the quasi-particle number is nonzero, as is given by Eq. (H.34); i.e., the number of the quasi-particles fluctuates.

H.3 Temporal Evolution of the Coherent State

H.3.1 Probability of Exciting the Phonon Field

In order to analyze the temporal evolution of the coherent state, the initial condition is assumed to be $|\psi\rangle = \tilde{a}_i^{\dagger}\,|0\rangle$, which means that the DP is generated at site i in the probe apex by injecting the propagating light at time $t = 0$. Since this initial state is not the eigenstate of the Hamiltonian $\hat{H}'$ of Eq. (H.9), phonons are excited by the DP–phonon interaction. For deriving the probability of this excitation, the creation operator $\tilde{a}_i^{\dagger}$ for the DP is expressed by the operator $\hat{\alpha}_i^{\dagger}$ for the DPP as

$$\tilde{a}_i^{\dagger} = \hat{\alpha}_i^{\dagger} \exp\left\{\sum_{p=1}^{N} \gamma_{ip}\left(\hat{\beta}_p^{\dagger} - \hat{\beta}_p\right)\right\} \tag{H.35a}$$

and therefore

$$|\psi\rangle = \tilde{a}_i^{\dagger}\,|0\rangle \equiv \hat{\alpha}_i^{\dagger}\,|\gamma\rangle\,, \tag{H.35b}$$

where Eq. (H.18) and the relation $\hat{c}_p - \hat{c}_p^{\dagger} = \hat{\beta}_p - \hat{\beta}_p^{\dagger}$ were used. In Eq. (H.35a), γ_{ip} is equal to χ_{ip}/Ω_p. The probability, P', that the phonons are still in the vacuum state at time t is given by

$$P' = \left| \langle\psi| \, e^{-i\frac{\hat{H}'}{\hbar}t} \, |\psi\rangle \right|^2. \tag{H.36}$$

It should be noted that the Hamiltonian $\hat{H}'$ on the right-hand side is given by Eq. (H.9), which is the one excluding the fourth term (hopping term) from the Hamiltonian of Eq. (4.44) in Chap. 4. Therefore, by applying the mean field approximation of Eq.(4.45) to the third term of Eq. (4.44), one derives

$$\hat{H}' = \sum_{i=1}^{N} \hbar\omega\hat{\alpha}_i^\dagger\hat{\alpha}_i + \sum_{p=1}^{N} \hbar\Omega_p\hat{\beta}_p^\dagger\hat{\beta}_p + \sum_{i=1}^{N}\sum_{j=1}^{N} \frac{\hbar\chi}{2}\langle x_j\rangle_i \frac{1}{N}\hat{\alpha}_i^\dagger\hat{\alpha}_i. \tag{H.37}$$

Inserting this into Eq. (H.36) gives the relation

$$\begin{aligned}
&\langle\psi| \, e^{-i\frac{\hat{H}'}{\hbar}t} \, |\psi\rangle \\
&= \langle\gamma| \, \hat{\alpha}_i \exp\left(-i\sum_{i=1}^{N} \omega\hat{\alpha}_i^\dagger\hat{\alpha}_i t - i\sum_{i=1}^{N}\sum_{j=1}^{N} \frac{\chi}{2}\langle x_j\rangle_i \frac{1}{N}\hat{\alpha}_i^\dagger\hat{\alpha}_i t \right) \\
&\times \exp\left(-i\sum_{p=1}^{N} \Omega_p\hat{\beta}_p^\dagger\hat{\beta}_p t \right) \hat{\alpha}_i^\dagger \, |\gamma\rangle \\
&= \langle\gamma| \, \hat{\alpha}_i \exp\left\{ -i\sum_{i=1}^{N} \left(\omega\hat{\alpha}_i^\dagger\hat{\alpha}_i t + \sum_{j=1}^{N} \frac{\chi}{2}\langle x_j\rangle_i \frac{1}{N}\hat{\alpha}_i^\dagger\hat{\alpha}_i t \right) \right\} \\
&\times \hat{\alpha}_i^\dagger \exp\left(-i\sum_{p=1}^{N} \Omega_p\hat{\beta}_p^\dagger\hat{\beta}_p t \right) |\gamma\rangle .
\end{aligned} \tag{H.38a}$$

In order to transform

$$\exp\left\{ -i\sum_{i=1}^{N} \left(\omega\hat{\alpha}_i^\dagger\hat{\alpha}_i t + \sum_{j=1}^{N} \frac{\chi}{2}\langle x_j\rangle_i \frac{1}{N}\hat{\alpha}_i^\dagger\hat{\alpha}_i t \right) \right\} \hat{\alpha}_i^\dagger \, |\gamma\rangle \tag{H.38b}$$

in the fourth row of Eq.(H.38a), a parameter κ is defined by

$$\kappa \equiv -i\left(\omega + \sum_{j=1}^{N} \frac{\chi}{2}\langle x_j\rangle_i \frac{1}{N} \right) t \equiv -i\left(\omega + \chi\langle x_i\rangle_i\right) t. \tag{H.39}$$

Using this parameter, Eq. (H.38b) is transformed to

$$\begin{aligned}
&\exp\left\{ -i\sum_{i=1}^{N} \left(\omega\hat{\alpha}_i^\dagger\hat{\alpha}_i t + \sum_{j=1}^{N} \frac{\chi}{2}\langle x_j\rangle_i \frac{1}{N}\hat{\alpha}_i^\dagger\hat{\alpha}_i t \right) \right\} \hat{\alpha}_i^\dagger \, |\gamma\rangle \\
&= \exp\left(\sum_{i=1}^{N} \kappa\hat{\alpha}_i^\dagger\hat{\alpha}_i t \right) \hat{\alpha}_i^\dagger \, |\gamma\rangle
\end{aligned} \tag{H.40}$$

$$= \left\{ 1 + \sum_{i=1}^{N} \kappa \hat{\alpha}_i^\dagger \hat{\alpha}_i t + \frac{1}{2!} \sum_{i=1}^{N} \left(\kappa \hat{\alpha}_i^\dagger \hat{\alpha}_i t \right)^2 + \cdots \right\} \hat{\alpha}_i^\dagger |\gamma\rangle$$

$$= \hat{\alpha}_i^\dagger \left\{ 1 + \kappa t + \frac{(\kappa t)^2}{2!} + \cdots \right\} |\gamma\rangle = \hat{\alpha}_i^\dagger e^{\kappa t} |\gamma\rangle ,$$

where the relations

$$\hat{\alpha}_i |\gamma\rangle = 0, \quad \langle\gamma| \hat{\alpha}_i^\dagger = 0, \quad \hat{\alpha}_i \hat{\alpha}_j^\dagger = \delta_{ij} + \hat{\alpha}_j^\dagger \hat{\alpha}_i \tag{H.41}$$

were used for deriving the third row from the second row. By using Eq. (H.40), Eq. (H.38a) is transformed to

$$\langle\psi| e^{-i\frac{\hat{H}'}{\hbar}t} |\psi\rangle \tag{H.42}$$

$$= \langle\gamma| \hat{\alpha}_i \hat{\alpha}_i^\dagger \exp\{-i(\omega + \chi\langle x_i\rangle_i)t\} \exp\left(-i \sum_{p=1}^{N} \Omega_p \hat{\beta}_p^\dagger \hat{\beta}_p t \right) |\gamma\rangle$$

$$= \langle\gamma| \left(1 - \hat{\alpha}_i^\dagger \hat{\alpha}_i \right) \exp\left(-i \sum_{p=1}^{N} \Omega_p \hat{\beta}_p^\dagger \hat{\beta}_p t \right) |\gamma\rangle$$

$$= \exp\{-i(\omega + \chi\langle x_i\rangle_i)t\} \langle\gamma| \exp\left(-i \sum_{p=1}^{N} \Omega_p \hat{\beta}_p^\dagger \hat{\beta}_p t \right) |\gamma\rangle .$$

When inserting this equation into Eq. (H.36), the exponential function at the head of the fourth row is neglected because its absolute value is unity. By rewriting other time-evolving terms as

$$f = \langle\gamma| \exp\left(-i \sum_{p=1}^{N} \Omega_p \hat{\beta}_p^\dagger \hat{\beta}_p t \right) |\gamma\rangle , \tag{H.43}$$

it represents the effect of phonon excitation triggered by the fluctuation of the number of quasi-particles. By differentiating it with respect to γ_{ip} $(= \chi_{ip}/\Omega_p)$, one obtains[4]

$$\frac{\partial f}{\partial \gamma_{ip}} = 2\left(e^{-i\Omega_p t} - 1 \right) \gamma_{ip} f . \tag{H.44}$$

---

[4] Since

$$|\gamma\rangle = \exp \sum_{p}^{N} \gamma_{ip} \left(\hat{\beta}_p^\dagger - \hat{\beta}_p \right) |0\rangle \tag{a}$$

holds by inserting $\hat{c}_p^\dagger - \hat{c}_p = \hat{\beta}_p^\dagger - \hat{\beta}_p$ into Eq. (H.24), inserting it into Eq. (H.43) derives

By integrating it with respect to γ_{ip} and summing up over $p = 1 - N$, the expression

(Footnote 4 continued)

$$f = \langle 0| \exp\left\{-\sum_{q=1}^{N} \gamma_{iq}\left(\hat{\beta}_q^{\dagger} - \hat{\beta}_q\right)\right\} \times \exp\left(-i\sum_{q=1}^{N} \Omega_q t \hat{\beta}_q^{\dagger}\hat{\beta}_q\right) \exp\left\{\sum_{q=1}^{N} \gamma_{iq}\left(\hat{\beta}_q^{\dagger} - \hat{\beta}_q\right)\right\} |0\rangle . \tag{b}$$

Here, the subscript p is replaced with q in order to avoid confusion in the following discussions. Differentiating Eq. (b) with respect to the coefficient γ_{ip} of the p-th term yields

$$\begin{aligned}
&\frac{\partial f}{\partial \gamma_{ip}} \\
&= -\langle 0| \exp\left\{-\sum_{q=1}^{N} \gamma_{iq}\left(\hat{\beta}_q^{\dagger} - \hat{\beta}_q\right)\right\} \left(\hat{\beta}_p^{\dagger} - \hat{\beta}_p\right) \exp\left(-i\sum_{q=1}^{N} \Omega_q t \hat{\beta}_q^{\dagger}\hat{\beta}_q\right) \\
&\times \exp\left\{\sum_{q=1}^{N} \gamma_{iq}\left(\hat{\beta}_q^{\dagger} - \hat{\beta}_q\right)\right\} |0\rangle \\
&+ \langle 0| \exp\left\{-\sum_{q=1}^{N} \gamma_{iq}\left(\hat{\beta}_q^{\dagger} - \hat{\beta}_q\right)\right\} \exp\left(-i\sum_{q=1}^{N} \Omega_q t \hat{\beta}_q^{\dagger}\hat{\beta}_q\right) \left(\hat{\beta}_p^{\dagger} - \hat{\beta}_p\right) \\
&\times \exp\left\{\sum_{q=1}^{N} \gamma_{iq}\left(\hat{\beta}_q^{\dagger} - \hat{\beta}_q\right)\right\} |0\rangle \\
&= -\langle \gamma| \left(\hat{\beta}_p^{\dagger} - \hat{\beta}_p\right) \exp\left(-i\sum_{q=1}^{N} \Omega_q t \hat{\beta}_q^{\dagger}\hat{\beta}_q\right) |\gamma\rangle \\
&+ \langle \gamma| \exp\left(-i\sum_{q=1}^{N} \Omega_q t \hat{\beta}_q^{\dagger}\hat{\beta}_q\right) \left(\hat{\beta}_p^{\dagger} - \hat{\beta}_p\right) |\gamma\rangle .
\end{aligned} \tag{c}$$

Here, by noting Eq. (4.38), it is found that Eq. (H.31) is effective even when $\hat{c}$ is replaced with $\hat{\beta}_p$, and resultantly, the relations

$$\hat{\beta}_p |\gamma\rangle = \gamma_{ip} |\gamma\rangle , \tag{d}$$

$$\langle \gamma| \hat{\beta}_p^{\dagger} = \langle \gamma| \gamma_{ip} \tag{e}$$

hold. By using them, in the first term in the fifth row of Eq. (c), the component having $\hat{\beta}_p^{\dagger}$ on the left of the exponential function is transformed to

$$\langle \gamma| \hat{\beta}_p^{\dagger} \exp\left(-i\sum_{q=1}^{N} \Omega_q t \hat{\beta}_q^{\dagger}\hat{\beta}_q\right) |\gamma\rangle = \langle \gamma| \gamma_{ip} \exp\left(-i\sum_{q=1}^{N} \Omega_q t \hat{\beta}_q^{\dagger}\hat{\beta}_q\right) |\gamma\rangle = \gamma_{ip} f . \tag{f}$$

In the second term in the seventh row, the component having $\hat{\beta}_p$ on the right of the exponential function is transformed to

$$\langle \gamma| \exp\left(-i\sum_{q=1}^{N} \Omega_q t \hat{\beta}_q^{\dagger}\hat{\beta}_q\right) \hat{\beta}_p |\gamma\rangle = \langle \gamma| \exp\left(-i\sum_{q=1}^{N} \Omega_q t \hat{\beta}_q^{\dagger}\hat{\beta}_q\right) \gamma_{ip} |\gamma\rangle = \gamma_{ip} f . \tag{g}$$

Next, in order to transform $\langle \gamma| \hat{\beta}_p \exp\left(-i\sum_{q=1}^{N} \Omega_q t \hat{\beta}_q^{\dagger}\hat{\beta}_q\right) |\gamma\rangle$ in the second term of the fifth row, it is noted that the following relation holds:

$$f = \exp\left\{\sum_{p=1}^{N} \gamma_{ip}^2 \left(e^{-i\Omega_p t} - 1\right)\right\} \tag{H.45}$$

is derived. By inserting this into Eq. (H.36), the probability of phonon excitation is expressed as

$$P = 1 - P' = 1 - |f|^2 = 1 - \exp\left\{\sum_{p=1}^{N} 2\gamma_{ip}^2 \left(\cos\Omega_p t - 1\right)\right\}. \tag{H.46}$$

Furthermore, the probability of exciting the specific mode p_0 of the phonon while other modes are in the vacuum state is given by

$$P_{p0} = \left[1 - \exp\left\{2\gamma_{ip0}^2 \left(\cos\Omega_{p0} t - 1\right)\right\}\right] \exp\left\{\sum_{p \neq p0}^{N} 2\gamma_{ip}^2 \left(\cos\Omega_p t - 1\right)\right\}. \tag{H.47}$$

(Footnote 4 continued)

$$\hat{\beta}_p \exp\left(-i\Omega_p t \hat{\beta}_p^\dagger \hat{\beta}_p\right) = \exp\left(-i\Omega_p t \hat{\beta}_p^\dagger \hat{\beta}_p\right) \hat{\beta}_p e^{-i\Omega_p t} \tag{h}$$

because $\left[\hat{\beta}_p, \hat{\beta}_p^\dagger\right] = 1$. Using Eq. (h), one derives

$$\begin{aligned} \hat{\beta}_p \exp\left(-i \sum_{q=1}^{N} \Omega_q t \hat{\beta}_q^\dagger \hat{\beta}_q\right) &= \exp\left(-i\Omega_p t \hat{\beta}_p^\dagger \hat{\beta}_p\right) \hat{\beta}_p e^{-i\Omega_p t} \exp\left(-i \sum_{q \neq p}^{N} \Omega_q t \hat{\beta}_q^\dagger \hat{\beta}_q\right) \\ &= \exp\left(-i \sum_{q=1}^{N} \Omega_q t \hat{\beta}_q^\dagger \hat{\beta}_q\right) \hat{\beta}_p \gamma_{ip} e^{-i\Omega_p t}, \end{aligned} \tag{i}$$

and thus, the relation

$$\begin{aligned} \langle\gamma| \hat{\beta}_p \exp\left(-i \sum_{q=1}^{N} \Omega_q t \hat{\beta}_q^\dagger \hat{\beta}_q\right) |\gamma\rangle &= \langle\gamma| \exp\left(-i \sum_{q=1}^{N} \Omega_q t \hat{\beta}_q^\dagger \hat{\beta}_q\right) \hat{\beta}_p e^{-i\Omega_p t} |\gamma\rangle \\ &= \langle\gamma| \exp\left(-i \sum_{q=1}^{N} \Omega_q t \hat{\beta}_q^\dagger \hat{\beta}_q\right) \gamma_{ip} e^{-i\Omega_p t} |\gamma\rangle = \gamma_{ip} e^{-i\Omega_p t} f \end{aligned} \tag{j}$$

is obtained, where Eq. (d) was used to replace the first row with the second row. The component $\langle\gamma| \exp\left(-i \sum_{q=1}^{N} \Omega_q t \hat{\beta}_q^\dagger \hat{\beta}_q\right) \hat{\beta}_p^\dagger |\gamma\rangle$ in the first term in the seventh row of Eq. (c) is also similarly transformed to

$$\langle\gamma| \exp\left(-i \sum_{q=1}^{N} \Omega_q t \hat{\beta}_q^\dagger \hat{\beta}_q\right) \hat{\beta}_p^\dagger |\gamma\rangle = \gamma_{ip} e^{-i\Omega_p t} f. \tag{k}$$

Finally, Eq. (H.44) is derived by inserting Eqs. (f)–(k) into the fifth and seventh rows of Eq. (c).

H.3.2 Fluctuations in the Number of Phonons

This section derives the magnitude of the fluctuations in the number of phonons when the phonons are in the coherent state. Since the value of the fluctuations is equal to the square root of the expectation value, as was given by Eq. (H.34), the expectation value of the number of phonons is derived in the following.

By using Eqs. (H.18) and (H.21), the phonon number operator $\hat{N}_p$ of mode p can be expressed in terms of the unitary-transformed operators as

$$\begin{aligned}\hat{N}_p &= \hat{c}_p^{\dagger}\hat{c}_p \\ &= \hat{\beta}_p^{\dagger}\hat{\beta}_p + \sum_{i=1}^{N}\sum_{j=1}^{N}\gamma_{ip}\gamma_{jp}\hat{\alpha}_i^{\dagger}\hat{\alpha}_i\hat{\alpha}_j^{\dagger}\hat{\alpha}_j - \sum_{i=1}^{N}\gamma_{jp}\hat{\alpha}_i^{\dagger}\hat{\alpha}_i\left(\hat{\beta}_p + \hat{\beta}_p^{\dagger}\right).\end{aligned} \tag{H.48}$$

Since time evolution of the number operator is expressed in Heisenberg representation as

$$\begin{aligned}\hat{N}_p(t) &= e^{i\frac{\hat{H}'}{\hbar}t}\hat{N}_p e^{-i\frac{\hat{H}'}{\hbar}t} \\ &= \hat{\beta}_p^{\dagger}\hat{\beta}_p + \sum_{i=1}^{N}\sum_{j=1}^{N}\gamma_{ip}\gamma_{jp}\hat{\alpha}_i^{\dagger}\hat{\alpha}_i\hat{\alpha}_j^{\dagger}\hat{\alpha}_j - \sum_{i=1}^{N}\gamma_{jp}\hat{\alpha}_i^{\dagger}\hat{\alpha}_i\left(e^{-i\Omega_p t}\hat{\beta}_p + e^{i\Omega_p t}\hat{\beta}_p^{\dagger}\right),\end{aligned} \tag{H.49}$$

the expectation value of the phonon number is derived as

$$\left\langle N_p(t)\right\rangle = \langle\psi|\,\hat{N}_p(t)\,|\gamma\rangle = 2\gamma_{ip}\left(1 - \cos\Omega_p t\right). \tag{H.50}$$

From Eq. (H.34), it is concluded that the value of the phonon number fluctuations is given by the square root of Eq. (H.50).

H.3.3 Eigenvalues for the One-Dimensional Lattice Without any Impurities

Since the equation of motion for the one-dimensional lattice without any impurities has been given by Eq. (4.3) of Chap. 4, this section derives its eigenvalue. For this derivation, the triple diagonal matrix is diagonalized, and its eigenvalues and eigenvectors are derived. For these purposes, the N-dimensional triple diagonal matrix

$$C = \begin{pmatrix} A & B & & \\ B & A & \cdot & \\ & \cdot & \cdot & B \\ & & B & A \end{pmatrix} \tag{H.51}$$

is considered, where A and B are constants. In order to calculate its determinant, the determinant of the n-dimensional triple diagonal matrix is written by f_n $(1 \leq n \leq N)$, and its recursion relations are derived by cofactor expansion:

$$f_n - Af_{n-1} + B^2 f_{n-2} = 0, \tag{H.52}$$

$$f_1 = A = \alpha + \beta, f_2 = A^2 - B^2 = \alpha^2 + \alpha\beta + \beta^2, \tag{H.53}$$

$$\alpha + \beta = A, \alpha + \beta = A. \tag{H.54}$$

From these relations, one obtains

$$f_n = \frac{1}{\beta - \alpha}\left(\beta^{n+1} - \alpha^{n+1}\right). \tag{H.55}$$

In order to derive the characteristic equation for the matrix C of Eq. (H.51), A is replaced with $A - x$ and we set $f_N = 0$, where x is the eigenvalue of the matrix C. That is, by solving

$$f_N = 0, \tag{H.56}$$

$$\alpha^{N+1} = \beta^{N+1}, \alpha = \beta \exp\left(2\pi i \frac{n}{N+1}\right) (1 \leq n \leq N), \tag{H.57}$$

and Eq. (H.52) simultaneously with respect to $A - x$, the eigenvalue x is obtained:

$$x_n = A + 2B \cos\left(\frac{n}{N+1}\pi\right). \tag{H.58}$$

Since the eigenvector $\boldsymbol{p}_n$ satisfies

$$\begin{pmatrix} A - x_n & B & & \\ B & A - x_n & B & \\ & \cdot & \cdot & \\ & & B & A - x_n \end{pmatrix} \begin{pmatrix} p_{1n} \\ p_{2n} \\ \cdot \\ p_{Nn} \end{pmatrix} = 0, \tag{H.59}$$

the recursion relations for the elements of the eigenvector are

$$p_{k,n} - 2\cos\left(\frac{n}{N+1}\pi\right) p_{k-1,n} + p_{k-2,n} = 0, \tag{H.60}$$

$$p_{2,n} = 2\cos\left(\frac{n}{N+1}\pi\right) p_{1,n}. \tag{H.61}$$

Solving Eqs. (H.60) and (H.61) simultaneously and imposing the nomalization condition

$$\sum_{k=1}^{N} p_{k,n}^2 = 1, \tag{H.62}$$

gives

$$p_{k,n} = \sqrt{\frac{2}{N+1}} \sin\left(\frac{kn}{N+1}\pi\right) (1 \leq k \leq N). \tag{H.63}$$

From these results, it is found that the eigenvectors are independent of the constants A and B in the matrix C. Furthermore, the matrix P, composed by arranging the eigenvectors in line, is confirmed to be an orthonormal matrix $\left(P^T = P^{-1}\right)$.

By inserting $A = 2$ and $B = -1$ into Eq. (H.51), the eigenvalue of Eq. (4.3) in Chap. 4, i.e., the square of the eigen angular frequency, is given by

$$\Omega_p^2 = \frac{k}{m}\left\{2 - 2\cos\left(\frac{p}{N+1}\pi\right)\right\} = 4\frac{k}{m}\sin^2\left[\frac{p}{2(N+1)}\pi\right], \tag{H.64}$$

whose square root corresponds to Eq. (4.21) in Chap. 4.

H.4 Diagonalization of the Hamiltonian Without DP–Phonon Coupling

In the case where the DP–phonon coupling constant χ is zero, the Hamiltonian can be written as

$$\hat{H} = \hbar\tilde{\boldsymbol{a}}^{\dagger} \begin{pmatrix} \omega & J & & \\ J & \omega & \cdot & \\ & \cdot & \cdot & J \\ & & J & \omega \end{pmatrix} \tilde{\boldsymbol{a}}, \tag{H.65}$$

$$\tilde{\boldsymbol{a}} = \begin{pmatrix} \tilde{a}_1 \\ \cdot \\ \cdot \\ \tilde{a}_N \end{pmatrix}, \tag{H.66}$$

and its eigenenergy is

$$\hbar\Omega_r = \hbar\omega + 2\hbar J \cos\left(\frac{r}{N+1}\pi\right). \tag{H.67}$$

On the other hand, the expectation value of the number of DPs, given by Eq. (4.56) in Chap. 4, is rewritten as

$$\langle N_i(t)\rangle_j = \langle \psi_j | \hat{N}_i(t) | \psi_j \rangle = \sum_{r=1}^{N}\sum_{s=1}^{N} Q_{ir}Q_{jr}Q_{is}Q_{js}\cos\{(\Omega_r - \Omega_s)t\} \tag{H.68}$$

$$\times \left(\frac{2}{N+1}\right)^2 \sum_{r=1}^{N}\sum_{s=1}^{N} \sin\left(\frac{jr}{N+1}\pi\right)\sin\left(\frac{ir}{N+1}\pi\right)\sin\left(\frac{js}{N+1}\pi\right)\sin\left(\frac{is}{N+1}\pi\right)$$

$$\times \cos\left\{2Jt\cos\left(\frac{r}{N+1}\pi\right) - 2Jt\cos\left(\frac{s}{N+1}\pi\right)\right\}$$

$$= \frac{1}{(N+1)^2}\left[\sum_{r=1}^{N}\left\{\cos\left(\frac{nr}{N+1}\pi\right) - \cos\left(\frac{mr}{N+1}\pi\right)\right\}\cos\left\{2Jt\cos\left(\frac{r}{N+1}\pi\right)\right\}\right]^2$$

$$+ \frac{1}{(N+1)^2}\left[\sum_{r=1}^{N}\left\{\cos\left(\frac{nr}{N+1}\pi\right) - \cos\left(\frac{mr}{N+1}\pi\right)\right\}\sin\left\{2Jt\cos\left(\frac{r}{N+1}\pi\right)\right\}\right]^2,$$

where the notations $m = j + i$ and $n = j - i$ were used.

Next, in order to grasp the behavior in the limit $N \to \infty$, by using $\theta_r = r\pi/(N+1)$ and replacing the summation in Eq. (H.68) with an integral, one obtains

$$\langle N_i(t)\rangle_j = \left\{\frac{1}{\pi}\int_0^{\pi} (\cos n\theta - \cos m\theta)\cos(2Jt\cos\theta)\,d\theta\right\}^2 + \left\{\frac{1}{\pi}\int_0^{\pi} (\cos n\theta - \cos m\theta)\sin(2Jt\cos\theta)\,d\theta\right\}^2. \tag{H.69}$$

By using the integral representation of the Bessel function $J_n(z)$ of the first kind

$$J_n(z) = \frac{1}{\pi i^n}\int_0^{\pi} e^{iz\cos\theta}\cos n\theta\,d\theta = \frac{1}{\pi i^n}\int_0^{\pi} \cos n\theta\{\cos(z\cos\theta) + i\sin(z\cos\theta)\}\,d\theta, \tag{H.70}$$

the expectation value in the limit $N \to \infty$ is obtained from Eq. (H.69) and is expressed as

$$\langle N_i(t)\rangle_j = \left\{J_{j-i}(2Jt) - (-1)^i J_{j+i}(2Jt)\right\}^2, \tag{H.71}$$

which corresponds to Eq. (4.57) in Chap. 4.

H.5 Expectation Value of the Displacement of Atoms

The expectation value of the displacement of atoms, as a result of DP–phonon coupling, has been presented by Eq. (4.43) in Chap. 4, i.e.,

$$\langle \hat{\boldsymbol{x}}_j \rangle_i = -\sum_{p=1}^{N} \frac{\hbar\chi P_{ip} P_{jp}}{\sqrt{m_i m_j}\,\Omega_p^2}. \tag{H.72}$$

By using

$$\Lambda = P^{-1}AP = P^{-1}\sqrt{M}^{-1}\Gamma\sqrt{M}^{-1}P, \tag{H.73a}$$

$$(\Lambda)_{pq} = \delta_{pq}\frac{\Omega_p^2}{k}, \tag{H.73b}$$

of Eqs. (4.7) and (4.8), the inverse matrix of

$$\Gamma = \sqrt{M}P\Lambda P^{-1}\sqrt{M} \tag{H.74a}$$

is derived as

$$\Gamma^{-1} = \sqrt{M}^{-1}P\Lambda^{-1}P^{-1}\sqrt{M}^{-1}, \tag{H.74b}$$

$$\left(\Gamma^{-1}\right)_{ij} = \frac{k}{\sqrt{m_i m_j}} \sum_{p=1}^{N} \frac{P_{ip}P_{jp}}{\Omega_p^2}. \tag{H.74c}$$

Then, the expectation value of the displacement can be rewritten by using these equations as

$$\langle \hat{\boldsymbol{x}}_j \rangle_i = -\frac{\hbar\chi}{k}(\Gamma^{-1})_{ij} \tag{H.75}$$

Since the values of the diagonal elements of the matrix Γ are 2 and those of the adjacent off-diagonal elements are -1, as given by Eq. (4.4) in Chap. 4, it is confirmed that Eq. (H.75) is independent of the effect of impurity atoms.

By denoting the matrix used for diagonalizing Γ by R and the diagonalized matrix by W, the inverse matrix Γ^{-1} can be readily expressed as

$$R^{-1}\Gamma R = W \tag{H.76a}$$

$$\left(\Gamma^{-1}\right)_{ij} = \sum_{n=1}^{N} R_{in}W_n^{-1}R_{nj}^{-1} = \frac{1}{N+1}\sum_{n=1}^{N} \frac{\sin\left(\frac{in}{N+1}\pi\right)\sin\left(\frac{jn}{N+1}\pi\right)}{1-\cos\left(\frac{n}{N+1}\pi\right)} \tag{H.76b}$$

which corresponds to Eq. (4.58) in Chap. 4.

Reference

1. Y. Tanaka, *Theoretical Models of Optical Near Fields Interacting with Localized Phonons* (Master's Thesis, Tokyo Institute Technology, 2007)

[illegible]

which can reproduce Eq. (4.[illegible]) in Chap. 4.

Reference

1. [illegible] Nanotechnology, 2007.

Index

A
Abrasive, 160, 166, 167
Absorption, 12, 17, 29, 61, 62, 71, 82–86
Absorption-edge, 148, 155
Acoustic phonon, 63
Active device, 111
Active layer, 207
Adiabatic approximation, 59–61
Adiabatic process, 85
Allowed, 41, 47
AND logic gate, 51, 54–56, 90, 97, 99, 111, 126, 127
Angular frequency, 12, 13
Annealing, 202–206, 208, 210–212
Annihilation operator, 9, 13, 16, 65, 74, 79, 260
Anti-binding excited state, 60
Anti-bunching, 118
Anti-commutation relation, 284
Anti-counterfeiting, 224, 226
Anti-electric field, 258
Anti-Hermitian operator, 15, 69, 301, 302
Anti-symmetric state, 38, 39, 50, 53, 54, 56, 89, 115, 116
Aperture, 1–3
Ar-ion milling, 97
Artifact-metrics, 226
Asynchronous architecture, 127
Asynchronous cellular automaton, 127
Atom, 59–61, 63, 65–68, 72, 73, 78, 80, 82, 248, 249
Atomos, 248
Atomic force microscope, 144–146, 148, 149, 152, 153, 163, 165
Atom-probe imaging, 204
Autonomous annihilation, 215, 233, 234
Autonomous control, 193, 194
Autonomous DPP generation, 171, 192
Autonomous formation, 237
Autonomous generation, 233, 234
Autonomy, 127, 128, 131, 132
Avalanche photodiode, 123
Azimuth angle, 27–29

B
Bandgap energy, 7, 13, 29, 191, 200
Band-pass filter, 36
Bare interaction operator, 20, 21, 23
Barrier layer, 94, 97
Base, 44, 49, 52
Basis, 49
Benard-Duraffourg inversion condition, 202
Bessel function, 77, 313
Bio-mimetic operation, 249
Biometrics, 226
Bit-flip, 120, 121, 134
Bloch function, 290
Bohr radius, 259
Boltzmann constant, 109, 119
Boltzmann distribution, 73
Born-Markov approximation, 44, 46
Born–Oppenheimer approximation, 59
Bose–Einstein distribution, 109
Boson, 13, 65, 68, 70, 275, 302
Boundary condition, 31
Break-over voltage, 203, 204
Bright state, 39
Buffer memory, 90, 115, 116

C
Capacitance, 119
Carrier confinement layer, 208
Cavity, 4–6, 8, 11, 12, 18, 20, 247, 248
Cavity loss, 71
Center of gravity, 41
Center of localization, 204
Center of mass, 288, 289, 291
Center of mass coordinate, 288

M. Ohtsu, *Dressed Photons*, Nano-Optics and Nanophotonics,
DOI: 10.1007/978-3-642-39569-7, © Springer-Verlag Berlin Heidelberg 2014

Central controller, 132
Characteristic equation, 311
Chemical etching, 161–166
Chemical–mechanical polishing, 160, 164
Chemical vapor deposition, 144, 169
Clean room, 159, 160
Cloud, 12, 31
Cluster, 238–241, 243, 244
CMOS logic gate, 120, 121, 124
Coarse-grained, 62, 67
Cofactor expansion, 310
Cohere, 62, 67, 72, 304, 305
Coherent phonon, 7, 71, 82, 249, 251
Coherent state, 71–74, 79, 80, 304, 305, 310
Collision broadening, 96
Commutation relation, 13, 16, 65, 68, 70
Complementary space, 21
Complete set, 99
Complex system, 120
Complimentary operator, 266
Conduction band, 3, 192, 197, 201, 202
Confinement potential, 288, 289
Conjugate, 255
Constructive interference, 148, 155
Continuous-wave, 95
Contrast, 92, 94, 95, 97, 108, 119, 124–126, 128, 131
Control terminal, 90
Conversion efficiency, 182, 183, 192, 199, 200
Cooperative phenomenon, 117
Coordinate representation, 5, 11
Core–shell structure, 232
Core–shell structured, 219, 220, 232
Correlation, 45
Coulomb interaction, 253, 254, 258
Coulomb interaction potential, 288, 289
Coulomb potential, 254
Coupled oscillation, 259, 260
Coupling coefficient, 20
Coupling constant, 68, 77, 81
Covert security, 224
Creation operator, 9, 13, 16, 17, 20, 23, 65, 68–71, 79, 260
Cross-correlation coefficient, 45, 118
Crystal ground state, 283, 288, 291, 293
Crystal lattice temperature, 85
Crystal vibration, 7
Current density, 254, 255
Cut-off wavelength, 7, 85, 191
CW coherent light, 209

D

Damage threshold, 164
Damaging, 157, 164
Dark state, 39
DC Stark effect, 196
De-excitation, 84–86, 172, 174, 197, 198
Defect, 66, 67
Degree of cleanness, 157
Delayed-feedback, 112
Delayed-feedback optical pulse generator, 90, 111, 113, 114
Delocalized mode, 66, 67, 73, 77
Democritus, 248
Density matrix, 42, 44, 48, 49, 52
Density matrix operator, 42, 49
Density of states, 45
Depletion layer, 192, 196, 202
Deposition, 137, 144–147, 161, 166, 168, 169
Deposition rate, 142, 143, 215, 216
Desktop machine, 158
Desorption, 236, 237, 239
Detailed information, 216, 218
Diagonal element, 45, 46
Diatomic molecule, 59
Dicke's super-radiance, 117
Dielectric constant, 20, 254
Differential external power conversion efficiency, 206
Diffraction, 8
Diffraction efficiency, 156
Diffraction grating, 155, 156, 165
Diffraction limit, 8, 146, 147
Digital-to-analog converter, 102, 103
Dirac δ function, 256
Directivity, 209
Directly modulated semiconductor laser, 209
Direct product, 22, 83, 85, 86, 138
Direct transition-type semiconductor, 200, 212
Dispersion relation, 247, 259–261
Displacement, 313, 314
Displacement operator function, 71
Dissociation, 59–62, 67, 73, 82, 137, 138, 141, 143–145, 161, 162
Dissociation energy, 60, 82
Double hetero-structure, 207
DP computer, 124
DP device, 37, 41–43, 47
DP-phonon interaction, 7
Dressed-photon device, 89, 117, 120, 121, 125
Dressed photon (DP), 1, 3–8, 9, 11, 18
Dressed-photon–phonon (DPP), 7, 9, 78, 82–87, 137, 142, 147, 148, 151, 160, 161, 164
Drift process, 243
δ-dyadic, 256
Dynamics, 47, 48

E

Effective energy, 11, 248
Effective interaction, 6, 219
Effective interaction energy, 18, 21, 22, 28–31, 33
Effective interaction operator, 269
Effective mass, 29, 35, 219
Effective mass approximation, 283
Effective operator, 266, 269
Efficiency of the energy transfer, 51, 54
Eigen angular frequency, 66
Eigenenergy, 16, 18, 21, 25, 29, 260–262
Eigenvalue equation, 261, 284, 285
Electric dipole-allowed, 7, 8, 85–87
Electric dipole approximation, 19
Electric dipole-forbidden, 6, 86–88, 89, 90, 104, 107, 116
Electric dipole-forbidden transition, 41, 248
Electric dipole moment, 13, 14, 29, 38, 39
Electric dipole operator, 20
Electric dipole transition, 6
Electric displacement operator, 14, 20
Electric displacement vector, 254, 255, 258
Electric field, 255
Electric quadrupole, 223, 227–229
Electrolysis, 200
Electromagnetic field, 11, 12, 18, 19, 31, 33
Electromagnetic mode, 4, 5, 11, 12
Electron, 11–14, 17, 35
Electron beam, 150, 152, 153
Electron-beam lithography, 97
Electron–hole pair, 5, 7, 9, 11–13, 17, 18, 20, 22, 23, 37, 42, 247–249, 259, 283, 287, 288, 290, 291
Electron–hole recombination, 3
Electron–phonon interaction, 3, 201
Electronic tag, 223
Electrostatic energy, 119
Elementary excitation, 259
Elliptically polarized, 184
Emission, 12, 17, 18
Energy conservation law, 12, 17, 22, 32
Energy consumption, 97, 118, 121–124, 132
Energy conversion, 171, 190, 192, 198–200
Energy dissipation, 40, 42, 44, 89–91, 119–123, 125, 221
Energy level, 11, 13, 89–92, 94–96, 102, 104, 105, 108–115, 117–119, 121, 122, 125–133
Energy transfer, 5, 6, 37, 39, 40, 42, 43, 46–48, 89, 90, 92, 94, 98, 99, 102, 104, 105, 108, 110, 111, 113–115, 118–120, 126–134, 215, 221, 223, 227, 231
Energy transmission length, 106–108, 128
Energy transmitter, 90, 104, 105, 107, 127
Energy up-conversion, 82–85, 173, 174, 176, 179, 182, 191, 192, 194, 196, 198–200, 202
Envelope function, 41, 90
Error rate, 119, 120
Etching time, 162
Exchange interaction, 253
Excitation, 60, 61, 72, 73, 77, 82, 84–86
Excitation probability, 72, 73
Excited state, 12, 22, 26, 32, 60, 82, 83, 85, 86, 137–140
Exciton, 37, 39–46, 48–50, 52, 53, 55, 89–92, 94–96, 99, 102–106, 108–119, 121, 125–127, 129–133, 259–262, 289
Exciton binding energy, 289
Exciton–polariton, 12, 18, 20–23, 25, 29, 30, 32, 260–262
Expectation value, 72, 74, 77, 79, 80
Exposure time, 148–150
External power conversion efficiency, 206
External quantum efficiency, 206

F

Fall time, 92, 93, 126
Fan-out, 99
Fermi energy level, 13
Fermion, 284
Fiber probe, 1, 2
Field operator, 283, 284, 288, 291
Figure of merit, 95
Filament current, 203, 204
Final state, 21, 22, 26, 32, 292
Finite temperature effect, 46
Finite-size effect, 81
Fluctuation, 12, 307, 310
Fluorescence, 171–176, 178, 180
Forbidden, 41
Forward bias voltage, 202–204
Forward-scattering, 123
Fourier transform, 14
Fractal nature, 233
Frank–Condon principle, 60
Frenkel exciton, 259
Frequency down-converter, 108
Frequency up-conversion, 109, 110, 112
Frequency up-converter, 108–112
Fresnel reflection, 123
Fresnel zone plate, 153–156
Full width at half-maximum (FWHM), 189, 197
Functional space, 21

G

Gain-saturation power, 211
Generic technology, 250, 251
Green gap, 208
Ground state, 60, 61, 83, 85, 86, 137–140, 161

H

Harmonic oscillator, 15, 64, 65
Heat bath, 6, 42–44, 46, 48, 49, 91, 108–110, 115, 119
Heisenberg representation, 76
Heisenberg's equation of motion, 275
Heisenberg uncertainty principle, 12
Hermitian conjugate, 9, 13
Hermitian conjugate operator, 266
Hermitian operator, 76, 266
Hierarchical hologram, 223–225
Hierarchical memory, 216–219
Hierarchy, 33, 35, 36, 215–217, 221, 223, 231, 248, 250
High-temperature phase, 230
Histogram, 233, 241
Hologram, 223–226
HOMO, 197, 198
Homo-structured pn-junction, 201, 208
Hopping, 302, 304, 306
Hopping constant, 68, 70, 75, 77, 78, 80
Hopping energy, 68
Hopping operator, 71, 79
Hough transform, 168, 169

I

Impurity, 66, 67, 72, 78, 80
Impurity atom, 314
Impurity site, 73, 78, 80, 81
Incoherent phonon, 73
Indirect transition-type semiconductor, 3, 200, 201, 204, 208, 211, 212
Information security, 216, 223, 224, 227, 229
Infrared excitation, 174, 181
Infrared radiation, 206
Inhomogeneous domain boundary, 202
Initial state, 21, 22, 32, 292
Injected current, 204, 206, 207, 211, 212
Inner product, 265
Inorganic semiconductor, 191, 198
In-phase, 71
Input interface, 90, 99, 121–124
Input terminal, 89, 90, 94, 95, 102, 107, 111, 118, 119
Integral Kernal, 292
Interaction energy, 92, 99, 102, 106, 115, 116, 121, 125, 126
Interaction Hamiltonian, 14, 139
Interaction range, 17, 19, 31, 35, 36
Interband transition, 3
Interband transition probability, 201
Intermediate state, 21, 22, 32, 33, 85–87, 178, 179, 181, 182, 184, 189, 190, 192, 198, 206, 292
Inter-material polarization, 227–229
Inter-nuclear distance, 59, 60
Inter-nuclear force, 59
Intra-material polarization, 227, 228
Inverse-Laplace transformation, 296, 297, 299, 300
Inverse matrix, 313, 314

J

Jahn–Teller distortion, 230
Joule-heat, 202

K

Key, 229

L

Lagrangian, 253–256
Laplace operator, 285
Laplace transform, 50, 53
Laplace transformation, 295, 296, 298, 299
Laser, 71, 195, 202, 209, 210
Lattice mismatching, 249
Lattice site, 260
Lattice vibration, 62, 63, 71
Law of inertia, 67
Light amplification, 211
Light emission, 172–174, 177–182, 186, 187, 195–198, 200, 206
Light emitting diode, 2, 200
Light exposure, 147, 157
Light harvesting antenna, 134
Light-matter fusion technology, 8, 250
Linearly independent, 276
Linearly polarized, 184, 224, 225
Lithography, 147, 148, 150, 152, 153, 157–160, 165
Localized mode, 66, 67, 72, 73, 75, 77, 80, 82
Localized site, 73
Lock, 229
Lock and key system, 229
Lognormal function, 194
Longitudinal motion, 63

Long-wavelength approximation, 6, 8, 40, 41, 292, 293
Lorentzian curve, 110, 112
Lowest energy state, 289
Low-temperature phase, 230
Lubricant film, 157
LUMO, 197, 198

M
Macroscopic system, 6, 18, 248
Magnetic dipole, 253
Magnetic flux density, 253, 254
Magnified transcription, 230, 231
Magnon, 259
Many-body system, 259
Many-particle problem, 283
Materials science and technology, 249
Mathematical science method, 249
Mathematical science model, 163
Matrix element, 50, 53, 291
Mean field approximation, 71, 75, 79, 80, 306
Memory access, 221, 223
Mesa-shaped, 97, 98, 123
Mesoscopic, 248
Metal–organic chemical vapor deposition, 233
Metamaterial, 4, 5
Minimal coupling Hamiltonian, 253
Mode, 11–13
Mode function, 14
Mode-locked laser, 209
Modulated, 5–7
Modulation, 5–7, 18, 31, 68, 82, 83, 172, 191
Modulation sideband, 5, 7
Molecular beam epitaxy, 97
Molecular vibration, 137, 139
Molecular vibrational state, 138–140
Molecule, 137–139, 141–147, 157, 161, 162, 164
Mole-fractional ratio, 169
Momenta, 3, 4, 201
Momentum, 62, 63, 65, 68, 255
Momentum conservation law, 201
Multi-mode, 7
Multi-photon excitation, 185
Multi-step excitation, 138, 176, 181, 185, 186, 191
Multiple exposure, 151–153
Multipolar Hamiltonian, 12, 253, 257, 258
Multipole, 253, 257

N
NAND logic gate, 99, 100
Nanomaterial, 3, 5–7, 11, 12, 17–20, 22, 25, 26, 29, 31, 32, 34–36
Nanometric system, 11, 18, 248
Nano-optical condenser, 90, 99, 101, 102, 121, 127
Nanoparticle, 144–147, 168, 169
Nanophotonics, 4, 5
Nanorod, 94, 110, 111, 117
Near-field optical interaction, 6, 18, 19, 89
Near-field optical microscope (NOM), 36
Negative resistance, 203, 204, 208
Non-adiabatic process, 61, 84
Non-degenerate excitation, 184, 189
Non-radiative relaxation rate, 44, 48, 49, 91, 105, 125, 126, 129
Non-resonant process, 32
Non-Von Neumann type computing system, 133
NOR logic gate, 99, 100
Normal coordinate, 64, 65
Normal distribution, 235
Normal mode, 259
Normal vibration, 259
NOT logic gate, 90, 95–99, 108, 121–124
Nucleus, 59, 60, 61
Number operator, 305, 310
Numerical aperture, 101
Nutation, 40, 51, 92, 104, 126

O
Observation probability, 77, 78
Occupation probability, 102, 103, 106, 109, 110, 113–116, 119, 121, 125–127, 129–133
Off-diagonal element, 45
Off-resonant, 92, 95, 119
One-exciton state, 49, 56
Optical absorption coefficient, 196
Optical amplifier, 111, 123
Optical cavity, 209, 210
Optical computing, 124
Optical fiber communication system, 190, 201, 206, 209
Optical frequency down-conversion, 171
Optical frequency up-conversion, 171, 172, 182–184, 186
Optical interaction, 5
Optical lithography, 1, 2
Optically active, 144
Optically inactive, 144, 150, 151
Optical nano-fountain, 101, 112
Optical near field, 5
Optical phonon, 63, 206, 259
Optical sum-frequency conversion, 187

Optical switch, 90, 92–96, 99, 111
Optoelectronic integrated circuit, 211
Orbital angular momentum, 285
Organic dye particle, 171, 191
Organic semiconductor, 191
OR logic gate, 99, 100
Orthogonal function, 290
Orthonormalized, 291
Orthonormal matrix, 64–66, 76, 81, 311
Oscillation threshold, 71, 209, 210
Output interface, 89, 121–124
Output terminal, 89, 90, 93–95, 99, 102, 106–108, 118, 122, 123, 125, 127
Overt security, 223

P

P space, 265, 266, 269
Parasitic capacitance, 209
Penetration length, 17
Phonon, 3, 4, 6, 7, 42–46, 48, 49, 59, 61, 62, 65–68, 70–73, 76, 77, 79, 80, 82–86, 91, 108–110, 112, 259
Phonon-assisted de-excitation, 249
Phonon-assisted excitation, 249
Phonon-assisted process, 137, 148, 154, 162, 167, 223
Phonon–phonon scattering, 71
Phonon scattering, 44, 204
Phosphorescence, 183
Photo-resist, 1–3, 147–153, 155, 157
Photochemical reaction, 221, 222
Photocurrent density, 195, 196
Photodetector, 190, 199, 205
Photodiode, 204, 205, 210, 211
Photo-induced phase transition, 230, 231
Photoluminescence, 146, 147, 169
Photo-mask, 1–3, 147–153, 155–158, 160, 165
Photon, 3–5, 7, 9, 11–14, 17, 18, 29
Photon correlation experiment, 118, 127
Photon energy, 3, 5, 7, 60, 61, 82, 85
Photonic crystal, 4, 5
Photonic device, 37
Photon number, 142, 143
Photon number density, 199, 200
Photosynthetic bacteria, 120
Photovoltaic device, 190–196, 198, 211
Pixel, 238, 241, 242
Planck's formula, 199
Plane wave, 14
Plasmon, 259
Plasmonics, 4, 5
Pn-junction, 192–196, 202, 204
Poisson distribution, 241
Polarization angle, 186
Polarization controller, 227–229
Polarization density, 254, 255, 257
Polarization state, 12, 27
Polaron, 259
Porous Si, 208
Power spectral analysis, 163
Power-Zienau-Woolley transformation, 254
Primary color, 173, 184
Probability density function, 71
Probe, 1–3
Projection operator, 18, 19, 21, 265, 266
Projection operator method, 248
Propagating light, 11, 12, 18, 29, 31, 34, 35
Pseudo footprint, 242–244
Pulsate, 111, 126
Pulse generator, 90, 111, 114, 117
Pump-probe spectroscopy, 164, 178, 181, 230

Q

Q space, 266, 272, 273
Qualitative innovation, 8
Quantitative innovation, 8
Quantization, 11, 18
Quantizing, 4
Quantum coherence, 42, 45
Quantum dot (QD), 6–8, 37–43, 46–54, 56
Quantum efficiency, 196, 199, 206
Quantum master equation, 42, 44, 49, 50, 52, 53
Quantum number, 41, 265
Quantum theory, 4, 5, 11, 12
Quantum well, 94, 95, 110, 111
Quasi Fermi energy, 202
Quasi-particle, 7, 9, 259, 304, 305, 307

R

Radiative relaxation rate, 31, 45, 46, 91, 93, 99, 106, 110, 113, 126, 128
Radical, 161, 162, 166
Raman signal, 164
Rare material, 208
Rate equation, 105
Real energy state, 181
Real photon, 12
Receiver, 121, 123
Recombination, 201
Reduced mass, 196
Redundancy, 124
Reference light, 184–189
Reflection, 104
Refractive index, 4

Relative coordinate, 288
Relaxation, 37, 40, 43, 44, 47, 48, 89, 90, 92, 96, 99, 104, 108, 112, 118–122, 127, 129, 177–182, 186–188, 191, 192, 198, 209
Relaxation oscillator, 208, 209, 211
Relaxation process, 12, 178
Relaxation rate, 91, 106
Resolution, 145, 147, 157, 159, 161, 162
Resonance condition, 47
Resonant, 90, 95, 96, 99, 102–104, 108–112, 115, 118, 119, 130
Resonant process, 32
Reverse bias voltage, 192–196, 198
Ridge waveguide, 209, 210
Rise time, 92, 93
Rotating wave approximation, 260
Rough information, 216, 222
Roughness, 160, 162, 163, 165–169
Route, 178–180, 182, 191, 192, 198

S
Scanning transmission electron microscope, 97, 98
Scattered light intensity, 163
Schrödinger equation, 265, 284, 286
Schwobel barrier, 166, 167
Second harmonic generation, 183
Self-organized criticality, 245
Sensitivity, 190, 205, 210
Shift, 196, 204
Sideband, 18, 68, 70, 82, 83, 138, 172, 191, 212
Signal light, 184–189
Signal processing rate, 124
Signal-to-noise ratio, 121, 190
Single-electron tunneling, 119, 120
Single mode, 304
Single photon, 118, 119, 121
Site, 63, 66–68, 70–72, 74–82, 290, 292
Size-dependent resonance, 33, 35, 215, 216, 239, 248, 249
Skew, 126
Skew resistance, 126, 127
Small-signal gain coefficient, 211
Smoothing, 160–163, 166
Soft X-ray, 153–156
Solar cell, 190, 199, 200, 211
Solar spectrum, 199
Sol–gel method, 169, 235
Spatial locality, 291
Spatial power spectral density, 163
Spatial resolution, 11
Spectral width, 146, 169
Spherical Bessel function, 286
Spherical harmonics, 313
Spherical wave, 31
Spin density wave, 285
Spontaneous emission, 84–87, 171, 172, 197, 201, 203, 212
Spontaneous emission process, 178, 180, 206, 207
Spring constant, 63, 66, 68
Sputtering, 166–169
Standard deviation, 106, 107, 305
1*s* state, 289
State-filling, 96
State-filling effect, 92, 96
State-filling time, 46
Stationary state, 240, 241
Steady state, 259, 262
Step-and-repeat, 157, 158
Stimulated emission, 84–87, 197, 202, 203, 209, 210
Stochastic model, 235–239, 241, 242, 244, 245
Stochastic science method, 249
Stokes shift, 197, 204
Stokes wavelength shift, 200
Streak camera, 190
Subsystem, 19, 20, 22, 29, 30, 32, 33, 35
Super-lattice structure, 208
Switching time, 95
Switch off, 91
Switch on, 92
Symmetric state, 38, 39, 50, 51, 115–117
Synchrotron radiation, 159

T
Tampering, 125, 126
Tamper-resistance, 125, 249
TE mode, 209
Thermal effect, 176
Thermal equilibrium, 73
Thermal equilibrium state, 83, 85, 86, 177–182, 191, 192
Thermal radiation, 199
Three-step excitation, 176, 178, 179, 181, 186, 189
Threshold, 149, 164, 221, 239, 242, 243, 245
Time delay, 111, 112, 114
Time inversion symmetry, 13
Traceable optical memory, 221, 222
Transition matrix element, 40, 293
Translational symmetry, 62
Transmission length, 104, 105, 107
Transmission loss, 104, 105
Triple diagonal matrix, 310

Tunneling, 5, 12, 17
Two-exciton state, 51, 52, 56
Two-step excitation, 178–180, 182, 186, 189, 191–193, 198, 202

U
Uncertainty principle, 32
Uncertainty relation, 4, 247
Unitarity, 261
Unit cell, 291
Unit operator, 273
Unitary operator, 301
Unitary transformation, 67, 69, 70, 79, 253, 261, 301
Unitary transform operator, 15
Up-conversion of optical energy, 171
Up-converting, 7
UV-cured resin, 98

V
Vacuum fluctuation, 32
Vacuum state, 22, 32, 37, 45, 71–73, 85, 283, 305, 306
Valence band, 3, 191, 192, 197, 201, 202
Vector potential, 253–255
Vibrational energy level, 60
Virtual photon, 5, 11, 12, 17, 32
Virtual process, 12, 32

W
Wannier exciton, 259
Wannier function, 290, 291
Wave function, 5, 11
Wavelength, 11, 12, 19, 31
Wavelength selectivity, 194–196, 199
Wave-number, 3, 4, 7, 62, 65, 201, 247, 248
Wave-vector, 13, 25, 253, 256
Well potential, 286
Wurtzite structure, 146

X
XOR logic gate, 49, 51, 55, 56, 99

Y
Yukawa function, 6, 18, 27, 31, 248

Z
Zero-point fluctuation, 12

MIX
Papier aus verantwortungsvollen Quellen
Paper from responsible sources
FSC® C105338

If you have any concerns about our products,
you can contact us on
ProductSafety@springernature.com

In case Publisher is established outside the EU,
the EU authorized representative is:
Springer Nature Customer Service Center GmbH
Europaplatz 3, 69115 Heidelberg, Germany

Printed by Libri Plureos GmbH
in Hamburg, Germany